THE ROLE OF DESIGN IN INTERNATIONAL COMPETITIVENESS

THE ROLE OF DESIGN IN INTERNATIONAL COMPETITIVENESS

D O Ughanwa and M J Baker

ROUTLEDGE
London and New York

First published in 1989 by Routledge
11 New Fetter Lane, London EC4P 4EE
29 West 35th Street, New York NY 10001

Printed and bound in Great Britain by
Mackays of Chatham PLC, Chatham, Kent

British Library Cataloguing in Publication Data

Ughanwa, D.O.
The role of design in international competitiveness.
1. Industries. Competitiveness. Role of design
I. Title II. Baker, M.J.
338.6′048

ISBN 0-415-00013-0

Library of Congress Cataloging in Publication Data

Ughanwa, D.O., 1945-
The role of design in international competitiveness.

Bibliography: p.
Includes index.
1. Design, Industrial—Great Britain. 2. Engineering design—Great Britain. 3. Great Britain—Industries. 4. Competition, International. 5. Exports—Great Britain. I. Baker, Michael John. II. Title.

TS171.4.U44 1988 745.2′0941 88-29718

ISBN 0-415-00013-0

Contents

Figures

Tables

Preface

Design is the seed corn of the future of any industrial society. It shapes the product, maximises value and minimises cost of new products to the benefit of both customers/users and manufacturers. It also differentiates new products, creates wants and satisfies needs, influences price and, not least, stimulates demand and sharpens competitiveness. The British Government has shown its awareness of all these salient values in setting up a funded consultancy scheme designed (a) to help British companies, in the short term, to get well designed products on the market quickly and (b) to persuade companies, through their own practical experience, that investment in design pays off — in fuller order-books and healthier profits, in the longer term.

But investment in design may not yield good dividends unless an organizational framework is created under which all facets of design can flourish. For example, our research vividly indicates that Britain does not lack in inventiveness, technological know-how or expertise yet the empirical evidence lucidly points to the inability of many British firms to marry these capabilities into a coherent whole to make the most of design. As maintained throughout this book, design is the concern of all in the organization and the key to successful, effective design is the integration of design, development, production, manufacturing, marketing and, indeed, all other relevant functional areas, with minimal boundaries.

This book is an attempt to provide an integrated approach aimed at solving design problems in order to enhance competitiveness in manufactured products. It develops a model under which various dimensions of design may be conveniently accommodated, effectively organized, efficiently coordinated and adequately managed to achieve success in international markets. Thus, it fills the gap which seems to be created by the isolated treatment of major aspects of design dimensions in the literature; and seeks to satisfy the need for a compact treatment of major aspects of a complex discipline, namely, design management and practice.

Each chapter treats a different topic and a consistent style of the use of conclusions based on both empirical and anecdotal evidence is maintained throughout the book. Therefore, each chapter may be viewed in its own right as an introduction to, and a reference work on, the subject.

All in all, the chapters constitute a handbook of design management and pactice which is intended to appeal to managers, practitioners of design management, researchers, lecturers, students and those interested or involved in design studies, in particular, the management of design/innovation and technology.

The book itself is based largely on the doctoral dissertation of Davidson Ughanwa, written under the supervision of Michael Baker at the University of Strathclyde between 1983–86, and developed further in a year as a Post Doctoral Research Fellow in support of a contract sponsored by the Design Council. As such the work owes much to the work of colleagues at the University of Strathclyde whose contributions are gratefully acknowledged. Most of all, however, the authors are indebted to the firms and senior managers who gave so freely of their time, experience and insight for without these this book would not have been possible.

Oyemeka Davidson Ughanwa
Michael J. Baker

Chapter one

The British disease: an examination of factors underlying the decline in British competitiveness in world markets

Introduction

Despite claims of an expanding economy, a close scrutiny of recent trade statistics in manufacturing goods indicates that imported manufactures have consistently eroded the home-market share of the UK producer. In the words of Lord Nelson (1984), 'Britain's manufacturing industry is today facing a most serious challenge. Its markets have been eroded, its competitiveness has been seriously weakened and its share of world manufacturing exports has shrunk; meanwhile imports of manufactured goods into the UK have increased . . .'

Over the last thirty years or so, government and independent bodies such as the National Economic Development Office (NEDO), the Design Council, the Council for National Academic Awards (CNAA), and the Boston Consulting Group (BCG) have undertaken a prolific number of in-depth studies into industries that were once the backbone of Britain's dominant manufacturing economy. The abundant publications on the results of these studies bear testimony to the rapid and competitive decline of such key industries. All of these tend to add to the growing concern throughout the UK about the uncertain future of these key industries.

Ironically, Britain's competitiveness is most seriously weakened in those areas where it led the world previously — shipbuilding, cars, motor cycles and bicycles — to name but some of the important examples. Two questions come to mind. What are the causes of this decline? Why have the UK's exports shrunk while its imports have increased?

The answers to these questions lie in the purpose of this chapter, namely, to examine the reasons for the continuing decline in the UK's international competitiveness.

The chapter contains four parts. The first takes a broad look at six industries in which Britain was once either a world leader or known to be competitively strong when compared with major foreign rivals, but which have consistently declined over the years and, in some cases, been completely decimated. They are: shipbuilding, motor cars, motor trucks, motor cycles, bicycles, and television.

The second part, which is an extension of the first, reviews a select tabulation of over forty cases spanning more than thirty years (1952–85) to explore additional reasons for the UK's weakening position in world trade.

However, this is not to say that all UK firms are non-competitive. Far from it. Some UK firms are highly competitive and this takes us to the third part of this chapter, which examines some of the firms that have successfully competed internationally.

In the final part we draw some conclusions that provide a framework for the succeeding analysis and research.

The decline in UK competitiveness in international markets: some empirical evidence

A firm can be described as competitive internationally when its total manufactured exports expressed as a percentage of its total sales revenue or output exceeds the equivalent exports or output of its rival firm(s) in foreign markets. Using this benchmark, a review of the research into the competitive decline of British industries confirms beyond doubt that even in industries in which Britain was a world leader, its competitiveness has been severely eroded if not entirely eliminated.

This section discusses six industries in which Britain was either once a world leader or was known to have been competitively strong. They are: shipbuilding, motor cars, motor trucks, motor cycles, bicycles, and television manufacture. While the evidence presented has been available for some time, it is believed that its concentration and discussion in a single analysis is essential to establish the nature and extent of the problem that faces British managers today. Thus, although many of the original studies are dated, their import is clear and largely confirmed by subsequent events.

The shipbuilding industry

In an extensive study of the 'Causes of the decline in British merchant shipbuilding and marine engineering' between 1890 and

1974, Albu (1976) found that the UK had consistently lagged behind its major overseas competitors.

Table 1.1 suggests that in 1890 Great Britain owned more than half of the world's mercantile fleet and built over 80 per cent of the new tonnage of merchant ships in the world that year. However, as Table 1.1 indicates, the UK's proportion of the world mercantile fleet had steadily fallen from about 52 per cent in 1890 to about 10 per cent in 1974. Similarly, the tonnage built over the same period consistently declined from about 81 per cent in 1890 to less than 4 per cent in 1974.

Table 1.2 shows the major innovations in ship design and construction, in marine propulsion machinery and their sources. It suggests that Britain led the world in ship design and construction and marine engineering up to the turn of the twentieth century. For example, of seventeen innovations between the nineteenth century and the first half of twentieth century, twelve (70.5 per cent) were solely of UK origin. Of the remaining five innovations, two seemed to originate simultaneously from the UK and USA and one from France and the UK. Only one innovation originated in the USA and Germany respectively.

The UK decline began in the second half of the twentieth century when Britain only accounted for three of the twelve innovations that occurred during that period. As Baker (1979d) observed, 'Since the start of the latter part of the 20th century, however, this position of pioneering near-monopoly has crumbled in the face of strong challenges from the world's other major shipbuilding nations'. This echoes Albu's (1976) finding that 'Britain's share of world shipbuilding has declined catastrophically

Table 1.1 Decline in British shipbuilding

	GRT owned	*Tonnage built*
1890	52.3%	81.5%
1900	50.7%	63.0%
1910	45.5%	58.5%
1920	41.5%	54.3%
1930	29.3%	57.2%
1950	21.5%	37.8%
1955	18.3%	27.7%
1960	16.6%	15.9%
1965	13.5%	8.8%
1970	11.4%	5.7%
1974	10.7%	3.6%

Source: Albu (1976) 'Causes of the decline in British merchant shipbuilding and marine engineering', *Omega* vol. 4, no. 5, p. 513.

Table 1.2 Main innovations and their origins

Period	*Innovation*	*Chief country of origin*
19th century (first half)	Iron hull	Britain
	Steel hull	Britain
	Separation of cargo carriers and passenger liners	Britain
	High-pressure steam engine	USA
19th century (second half)	Oil tanker	Britain
	Use of model tanks	Britain
	Sub-division of hull and cellular bottom	Britain
	Screw propeller	UK & USA
	Compound engines	Britain
	Water-tube boiler	UK & France
	Powered steering gear	Britain
20th century (first half)	Steam turbine	Britain
	Geared steam turbine	Britain
	Diesel engine	Germany
	Longitudinal framing (Iskemood system)	Britain
	All-welded hull	Britain (adopted first in USA & Germany)
	Tilting pad thrust block	UK & USA
20th century	Gas turbine (not yet adopted in merchant ships)	Britain
	Very large oil tanker	Japan
	Standard ship built in prefabricated units under cover	USA
	Tenth scale drawings and automatic steel cutting	Germany
	Specialised ship (including gas carriers)	USA
	Electronically controlled and automatic engine room	Japan and Sweden
	Advanced steam data for turbines — use of electrical plant manufacturers' designs	USA
	Very large single-shaft diesel engines	Denmark & Switzerland
	Stern oil seals	Germany
	Withdrawable stern gear (none so far fitted by UK shipbuilders)	Britain — followed by USA & Sweden
	Keyless bore propeller	Britain
	Large ducted propeller	Japan & Norway

Source: Albu (1976) 'Causes of the decline in British merchant shipbuilding and marine engineering', *Omega* vol. 4, no. 5, p. 515.

during the twentieth century.' He pointed out that 'successive committees of enquiry (the DSIR Report (1960), the Patton Report (1962), the Geddes Report (1966), and the Booz, Allen and Hamilton Report (1972)), have attributed this [i.e. the UK's catastrophic decline] to a loss of competitiveness deriving from poor production methods and laggardly innovation.' For example, he cited British yards' failure 'to take advantage of the enormous expansion in world tonnage which took place after the war, from 69 million tons in 1939 to 222 million tons by 1970'.

As a result of this 'increasing world shipbuilding capacity', he continued, '... by 1960 Britain had been overtaken by Japan with a 40 per cent share and by the 1970s by Germany with 20 per cent, Sweden 8 per cent and Holland over 5 per cent. Today there are many new countries developing shipbuilding capacity; among them Spain (now among the two or three largest in the world), Brazil, Korea and Portugal.' He supported his argument by pointing out that, 'In 1974, 63 liquified gas and chemical carriers were built; 27 in Japan, 9 in Norway, 8 each in the Netherlands and West Germany and none in the UK.'

In addition, Albu (1976) catalogued the following reasons for the UK's apparent lack of competitiveness in the shipbuilding industry:

1. Poor management and inflexible labour that becomes an increasing handicap in a period of rapid technological change.
2. Decline in technical competitiveness.
3. Inadequate technology to compete with rivals, such as Japan, in ship design and construction, for example, the reduction of thickness of steel plate by the use of high-tensile steel leading to reduction in weight and cost.
4. According to the original version of the Department of Scientific and Industrial Research Report (DSIR), the failure rate of diesel engines, whatever the design, built in British engine works was much higher than that of their competitors.
5. Typically, the British industry, having started an innovation, failed to develop it to its logical conclusion, and even when the UK shipbuilders had the advantage of building the first of a new type they failed to capitalize on this lead.
6. Most British builders failed to appreciate the cost advantages of standardization; such as the elimination of teething troubles, simplification of production methods and

the spreading of the overhead costs of planning and controlling production.

7. The design effort tended to be dictated by the short-term demands of the customer and not directed to long-term developments.
8. In Japanese shipyards nearly every manager and engineer is a university graduate. In the UK, only very few of those employed in the yards are graduates. Table 1.3 illustrates this. However, the latest statistics show that the number of UK engineers qualifying annually has fallen, while in Japan it has increased. According to Barratt (1984c), 'Japan produces 70,000 engineering graduates each year, against the UK's 7,000.' Nevertheless, Table 1.3 appears to suggest that Great Britain is proportionately better than the USA when population is taken into account. For instance, whereas the population of Great Britain was 55.93 million in 1976, that of the USA was 215.12 million (about four times greater) for the same period; the US graduate engineers' production ratio was marginally worse than that of the UK. That is, the US produced one engineer per 6,000 population as against UK's one engineer to about 5,000 population.
9. The industry's long record of bad industrial relations is well known and it has led to low productivity and a decline in quality.

Table 1.3 Number of engineers qualifying annually

	University or equivalent	*Non-university*
USA (1976)[a]	37,970[1]	25,089[2]
France (1975)[b]	11,205[3]	12,778[4]
West Germany (1976)[c]	3,960[5]	11,830[6]
Great Britain (1976)[d]	11,025[7]	6,594[8]
Japan (1973)[e]	62,961	8,235[9]

Source: Albu (1980) 'British attitude to engineering education: a historical perspective', in K. Pavritt (ed.) *Technological innovation and British economic performance*, Macmillan, London, pp. 67–87. a) Engineers Joint Council, New York. b) Engineers training in France: Service Scientifique Ambassade de France à Londres; Service des Etudes Informatiques et Statistiques, Secretariat de l'Etat aux Universitaires. c) Statistisches Bundesamt, Bundesrepublik Deutschland. d) DES Statistics. University and Polytechnic include some London external degrees obtained by part-time education. e) Japanese Ministry of Education: Report on Education (1974).

Notes: 1. First degrees in engineering 2. Two-year degrees in engineering and industrial technology 3. Diplomates of Grandes Ecoles 4. Diplomates of Instituts Universitaries (1973) 5. Diplomates of THS excluding architects 6. Graduates of IS (1970) (15,700 in 1973) 7. University and Polytechnic First Degrees 8. HNC and HND (of which 1,803 full-time HND) 9. Short University Course.

10. Owners complain of long delays in delivery and lack of interest in quality control, which is confirmed by the classification societies (design organizations responsible for specifying standard designs for ships, e.g. Parsons and Marine Turbine Research and Development Association set up in 1944, Pamatrada standard designs, and the Yarrow Admiralty Research Department (YARD) formed in 1946).
11. The inability of management to foresee the technological changes that were coming.
12. Management conservatism.
13. Resistance to change.
14. Unwillingness of better qualified and more entrepreneurial engineers to enter the industry or if they do, soon to leave it for classification societies or the ship brokers so reinforcing its decline.
15. Misplacement of priority on R & D expenditure, for example, management directed funds into areas that would not give the industry particular competitive advantages.
16. The total amount spent on R & D was insufficient.
17. Little or no 'in-house' development carried out.
18. Lack of co-operation between ship owners and shipbuilders on research programmes.
19. No research was conducted on production problems.
20. Inability of management to implement the recommendations of committee reports, for example, the recommendation by the Rochdale Committee (1970) that the 'existing R & D facilities be integrated to be supported by the shipping and shipbuilding industries' was never carried out.
21. Lack of government support, for instance, for building ships incorporating risky innovations as was done in Japan. In addition, the government has not assumed full responsibility for the industries' future policies.
22. Machinery made to the designs of British Marine Engineering (as criticized by DSIR Report, 1960) had poorer fuel consumption than other (i.e. competitors') designs.
23. Turbines built in Britain suffered many more breakdowns than those built elsewhere.

Based on the above weaknesses in the British shipbuilding industry, Albu (1976) concluded: 'It is certain that no action to revive the industry can have any hopes of success unless it provides a framework in which all its deficiencies: poor management, inflexible labour and technical backwardness, can be dealt with and this must involve an infusion of new managerial and technological blood and a radical change in trade union attitudes.'

Albu's work is particularly important to this study. First, as a trading nation, shipping and shipbuilding are of strategic importance to Great Britain and any laggardness in this area is likely to affect the UK's economy and probably its general standard of living. Second, Albu's (1976) findings suggest that the UK lags behind its major competitors mainly on non-price factors that are largely influenced by design.

A recurring theme of this book is that 'design shapes the product'. Since the British shipbuilding industry has an 'inadequate technology to compete with its rivals' such as Japan, in ship design and construction, and since also the UK's failure rate of diesel engines is much higher than that of its competitors, it could reasonably be argued that the major cause of lack of competitiveness in the UK shipbuilding industry is weak design.

Also, the design effort in the shipbuilding industry referred to earlier, 'has tended to be dictated by the short-term demands of the customer and not directed to long-term development' — an evidence of a lack of foresight that we (Baker, 1979c) have described elsewhere as 'pernicious and debilitating' in the sense that such a strategy tends to 'encourage short-term expediency at the expense of long-term prosperity'.

The motor-car industry

The available evidence indicates conclusively that the UK lags behind its major competitors in the motor-car industry.

Table 1.4 Production and export figures in the motor industry (in thousands of vehicles)

	Production			*Export*		
	1975	1978	1982	1975	1978	1982
UK	1268	1223	888	516	466	313
Belgium	792[a]	1053[a]	950[a]	735	1002	903
France	2546	3111	2777	1769	1579	1464
W. Germany	2907	3890	3761	1475	1904	2194
Italy	1348	1508	1297	661	640	437
Japan	4568	5748	6882	1827	2819	3770
USA	6717	9175	5073	154	248	84

Source: Nelson of Stafford (1984) 'The First Lord Nelson of Stafford Lecture', IEE Proceedings, vol. 131, no. 8, November, p. 627.
Note: a. Production and assembly.

Table 1.4 suggests that the UK's performance in the motor-car industry is rapidly diminishing both in terms of production and export shares when compared to that of its major competitors. For example, whereas production fell from 1.2 million cars in 1975 to 0.8 million cars in 1982 (a reduction of about 30 per cent), exports fell from 516,000 to 313,000 cars (a decline of about 30 per cent) for the corresponding period. Compare this to Japan's increase in production of 51 per cent and rapid export growth of 106 per cent, for the same period. Both France and West Germany have also performed better than the UK in both production and export.

The US decline in car exports may be explained by the fact that the major US car manufacturers such as Ford and General Motors have a number of large subsidiaries sited in Western Europe (UK, West Germany, Spain, etc.). The fact that typical American cars are often too large to compete in countries where parking space appears to be scarce or where fuel is costly to buy, also adds to the explanation.

The 'alarming' increase in Japanese car production and export is worth stressing, especially the increase in Japanese car exports to North and South America and Western Europe. For example, in 1962 Japanese car exports to both North and South America was only 4,000, but by 1980 exports had risen to 2 million in under twenty years.

Similarly, the export of Japanese cars to Western Europe started in 1962 with just 1,000 cars, but by 1980 it had soared dramatically to one million cars, exactly half of the number of cars exported to North and South America — a worrying figure when one compares the size of the Western European market with that of North and South America combined. According to an Economist Intelligence Unit Report (1980) 'In 1979, Japan produced 6.1 million passenger cars of which 3.1 million were sold overseas. North America and Europe were the two largest importers.' These increasing exports of Japanese cars to Western Europe led the leading West European car and truck manufacturers — Britain, West Germany, etc., to jointly lay a formal complaint to the European Commission about the use of West European markets as a 'dumping ground' for Japanese cars.

In the UK itself, car workers at Dagenham wrote to the Trades Union Congress (TUC) 'calling for a total ban on Japanese cars' (Cooper, 1980). According to Rines (1980), 'For the unions in Detroit and Dagenham, the Japanese marketing attitude of "what's mine is mine, and what's yours is mine as well" can no longer be tolerated.' He asserts, 'Over the last couple of months the scale of Japanese imports in the UK has certainly stretched the supposed limit of 10 per cent. What really irks is that over the same period, sales of foreign cars in Japan have fallen by 24 per cent.'

In 1975, the UK authorities were 'forced' to negotiate 'unofficially' with the Japanese car producers to hold their car imports into the UK constant for a period of time. As Roy, Walker, and Walsh (1983) put it, 'Much has been made of the Japanese car imports to the UK but, following an unofficial agreement with the Japanese Automobile Manufacturers Association in 1975, the level of imports has been held steady at around 11 per cent for five years.' The question is, what happens after five years? For how long can this 'wait-for-a-while' negotiation go on?

To the Japanese, this agreement is a 'favour' to allow British Leyland (BL) a breathing space to recover or 'put its house in order'; but did BL take advantage of this to recover? The answer is negative. According to Roy *et al.* (1983), 'More seriously still BL continues to slip further into second division volume car producers – 0.65 million units in 1977. Conventional wisdom indicates that around 1.0 million units per annum are necessary to give economies of scale, necessary to challenge the Japanese competitors, at least, in the home market. As Roy and his colleagues indicate, 'Within this general downward trend, the UK (car) industry is declining faster than expected and import penetration is 12.6 per cent higher than the most pessimistic prediction of the report on the industry by the Central Policy Review Staff (1975).' This downward trend can be explained by the rapidly increasing import penetration of Japanese cars into Western Europe.

In the US, a similar complaint was filed with the International Trade Commission in Washington in a bid to get a temporary ban placed on Japanese imports. As Rines (1980) puts it, 'And in the States, imports from Japan have scorched ahead to give Japan a 22.6 per cent market share.' He argues that 'There are three common features linking the US and UK car markets these days. In both sales have slumped. In both the share taken by Japanese manufacturers is on the up, and in both tougher action against the likes of Honda, Datsun and Toyota is being seen as the way to stop the rot.'

Rines (1980) also suggests other methods of stopping the rot, such as a complete barricade of the European markets (like Ford, USA), or acting politically, by fixing 'high tariff' or adopting 'stringent quotas on Japanese car imports'. However, he argues that confusion still exists among European car manufacturers as to whether a complete 'barricade' will enable them 'to make up lost ground, or whether open competition will produce faster change'.

Combining these problems apparently caused by the 'alarming' increase of Japanese imports into the UK, Western Europe, and the US, one may ask, what is the most effective method of minimizing the Japanese imports:

- take legal action to stop the Japanese imports,
- negotiate with the Japanese to hold imports constant for a period of time,
- fix high tariffs on Japanese imports,
- adopt a stringent quota system on Japanese imports, or
- barricade the import of Japanese cars?

At best, all of these five methods may provide an effective short-term solution to the problem but an effective long-term economic solution still needs to be found. This is because the solution to the rapid increase of Japanese car imports into the UK or any other country lies not in legal wrangling, or negotiating favourable trade terms, nor even in fixing high tariffs or adopting a stringent quota system or excluding Japanese cars, but in matching their technical quality and design excellence. This fact has been admitted by the Ford executives themselves.

As Rines (1980) puts it, 'Having toured the country's car plants, Ford's top management certainly needs no convincing about the lead Japan now enjoys at the top of the productivity league ... Ford executives returned from the visit shaken and unnerved by what they had seen of the Japanese worker's dedication to his job.' He then concludes, 'It can be argued, as the Japanese undoubtedly do, that the reasons for the success of the Eastern car men is the excellence of their products', emphasizing the importance of new-product design in international competitiveness.

In order to compete and match the technical quality and design excellence of the Japanese cars, therefore, only two major alternative strategies appear to be available:

(1) Set up collaborative arrangements, or
(2) Improve our own technology.

Failing this we can either try and close the door (unlikely) or capitulate (likely).

Collaborative arrangement between UK and Japanese car manufacturers provide the opportunity to cross-fertilize ideas in the areas of new-product development, design, and manufacturing, as well as to co-operate to produce sophisticated products for world markets. Some collaborative arrangements in design of new products are already under way between Austin Rover and Honda, Bedford and Pontiac, as well as between GM's Vauxhall (UK) and Isuzu (Japan).

Improving one's own technology covers the whole technological spectrum — radical breakthrough innovation, major innovation, incremental innovation, and improvements. All these involve adequate spending on new-product design and development by the manufacturing firm, and a continuous search for new ideas including transfer of technology — be it inter- or intra-firm transfer.

Perhaps an illustration of what is meant by 'adequate spending' may help to make the point. Rhys (1984) discovered that the range of expenditure needed for a modern integrated assembly plant of optimum size such as the Daimler-Benz plant at Woerth in West Germany, is between £150-£300 million. In the late 1970s, British Leyland spent only £32 million on its new assembly hall in Lancashire. Similarly, in France, Renault was given £500 million by the state for the development of a family of just four diesel engines, but between 1975–83, British Leyland was allocated £350 million to finance everything — new plant, vehicles, components, research, development, etc., (Rhys, 1984).

Table 1.5 suggests that the setting up of a modern assembly plant that is comparable to a continental standard and the development of one gear box have, in fact, 'swallowed up' the £350 million allocated to BL by the government (for use over a period of eight years). Thus, other vital costs such as costs of developing a new engine (£150-£200 million), research and design (£40 million), etc., appear not to be considered by government at all. Obviously BL's research programme will be constrained by this

Table 1.5 Capital cost of facilities (1982 prices)

Cost of sub-assembly plant	
all new truck	£120 million–£200 million
truck with existing components	£50 million
new cab	£60 million
new axles	£80 million
new engine	£150 million–£200 million
new gearbox	£50 million–£100 million
Cost of assembly plant	£150 million–£300 million
Cost of sub-assembly plant	£150 million
Research and design centre	£40 million
Company-wide investment on new CV range (using existing components)	£350 million

Source: Rhys (1984) 'Heavy commercial vehicles: a decade of change', *National Westminster Bank Quarterly Review,* August, p. 26.

poor funding, and this will have contributed to BL's non-competitiveness in the motor car industry.

Banning the importation of Japanese cars into the UK is a third possible alternative that may be used as a solution of last resort, that is, when neither of the two main solutions yields a fruitful result. Under this strategy, the UK customers may have to be persuaded to buy home-made cars on the grounds of chauvinism, backed up probably by a massive advertising campaign (including public relations), availability of cars and parts, and improvements in design and services (after-sales and delivery). Persuading customers to buy may be a very difficult task, particularly when the foreign product is believed to be better.

Further, this alternative appears to be another way of inviting the Japanese to shut their doors to exports from the UK, which is unlikely to be to the UK's long-run benefit. For example, John Nott (1980) (former Secretary of Defence) has made it abundantly clear on numerous occasions that he believes that 'for a nation which depends on exports, to impose import controls would be *suicide*'. In addition, the UK has a lot to gain technologically and otherwise by keeping its market open to Japanese products and vice versa. As Levacic (1984) puts it, 'So it is fair to say that the arrival of Japanese manufacturers helps to re-invigorate the domestic industry.'

Comparing all the alternatives, it appears that 'collaborative arrangement' offers the most effective solution to the rapidly increasing import penetration of Japanese cars into the UK. It is often said that 'two heads are better than one'. A cross-fertilization of ideas between UK and Japanese experts appears to offer an opportunity for developing sophisticated cars in response to the increasingly sophisticated tastes of customers.

According to Eddie Peppal, (1984), Styling Director, International Automotive Design 'It is estimated that by the year 2000 there will be a further 500,000 light vans and up to 75,000 more heavy goods vehicles on Britain's roads.' Consequently, 'future cabs will contain a variety of electronics-based instrumentation that will reduce the amount of paperwork that drivers have to undertake. Simple computer cassettes could be used to guide drivers on journeys across Britain and Europe, and every convenience — from sleeping accommodation to facilities for preparing hot snacks — will be built in the cab.' All these are anticipated future needs of drivers, which in turn, will make motor vehicles sell and thus increase vehicle exports.

It follows, therefore, that electronics will create a revolution in the drivers' environment. Since Japan is an 'undisputed world champion' in the electronics industry, it can be argued that a collaborative arrangement between the UK and Japan in motor-car manufacturing will be rewarding, particularly in achieving the above anticipated projects, as well as helping to close the technological gap (at least in the electronics industry) between the UK and Japan. In this way, one hopes British-made cars will be as good as or better than the Japanese cars, particularly when electronics technology is blended with the traditional British 'meticulous workmanship' as epitomized by Rolls Royce. For example, Nagashima (1970) found that '"made in England" still carried the traditional image of excellence and maintains a strong prestige value in Japan'. He stated: 'The Japanese respondents in this study regarded English products as expensive and as luxury items, but at the same time considered them very reliable and handmade in the old tradition of careful and meticulous workmanship.'

The second reason for suggesting 'collaborative arrangements' is that the 'improving own technology' alternative can also be achieved through these means. For example, ICL used Fujitsu's microchip technology to develop 'the most technologically advanced main-frame computer from the UK-based manufacturer'.

In conclusion, it would seem that collaborative arrangements with Japanese car manufacturers rather than legal wrangles to restrict imports, raising tariffs, applying stringent quotas, or barricading the importation of Japanese cars, offer the best alternative for not only minimizing increasing Japanese car imports into the UK, but also matching and excelling the Japanese cars in terms of quality.

The motor-truck industry

Table 1.6 shows a breakdown of trucks output in the categories over 6 and 15 tonnes gross weight for the period 1978–80. Judging from the context of a country's 'wholly owned' truck company, it can be seen that British Leyland's (BL) truck company compared unfavourably with their major foreign competitors. First, BL appears to compete in only two of the nine segments of the trucks market. Second, BL is conspicuously absent in two major sectors (6 tonnes (large) and 15 tonnes (large)) with high-volume production. Third, whereas BL was placed in the seventh position in the medium 6-tonne segment, its performance was very similar to that of White, Nissan, and Isuzu in the small-medium, 15-tonne, low-volume range.

The fact that BL is relegated to a medium and small-range motor-

Table 1.6 How companies compare 1978–80

'Typical' output of trucks over 6.0 tonnes gross weight & buses		*'Typical' output of trucks over 15 tonnes gross weight*	
Large		Large	
Ford (US and UK)	160,000	Daimler-Benz	70,000
General Motors Corp. (US and UK)	130,000	Large–medium	
Daimler-Benz	150,000	Iveco	40,000
Iveco	80,000	Mack	35,000
International Harvester	85,000	International Harvester	35,000
Medium		Medium	
Hino	36,000	Volvo	25,000
Renault	40,000	Hino	16,000
Mitsubishi	25,000	Renault	20,000
Volvo	30,000	Scania	18,000
Scania	25,000	General Motors Corp (US)	25,000
Nissan Diesel	35,000	Paccar	22,000
Isuzu	35,000	Ford (US)	25,000
Mack	35,000	MAN	16,000
Leyland	28,000		
MAN	22,000	Small–medium	
Paccar	22,000	Mitsubishi	11,000
		White	12,000
Small–medium		Nissan Diesel	12,000
Freightliner	10,000	Leyland	12,000
Dodge (UK–Spain)	15,000	Isuzu	12,000
DAF	15,000	DAF	10,000
Enasa	10,000	Freightliner	10,000
		Ford (UK)	7,000
Small		General Motors Corp (UK)	9,000
Sonacome	6,000	Dodge	7,000
ERF	3,000	Enasa	7,000
Seddon Atkinson	3,000	Bedford	7,000
Foden	1,500		
Others	(1,000–2,500)	Small	
		ERF	3,000
		Seddon Atkinson	3,000
		Foden	1,500
		Others e.g. Hestair Mowarg, Faun Bellow	500

Source: Rhys (1984) 'Heavy commercial vehicles: a decade of change', *National Westminster Bank Quarterly Review*, August, p. 29.

Note: This table does not include commercial vehicle production by licensees of the firms listed here (e.g. MANs licensees in Hungary and Romania) or various Comecon producers such as the factories supplying Belaz vehicles in the USSR or Ikarus buses in Hungary. Neither does it include the sixty-five or so firms making less than 1,000 vehicles a year.

truck manufacturer may be a fair reflection of its inability to compete in the up-market (6 and 15 tonnes (large) segments), which are assumed to be characterized by technical sophistication, design excellence, and reliability. Thus, BL appears to be technically backward when compared to its major competitors in the motor-truck industry.

Some problems prompted the reorganization of BL, which resulted in the split-off of Jaguar as a separate company from BL in April 1980, and the divestment in 1986/7 of BL's unprofitable truck division to the Dutch truck company — DAF. However, the divestment deal still guarantees BL a 40 per cent share in the new company (DAF-BL Truck).

The motor-cycle industry

Table 1.7 shows the UK's performance in the world motor-cycle market for the period 1968 to 1974. Of concern is the sharp decline in the UK's market share of motor cycles between 1968 and 1974, despite a considerable increase in its market size (in volume terms) for the same period. For example, the UK's market share took a plunge from 34 per cent in 1968 to an all time low of 3 per cent in 1974. Similarly, its share of the North American market fell from 11 per cent in 1968 to just 1 per cent in 1974, while its share of the European market is nearly wiped out, from 2 per cent in 1968 to less than 1 per cent in 1974.

The analysis of the industry by the Boston Consulting Group showed that 'Between 1960 and 1974, UK production of motor cycles and mopeds declined by a factor of 8 from 160,000 units to 20,000 units in a period of consistently expanding markets' (Baker, 1979c). This appears to confirm the NEDO Report (1977) which states that 'The UK tends to lose share of world trade fastest when trade is expanding most rapidly ...'

Table 1.7 British market performance: 1968–74

	Market size		*Market growth*	*British market shares*	
	1968 (000 units)	*1974 (000 units)*	*1968–74 % p.a.*	*1968 %*	*1974 %*
North America	458	1,006	15	11	1
UK	38	91	16	34	3
Europe	100	290	19	2	<1

Source: Baker, M.J. (1979c) 'Export myopia', *The Quarterly Review of Marketing*, vol. 4, no. 3, Spring, p. 7.

In marked contrast, the Japanese share of this market rose from 4 per cent in 1960 to a record 72 per cent in 1973 and they have continued to dominate the UK motor cycle and moped market. As Spandler (1980) points out, 'Sales of mopeds for the first five months of 1980 are up 57 per cent on the same period last year with the Japanese taking 75 per cent of the market.' He reveals that 'Honda is the brand leader (in the UK market) with 48 per cent of all motorcycle sales and is also the biggest spender on advertising.' He adds: 'The market for small bikes is growing rapidly, and Honda has a great reputation for reliability, backed by the biggest dealer network in the UK.'

The major challengers to Honda's dominance in the UK market are also Japanese manufacturers, such as Suzuki, with about 30 per cent share of the UK moped market. The only UK manufacturer of mopeds — Norton Villiers Triumph with its BSA model, for example, has only an 8 per cent share of the home market. Unfortunately, this company still imports all parts used for its manufacture from Italy, virtually ruling out any effective competition with the Japanese producers as far as original home design is concerned.

As Spandler (1980) reveals, 'it is the new Italian moped manufacturer (Garelli), holding 4 per cent of the UK market, which is currently making elaborate plans to challenge the Japanese domination of the UK moped market, not Norton Villiers Triumph, which has twice Garelli's share. We should not be surprised, therefore, if Garelli overtook Norton Villiers Triumph in the UK market, as a result of the latter's failure to defend its market share.'

The bicycle industry

Like motor cycles, the bicycle industry has also declined drastically. An extensive survey conducted by the Open University Design Innovation Group (1984) reveals that 'At the height of the 'bicycle boom' towards the end of the 19th century, there were some 830 cycle manufacturers in Britain. Today the British cycle industry is dominated in terms of size, market share and design capability by one firm: TI-Raleigh ... the remainder of the UK cycle industry comprises about a dozen small firms ...'

On the export front, the Design Innovation Group (1983) states that 'Britain has traditionally been a major exporter of bicycles and components, particularly to developing countries, and British-made machines have dominated in the home market. However, since the 1960s the volume of bicycles exported has stagnated or declined while imports (mainly from Italy, West Germany and

Austria have rapidly increased to meet a growing UK cycle market.' The Group points out that 'In 1968, imports represented only 2 per cent of home sales by numbers; by 1978 imports had captured over 30 per cent of the UK market which itself had nearly doubled.'

Why did UK exports stagnate? Why did its domination of the home market decline? Why have foreign imports rapidly increased? All these are questions to which answers must be found.

According to the Group (1983), 'the increasing import penetration of bicycles has largely been due to price competition at the cheap end of the market and aggressive sales methods.' They point out that 'Small-wheel adult and childrens' cycles with welded frames and low-cost components are being sold in large numbers by direct mail at prices often below UK manufacturing costs.'

The Group maintains that 'Import penetration of components (e.g. saddles, brakes) is being achieved by fierce price competition at the volume end of the market and by technical innovation and quality at the more sophisticated end. In particular, because no UK manufacturers of lightweight alloy components survived the slump in bicycle sales of the 1950s, virtually all parts (gears, chainsets, brakes, etc.) used on lightweight sports models are imported — mainly from Italy, France, and Japan.' This suggests that the UK lags behind its major competitors not only in price, but also in technical sophistication and quality, all of which are influenced by design. In addition, the UK appears to lag behind in in-house development and adequate spending on capital investment (hence the UK imports virtually all bicycle components from abroad); process innovation; technical specification; and value for money (hence the rapid increase in demand for foreign-built bicycles that seem to have more value-satisfying characteristics).

In his study, Roy (1984) found that 'The very rapid penetration of (bicycle) imports into Britain (from 4 per cent of home sales in 1970 to nearly 40 per cent in 1980) has been based on low-price small-wheel models with 'U' frames produced by automated welding methods in West Germany or by low-cost labour in Italy, Eastern Europe and the Far East.' He feared however, that 'British (cycle) firms are unlikely to be able to compete against imports in this sector without major investment in process innovation and greater attention to design for economic manufacture.' Clearly the British bicycle industry is non-competitive in design, process innovation (automation), major investment, and the ability to link new-product designs to ease of manufacture.

Roy (1984) argued that the UK cycle manufacturers have neither competed effectively on price nor quality. He contended that 'almost all the components used, other than those needed to build the frame, were imported from Germany, Italy, France, and especially from Japan, whose cycle component industry has established a world-wide reputation for quality, innovation and value for money.' He asserted: 'The increasing use in the British cycle industry of imported components and almost total dependence on imported lightweight alloy components, is widely regretted: "It's the saddest part of the job" was a comment echoed by several interviewees.' This again demonstrates the UK's laggardness in product quality, value for money, technical innovation, and timely and adequate investment when compared to its overseas competitors.

It is unfortunate that most of these countries that have now excelled Britain in bicycle manufacture have largely drawn on original British-design concepts. Roy (1984) points out, 'The most radical innovations in cycle design and technology have tended in recent years to be developed abroad, although deriving from earlier British design ideas which did not gain commercial success', implying that UK cycle-producers are inventive but not innovative. According to Roy, 'Although many of the inventions and innovations on which the cycle industry is based were originally developed in Britain, the (bicycle) industry today is generally very cautious towards new ideas and novel designs.' This view is in line with the conclusion drawn by the Open University Design Innovation Group (1983) which states: 'In general, the UK (cycle) industry tends to be very sceptical of radical innovations preferring to concentrate its efforts on market-oriented design changes.'

However, Roy (1984) appears to have some reservations about the 'market-oriented design' approach. It is his contention that 'While the British cycle industry is good at a market-oriented approach to product design and innovation, there is a danger that it will ignore more radical and technically innovative ideas which might form the basis for overseas competition in the longer term.'

TI-Raleigh is cited as a typical example of the UK bicycle-manufacturing firm that 'appears to have adopted a strategy towards innovation that involves letting inventor/entrepreneurs or foreign manufacturers develop and test the market for novel designs, and producing its own versions if an innovation proves sufficiently successful'. One may be tempted to ask, when is an innovation sufficiently successful? Can a progressive firm afford to wait until its rival's innovation is sufficiently successful?

It is regrettable that TI-Raleigh — the only major UK bicycle manufacturer, with all its vast technical and financial resources, coupled with its international reputation for 'quality and value', seems now to be reacting to, rather than anticipating, competitors' actions in world markets. It can be argued that reacting to rivals' innovative activities in the marketplace is not an effective competitive strategy and does not promote technical progressiveness (see Figure 1.1). A reactive strategy or 'strategy of the fast second' consists of waiting to see what your competitor will do first then moving into the market quickly if his strategy appears successful. Levitt (1966) calls this a '"Watchful waiting strategy" or "The Used Apple Policy".' 'According to this policy', he argues, 'a company consciously and carefully adopts the practice of never pioneering a new product. It says, in effect, "You don't have to get the first bite on the apple to make out. The second and third bite is good enough".' This clearly describes TI-Raleigh's behaviour or 'fast second' strategy in the bicycle market. For example, following the impact and success of Alex Moulton's small-wheel bicycle (winner of a Design Council Award in 1964), Raleigh quickly introduced its own rival range. Similarly, following the successes of two American bicycle models — 'high riser' cycles (popular amongst US teenagers) and the 'BMX' range introduced in 1982 (used for US bicycle motocross sports), Raleigh reacted by introducing its own versions of 'high riser' and 'BMX', respectively.

It is not known why TI-Raleigh has opted for this 'follow-the-leader' policy. One reason may be to avert the risk of introducing a new product or being first to market. However, Levitt (1960) has strongly argued that 'While the innovator faces the risk of his product not finding a ready market, the would-be imitator (such as Raleigh) faces the equally palpable risk of reaching the market

Figure 1.1 Leader and follower types

LEADERS	FOLLOWERS
BROAD-SPAN INNOVATORS lead by introducing new products into a number of related markets	REACTORS introduce new products in response to competitive pressures
NARROW-SPAN INNOVATORS lead by introducing new products into particular market segments	DEFENDERS safeguard existing products mainly via process innovations, i.e. cutting manufacturing costs

Source: Johne (1983) 'How to lead by innovation', *Management Today*, September, p. 91

when it is already glutted with many competitors — and often rapaciously price-cutting competitors at that.'

Although TI-Raleigh has managed to survive through this 'watchful waiting' strategy, the result of such 'opportunism' appears to have adversely affected its share of the bicycle market. According to the report by the Director of Fair Trading (1981), TI-Raleigh's UK market share fell from 57 per cent in 1974 to 46 per cent in 1979, largely because of the rapidly increasing import penetration into the UK.

One can argue that a 'follower' or 'reactive' strategy, particularly in a manufacturing industry, does not encourage firms to be progressive and think about products for tomorrow. Rather, they appear to think in retrospect of past products (i.e. products already developed and introduced by their competitors). A customer who has seen or tested the existing product always looks forward to a future product that will be different or more exciting than the existing product, hence it is argued that once a (customer's) need is satisfied, a higher level of need seems to emerge. If this is so, then would-be sellers should always address themselves to future products (products different from existing products), particularly when one considers that we (customers and innovators alike) will spend the rest of our lives in the future.

One can reasonably conclude that the UK lags behind its major competitors in the bicycle industry because of a lack of competitiveness in good design, technical quality, in-house development, R & D expenditure, and price.

The television industry

Given that television was the creation of British inventor John Logie Baird, it is pertinent to inquire why so many TV sets on sale in Britain are of foreign origin. The overall concern of this section, therefore, is to find out why the UK market appears to be more vulnerable to foreign TV sets than those of its major competitors. It is hoped the reasons that will emerge will help to throw more light onto the UK's non-competitiveness in international markets.

Table 1.8 shows the output of TV sets in Britain and France, and indicates clearly the wide gap that appears to exist between them. For example, in 1953, France produced only 5 per cent of the total British output. However, by 1966, France has caught up with the UK, which seems to suggest that the French production methods, among other things, have become more efficient than those of the UK.

In terms of average production per worker in the TV industry,

Table 1.8 Production of TV sets in Britain and France, in 1000s (1953–82)

Year	BRITAIN All sets	BRITAIN Mono-chrome	BRITAIN Colour	FRANCE All sets	FRANCE Mono-chrome	FRANCE Colour
1953	1,150	1,150		59	89	
1959	2,865	2,865		510	510	
1960	2,140	2,140		655	655	
1961	1,255	1,255		822	822	
1962	1,465	1,465		991	991	
1963	1,600	1,660		1,152	1,152	
1964	2,185	2,185		1,332	1,332	
1965	1,590	1,590		1,250	1,250	
1966	1,400	1,400		1,360	1,360	
1967	1,270	1,240	30	1,324	1,304	20
1968	1,960	1,820	140	1,457	1,400	57
1969	1,900	1,760	140	1,416	1,311	105
1970	2,215	1,725	490	1,397	1,209	188
1971	2,390	1,560	830	1,469	1,169	300
1972	3,030	1,550	1,480	1,526	1,106	420
1973	3,105	990	2,115	1,706	1,125	581
1974	2,675	695	1,980	1,616	1,022	614
1975	2,180	505	1,675	1,661	926	735
1976	2,070	530	1,540	1,780	825	955
1977	2,295	680	1,615	1,784	653	1,131
1978	2,410	695	1,715	1,887	533	1,354
1979	2,450	775	1,675	1,884	492	1,352
1980	2,359	588	1,771	1,906	507	1,399
1981	2,307	356	1,951	1,895	449	1,446
1982	2,281	141	2,140	2,090	445	1,645

Source: Levacic (1984) 'Do mercantilist industrial policies work? A comparison of British and French TV manufacturing', *National Westminster Bank Quarterly Review*, May, p. 54.

the UK also appears to lag behind its major competitors. For example, Reitsperger (1982) found that the 'average number of sets produced per day per production worker employed (1979) showed 1.7 each for Matsushita and Sony (both of Japan), 1.2 for ITT (USA), and 0.65 for Thorn (UK).' This lower output suggests deficiency in British production technique, inefficient management (including worker motivation), and so on. In that higher productivity often leads to economies of scale and an increase in profits, it also creates the opportunity for further investment to improve the design, quality, and reliability of TV sets.

Comparing the percentage of TV penetration in both the UK and France, between 1953 and 1980 (see Table 1.9), Levacic (1984) found that the UK market was more vulnerable to the penetration of foreign TVs than the French market. This may be

Table 1.9 Percentage TV penetration in Britain and France[a]

Year	*All sets* (%)	BRITAIN *Mono-chrome* (%)	*Colour* (%)	*All sets* (%)	FRANCE *Mono-chrome* (%)	*Colour* (%)
1953	21	21	—	1	1	—
1963	87	87	—	33	33	—
1966	92	92	—	52	52	—
1967	93	93	0.5	58	58	0.1
1968	95	94	1	63	62	0.3
1969	96	94	2	68	67	1
1970	98	94	4	72	70	2
1971	101	92	9	76	72	4
1972	104	86	18	80	74	6
1973	111	80	31	83	74	9
1974	115	73	42	86	73	13
1975	116	67	49	88	71	17
1976	118	62	56	90	68	22
1977	120	57	63	92	65	27
1978	122	53	69	94	60	34
1979	128	53	75	96	56	40
1980	133	54	79	96	50	46

Source: Levacic (1984) 'Do mercantilist industrial policies work? A comparison of British and French TV manufacturing', *National Westminster Bank Quarterly Review*, May, p. 55.

Note: a. Measured as number of sets in use — number of households × 100.

because the French market was protected, which would account for the growth of the domestic industry. This may be inferred from Table 1.10, which shows that the retail price of a 22-inch colour-TV set was lower in Britain than in France, suggesting a more competitive market.

According to Levacic (1984), 'There are now six major Japanese TV manufacturers and one Taiwanese company in Britain.' Among them are: Mitsubishi, Hitachi, Toshiba, Sanyo, Tatung, Sony, and National Panasonic. The first five were purposely encouraged by British government under the auspices of NEDO to set up business in the UK in order to 'utilise TV plants which the existing British-based manufacturers — including ITT and Philips — could not run profitably', a clear evidence of lack of competitiveness in technology and innovation on the part of the UK TV manufacturers. This appears to be in line with the Boston Consulting Group's (1984) survey, which compares TV manufacture in Britain, West Germany, Korea, and Japan. The result shows that the 'European manufacturing costs exceeded those in Japan by about 25 per cent, despite lower wages in Britain.'

It is pertinent to note that the import penetration of TV sets into

Table 1.10 Retail price of a colour TV set at current prices (£s)

	1968	*1973*	*1977*	*1979*	*1980*	*1983*
Britain	300	220	260	300	337	290
France	416	358	400	490	403	350

Source: Levicic (1984) 'Do mercantilist industrial policies work? A comparison of British and French TV manufacturing', *National Westminster Bank Quarterly Review*, May, p. 56.

Note: Estimated average price of a 22 inch colour TV.

the UK has steadily increased in spite of the fact that 'the PAL licence permits a manufacturer to export no more than 50 per cent of its output', which appears to be a good reason why the UK's import figure should have been lower than that of France. Unfortunately, it is not. (Britain is currently using the PAL colour system, manufactured under licence from AEG-Telefunken of West Germany).

There appears to be some fear that the UK market may be used as another dumping ground for the 20-inch screen range of foreign TV sets, given that Britain fails to improve the technical design quality and reliability of its TV sets before the 'exclusion clause' contained in the PAL licence agreement is abrogated by 1990. As Levacic (1984) explains, 'The PAL licence prevented any country which does not use the PAL system from exporting PAL sets of 20-inch screen size and above until the mid-1980s.' Clearly, Japan and other major competitors may be anxiously waiting for the lifting of such a barrier to exploit the opportunity. Already, the Boston Consulting Group (1978) has reported that the reason Britain lags behind Japan, West Germany, etc., in TV manufacture is because of the UK's 'inability to exploit economies of scale and "x-inefficiency" (cost in excess of the minimum level attainable at any scale of output).' By contrast, the Japanese are renowned for their outstanding ability at installing sophisticated automated assembly for reducing 'x-inefficiency', such as reducing the number of components per set, minimizing faults per set, reducing calls per set (i.e. the number of times that a set when put into operation in a customer's house requires attention from a service engineer), improving performance, quality, and reliability, as well as increasing productivity.

The gap between the two countries is apparent in the Boston Consulting Group chart (Fig. 1.2) NEDO (1983), which shows the difference in reliability that appears to exist between TV sets made in Britain and those built in Japan in 1977. As can be seen, the Japanese call rate lies between 10 and 20 per cent, whereas that of

Figure 1.2 Field call rate comparisons for television receivers (1977)

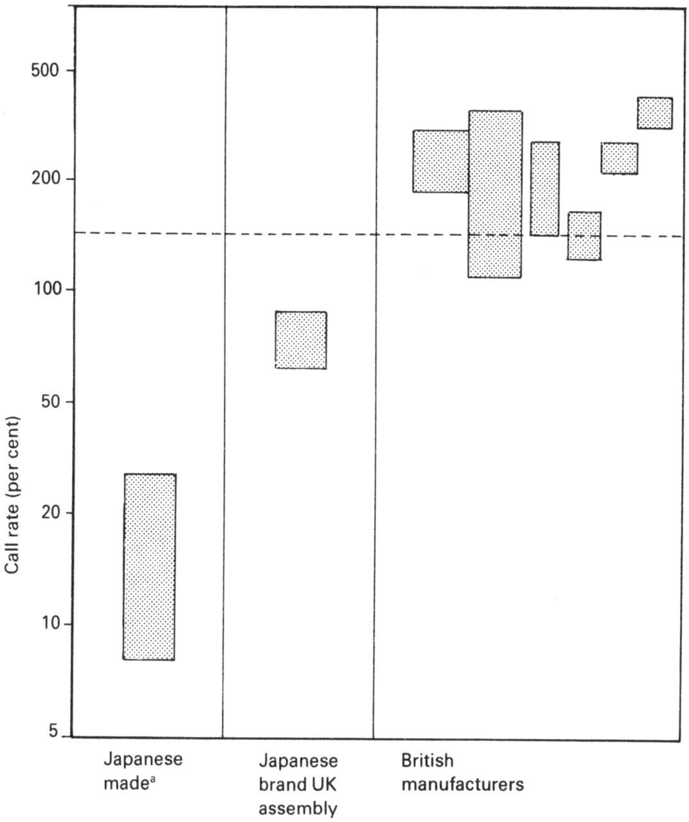

Note: a. mainly small screens.

Source: BCG Report (1978) November. (Reproduced in NEDO (1983) report.)

the UK lies in the region of 150 to 350 per cent. NEDO explains that a 10 per cent call rate means that in any one year, ten out of a hundred TV sets would require a call, suggesting that on average, one set requires attention once every 10 years. In other words, whereas one Japanese TV set requires attention once every ten years, the British rival set requires two calls per annum on average, indicating that the Japanese sets are more reliable than their UK rival brands.

Consequently, NEDO proceeded to examine the relationship between production rejects (faults per set) and first-year call rate (calls per set) for all sets made in the UK and Japan between 1977 and 1982 (Figs 1.3 and 1.4). The result indicates that the lower the production rejects, the lesser the number of calls per set during the first 12 months of use, and the higher the reliability and quality of the TV set. Figure 1.3 demonstrates that the UK TV manufacturers are rapidly closing the quality and reliability gap that appears to exist between them and their Japanese counterparts. For instance, in 1977, the UK's call rate per set was 5 while its reject level was at about 2.5; but by 1979, the first-year call-rate figure had fallen from 5 to 2 (60 per cent), whilst its rejects were reduced by about 1.5 to 1. The 1982 statistics were encouraging, displaying some significant improvement in production quality and reliability, both countries closely lying between the range of 0 and 1 (Fig. 1.3). In fact, Figure 1.5 illustrates more clearly, with bar charts, the proportion of TV sets built in the UK, Japan and 'other'

Figure 1.3 In-field reliability and production quality

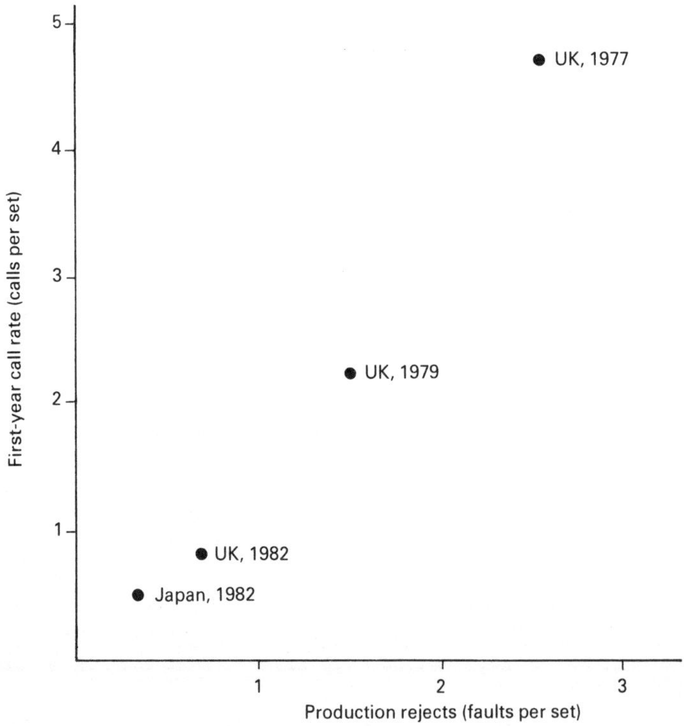

Source: NEDO (1983) 'Standards, quality and competitiveness', NEDO, London, May, p. 16

Figure 1.4 Relationship between reliability and production quality, UK-made colour television sets (1977–82)

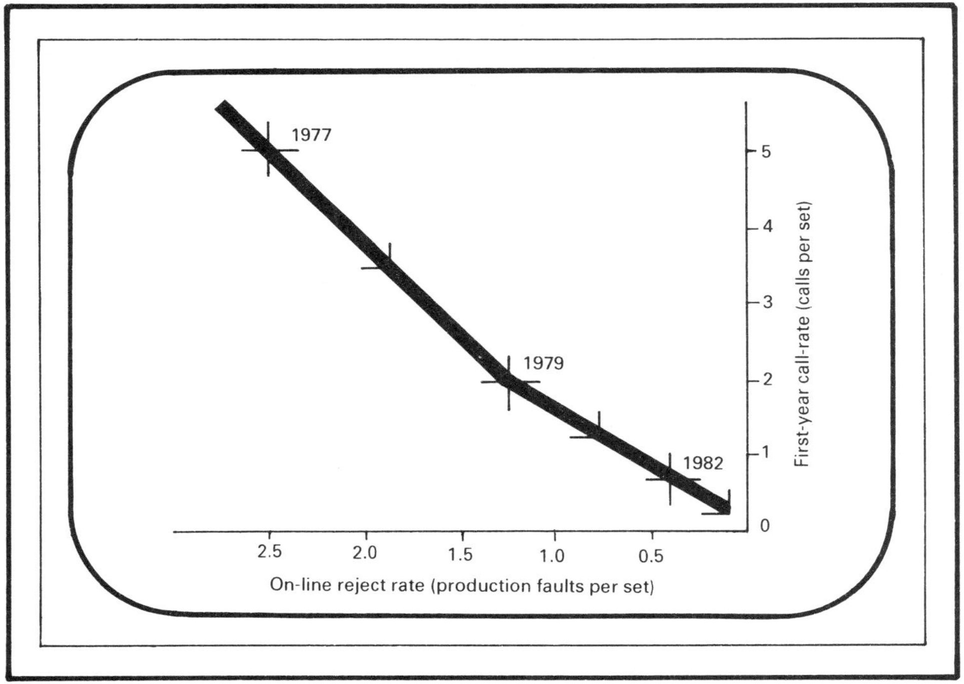

Source: Consumer Electronics Economic Development Committee (1983) 'Fact', NEDO, London

Figure 1.5 Survey results: proportion of sets with no repairs in the last twelve months

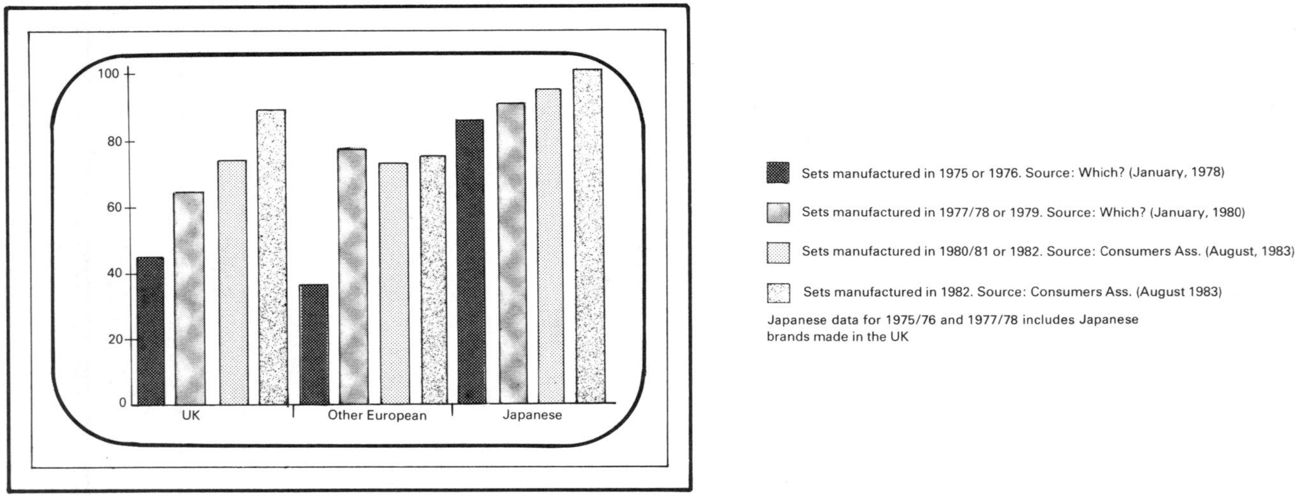

Source: Consumer Electronics Economic Development Committee (1983) 'Fact', NEDO, London, November

European countries, with no repairs for the first 12 months of being put into use, between 1975 and 1982. It can be seen that Britain has dramatically improved the quality and reliability of its sets from about 40 per cent in 1975/6 to an impressive 93 per cent in 1983 — a standard that beats its 'other' European counterparts, whose proportion of 'no repairs' stands at about 75 per cent on average; and compares favourably with Japan, which still maintains its lead during the corresponding periods.

The Economic Development Committee puts down the reasons for this remarkable UK achievement to improvement in: manufacturing methods, circuit design, component quality, which halved the number of components per set, and the adoption of a combined production technique — the Japanese practice of 'quality circles' and the American 'parts per million' (PPM) approach to quality (which stipulates a zero-defect objective, close collaboration between supplier and customer, inter-company liaison at several levels, and the idea that quality is everyone's business).

One important reason that appears to be neglected is the non-interventionist and non-protectionist industrial policy of Great Britain which, although it has helped to encourage import penetration, has also stimulated competition. In the words of Griffiths (Chairman, of the Electronic Consumer Goods Sector Working Party) (1983), 'I should like to say that the presence of Japanese companies in the UK has been positively beneficial to this process providing both commercial pressure and technical assistance.' Similarly, the Consumer Electronics Economic Development Committee (1983) comments: 'So competition provided one spur to improved quality.'

This review of the UK's performance in the TV industry indicates that made-in-UK TV sets suffered some setbacks in terms of quality reliability and efficient manufacturing methods between 1975 and 1979, when compared with the Japanese and major European rival brands (see Figs 1.2, 1.3 and 1.5). As NEDO (1983a) describes it, 'It was an accepted fact of life that sets occasionally broke down .... A survey of sets made in 1975/76 found that fewer than 50% of sets made in Britain or continental Europe survived their first year without needing at least one repair, but only 12% of Japanese brands needed repairing.'

However, the 1982 statistics show that the UK has dramatically narrowed this gap and has so significantly improved the quality and reliability of its TV sets that it now leads its major European competitors, such as France and West Germany (Figs 1.4 and 1.5). This manifests itself in the Consumer Electronics Committee

(1983) report, which states that 'In order to make reliable TV sets manufacturers must essentially get three things right. They must use reliable components; they must use as few components as possible; and they must employ the best production methods for assembling the components into the finished sets. UK set makers have done all three in recent years.' For example, the Committee (1983) also points out that 'There has been a dramatic improvement in quality of UK-made TV components; fewer components are used; automated production methods are used throughout the industry.'

Nevertheless, Figures 1.3 and 1.5 strongly suggest that British TV sets still compare less favourably with their Japanese rival brands in terms of quality, reliability, value for money, and efficiency in production techniques. If so, then more still requires to be done by UK TV manufacturers in this very important market to ensure that these design attributes (quality, reliability, etc.), which currently determine the competitiveness of TV sets in world markets are improved upon.

It is clear from the analysis of the performance of these six industries that their consistent decline over the years is largely a function of their inability to compete with their major foreign competitors on design dimensions — reliability, performance in operation, technical sophistication, production quality, quality of after-sales service, and so forth, all of which points to the significant role of design in international competitiveness.

However, the evidence of the factors underlying the decline in British competitiveness contained in these 'case histories' is but a small part of that available. In the Appendix to this chapter we have compiled a summary of over forty studies/analyses, spanning the period 1952–85, that indicates clearly that the symptoms and causes of the UK's steady decline in world trade are multifarious and multifaceted.

Nevertheless, the similarity and incidence of the reasons advanced in the Appendix and indeed other evidence in this book strongly indicates that the actual causes of the rise and fall of many British industries are far beyond the reasons often put forward in explanation. Such reasons as protectionist measures adopted by host governments, cheap foreign raw materials/labour, and the like, appear to be incompatible with our findings. At best, such reasons provide insight into some of the causes; at worst, the root causes are yet to be found.

Analysis of the Appendix enables one to draw up a short list of important and probably the most widely mentioned characteristics, which may be categorised under four main headings, namely,

'Price,' 'Design', 'Non-Price', and 'Design-related management factors'.

Price factors

The UK seems to be placed at a price disadvantage in three main areas. First, is the price disadvantage due to higher unit value of foreign imports, resulting from technical sophistication, 'better overall performance, superior overall design', reliability, efficiency, and higher productivity. Second, is the price disadvantage resulting from the low unit value of UK exports, mainly due to lack of technical quality, poor design, unreliability, etc. Finally, while UK unit prices are sometimes higher than those of its competitors, these prices do not seem to be compatible with quality. Such higher prices are mainly caused by higher unit costs of production due to higher costs of labour and raw materials, and inefficient manufacturing process.

As a result of these flaws, the UK cannot compete effectively on the price factor in international markets.

Design factors

Lack of good design

This is an important factor widely mentioned in different forms. It ranges from lack of value for money, lack of technical quality through unreliability, absence of aesthetic appeals, poor packaging, design not up-to-date, to use of inferior materials.

According to Walsh and Roy (1983): 'By "good design", firms mean attributes that increase value; the product may be of higher quality than competing products (e.g. stronger, longer lasting, better looking, more reliable or made to higher specifications); it may allow the user to do something previously impossible; it may reduce the customer's costs; or may meet user needs better than competing products.' This suggests that for a product to be competitive in the market place, it must be of good design in the first place, that is, it must meet the customer requirements/specifications.

However, a 'good' design may end up in a firm's 'cupboard' if it is not designed with 'ease of manufacture' and 'marketability' in mind and this is one major aspect in which the UK seems to be markedly lacking when compared with its major foreign competitors.

Inability to link design to ease of manufacture and marketing

This factor is important because in many a case, excellent designs have failed because designers themselves have failed to design new products with 'ease of manufacture' in mind. In other words, they seem to believe that their duty is to design a new product and then pass it to the production engineers whose responsibility it is to oversee the problem of manufacturability.

In the same vein, production engineers appear to be less concerned with the marketability or saleability of whatever they produce. To them, whether or not the product sells, is the 'headache' of the marketing manager and his salesforce. That is to say, it is the business of the sales person to ensure that the product is sold whether or not the product is badly designed and/or badly manufactured.

This attitude of not relating design to 'ease of manufacture' and 'marketability' by many UK manufacturers has been well documented in many studies, Pilditch (1978), Rothwell, Schott, and Gardiner (1983), and Osola (1984) see Appendix 1, p. 44 for detail.

Last but not least, Roy and Bruce (1984) asked respondents: 'Are your products designed specifically with ease of manufacture in mind? If yes, how important is this relative to other design considerations?'. To this, the Senior Production Manager of Dawes Cycles Ltd responded: '*Bikes not designed specifically with ease of manufacture in mind — this is a secondary consideration —* the prime one is to get a high-quality design that will sell. . . .' How can 'a high-quality design' sell when it is not easily producible? This is because 'a high-quality design' can be turned into a poor-quality product if it is badly manufactured, the reason being that such design is not initially related to ease of manufacture. Consequent upon this, the product may not sell.

Non-price factors

Inadequate delivery

About one in every ten studies perceives 'inadequate delivery' as a factor underlying the UK's lack of competitiveness when compared with its major rivals. It would appear that many UK manufacturers tend to forget that what customers buy is not only the physical product, but a package of new product attributes that provides satisfaction or value for money to the customer. This package includes 'efficient delivery', which most UK producers seem to lack.

As revealed in various studies, the UK lags behind its major competitors because it has longer lead times, does not meet delivery dates, often sells on delivered terms that have no basis other than 'tradition', is unwilling to sell on a delivered basis in line with its continental counterparts, and finally, is unable to calculate accurately 'delivered costs'. Thus, inadequate delivery tends to detract from the competitiveness of a product for which a delivery service is required.

Inefficient after-sales service

Inefficient after-sales service is also mentioned often as a factor that tends to undermine the UK's competitiveness in international markets, particularly in the heavy mechanical/engineering machinery sector, such as agricultural and textile machinery. Because of the sophisticated nature of such machines, users are often ill-equipped and sometimes ill-trained to cope with their routine maintenance and/or repairs, when they break down. For such machines, after-sales service is often among top buying-decision factors. Thus the inability of many UK suppliers to offer efficient and prompt after-sales service often discourages purchase and thus retards competitiveness in the export trade.

Rollason (1971), Department of Health (1976), Department of Trade and Industry (1978a), Michell (1979) have all found, among other things, that efficient after-sales service largely contributes to the competitiveness of manufactured exports.

Marketing factors

Weakness in marketing and marketing research and inability to adjust to specific markets and customer/user needs, are among the marketing factors underlying the UK's apparent lack of competitiveness in world trade. This is particularly important because available evidence suggests that the UK has either lost out to its competitors in certain markets (such as Third-World markets previously dominated by Great Britain but now being rapidly lost to Japan, West Germany, etc.), has not gained a strong foothold in some competitors' markets (such as France and West Germany), or has not been able to enter other markets (such as Japan), because of her inability to carry out in-depth studies of these markets. Consequent upon this, the UK product cannot adjust to specific market or customer/user needs. In the words of Sir Hugh Cortazzi (1984), 'The trouble with you British is that you don't try hard enough. The Japanese market is wide open, but you don't study the market in depth, you don't adjust your products to our needs ...'

It could, therefore, be argued that most UK manufacturers appear to be selling-orientated rather than marketing-orientated. Moreover, they tend to assume that 'what was good for our forefathers is also good for us', and so ignore the need for up-to-date research into the varying needs and wants of customers/users in specific markets (including the home market).

Perhaps, some lessons may be drawn from a similar assumption that nearly spelled disaster for Henry Ford. First, Ford seemed to assume that customer tastes/preferences were constant. According to Levitt (1960) 'Detroit (i.e. Ford) was not persuaded that he (the customer) wanted anything different from what he had been getting until it lost millions of customers to other small car manufacturers', that is, until customers switched loyalty to better differentiated rival brands; and potential customers who often depend on opinion leadership, may not buy at all, thus, leading to loss of sales and market share. Second, Ford appeared to have paid 'stepchild' attention to other areas of unsatisfied customer needs, such as repair and maintenance in spite of dissatisfaction with Ford's servicing arrangement repeatedly expressed by customers, suggestive of the fact that Ford was mainly product-orientated, not customer-driven. For instance, Levitt (1960) found that of 7,000 Ford dealers, only fifty-seven provided night-maintenance service.

Finally, Levitt described Henry Ford as 'the most senseless marketer in American history ... because he refused to give the customer anything but a black car'. Although no further evidence appears to be given in support of this assertion, more to the point seems to be that Ford used black paint because it was the only colour that dried quickly enough to keep his production line moving. This was essential for Ford to sustain his mass-production strategy, which yielded massive economies of scale. In turn, Ford was able to reduce the selling price per car to only $280 — well within range of the affordable income of average wage earners. His mistake was not to modify this strategy when it became clear that some customers would pay more for a differentiated product.

Management factors

Unwillingness to change attitudes

Unwillingness to change attitudes features prominently in the league table of reasons for the UK's lack of competitiveness. For example, attitudes to investment, such as inadequate spending on

R & D, as discussed earlier. 'Sticking to traditions' and thus unwilling to change 'old' management practices (such as selling on delivered terms that have no basis other than 'tradition' (see Appendix) is another example of the UK's unwillingness to change attitudes.

Other examples abound. For instance, 'sticking to the tradition of according much higher priority to pure sciences and mathematics than the applied or practical arts', as illustrated by Keith Lucas's (1980) Report (Appendix). According to Fores (1978), 'part of our manufacturing malaise is due to the propagation by scientists and economists of the false idea that innovation in manufacture arises from the application of new scientific knowledge'.

Moreover, in a number of cases, UK management has been found to be unwilling to heed expert advice. For example, what is now known as the 'Metro' car was redesigned about eight times because of the 'conservative' attitude of British Leyland's management. Initially, British Leyland management and their engineers stuck to 'tradition', being more concerned with functional performance rather than aesthetic appeals until Bache (the Styling Director of BL Cars and a specialist in aesthetics) was able to persuade the management to change their attitude towards the overall Metro design. Consequently, a 'lot of production engineering which had already been done on the ADD 88 model' had to be started from scratch following the unfavourable comments made by customers during a 'car clinic'. According to Manasian (1980), 'the changes that were essential were going to be very expensive'. Indeed it was. For instance, Bache, who has been known for his 'reputation for being hostile to outside design consultants', was forced to call in Pininfarina (an Italian firm long associated with BMC) to design a new exterior shell for the new Metro car.

Overall, BL invested a total of £285 million on the 'Metro project' alone. It is estimated that this sum might be much higher, considering the costs of redesign (including the hiring of foreign consultants). For example, Manasian (1980) found that the Metro engine modifications alone cost £30 million.

Unwillingness to learn foreign languages

The unwillingness to learn foreign languages is a factor mentioned by a few but strongly stressed. It is a factor presently overlooked but which, it is predicted, will have a marked impact on British business, particularly in the major rival countries, in the 1990s, if necessary steps are not taken to improve the language situation.

With modern technology becoming more sophisticated and

products much more so, 'sheltered markets' rapidly disappearing, and world markets increasingly becoming more competitive, the UK can no longer afford to concentrate on the home market (which is currently being aggressively attacked by foreign competitors — Japan, West Germany, the USA, France, etc.). British business executives have to go out and talk business with their foreign counterparts as well as key foreign buyers in their own languages. This cannot be done effectively through an interpreter, but only via a direct dialogue between and among executives themselves in the foreign language concerned, to ensure that every detail is discussed and understood by both parties.

It is evident from previous studies that the British, particularly British business people, do not seem to be willing to learn foreign languages. For example, Aitchison (1978) found that 'Reading fluently in English is astonishingly high, ranging from 90 per cent in Scandinavia and Holland to about 40 per cent in Italy and Spain.' He maintained that 'Taking the nine continental countries (i.e. excluding Britain), no less than 55 per cent of executives read English fluently, or fairly, with 26 per cent reading fluently and 29 per cent fair.' However, Aitchison regretted that 'The British do not return the compliment.' He pointed out that 'Although 32 per cent read French, only 9 per cent read French fluently and of the 12 per cent reading German only 3 per cent read fluently [see Table 1.11]. So not only are the levels of reading ability of the British low, but the average level of competence is lower than for Europeans.' He said that 'Of even greater significance is the data on business language' (as shown in Table 1.12). In a question 'Which languages do you use frequently in your business life?', Aitchson found that 'In every country English is the most important second business language, after the native language (or languages in bilingual countries). On the continent 31 per cent of executives use English frequently in business.' He asserted: 'It is figures like these that help to explain, if not excuse, the low level of language capability of the British. The Dutch, the linguists of Europe, can read on average 2.2 languages in addition to their native language, whereas the British can manage only 0.4.'

Management myopia

Available evidence suggests that one of the major causes of decline in the UK industry is what is termed here as management myopia (an apparent lack of foresight on the part of British management, particularly on forward planning).

Albu (1976) found that one of the major causes of the British

Table 1.11 Reading of foreign languages (reading fluently or fair (scales 2 and 3))

	English %	*French* %	*German* %
Sweden	86	13	63
Denmark	88	20	78
Netherlands	88	48	88
Germany	57	20	100
Switzerland	55	76	84
Belgium	64	93	43
France	54	100	15
Italy	39	58	15
Spain	40	59	5
Britain	100	32	12
Total	65	51	43

Source: Aitchison (1978) 'Europe's International Top People', *Marketing*, December, p. 41.

Table 1.12 Languages used frequently in business

	English %	*French* %	*German* %
Sweden	36	4	17
Denmark	45	4	32
Netherlands	50	17	34
Germany	39	12	99
Switzerland	45	73	17
Belgium	30	73	17
France	22	94	5
Italy	23	20	8
Spain	21	17	2
Britain	100	9	4
Total	45	32	31

Source: Aitchinson (1978) 'Europe's International Top People', *Marketing*, December, p. 41.

decline in the shipbuilding industry was 'the inability of management to foresee the technological changes which were coming'.

In an influential study, Carter and Williams (1957) found that 'In general, the opportunities for investment have been taken for granted (by management) or assumed to be something given by chance or by causes external to the economic system or mysteriously created by the beneficent interplay of competition.'

The fact that management take investment opportunities for granted, and believe in chance and mystery seems to lead to the conclusion that UK management lack forward planning — an evidence of myopia (short-sightedness). Thus Baker (1979c) blames the UK's poor performance in world export trade on 'policy makers and senior managers' for encouraging 'short-term expediency at the expense of long-term prosperity ... or for having the natural proclivity to live for today and ignore tomorrow in the expectation that if it does come our best forecast is that it will be like today.' In other words, management seem to assume that the market environment is static. As a result, they tend to ignore environmental changes as well as long-term planning.

Again, in another study, 'Innovation — Key to Success', Baker (1982) found that the decline and fall in UK industry was largely due to management's 'pursuit of markets to suit technologically obsolescent products to the neglect of product and process innovation and the creation of new markets'. Again, this suggests, a lack of foresight on the part of management for concentrating on out-of-date products at the time when their major competitors (West Germany, Japan, etc.) were 'modernizing' their products and processes.

Finally, commenting specifically on the Japanese 'invasion' of the UK 'video-cassette recorders and digital audio-discs' market, Sir Hugh Cortazzi (1984) asked: 'Why for instance did British manufacturers not foresee the demand for video-cassette recorders and digital audio-discs and start to meet the home demand before the Japanese could step in?'.

There are many more examples, but the above instances have made the point that many UK industries appear to be myopic as far as planning ahead is concerned. They tend to prefer short-term plans at the expense of long-term strategies.

Attempts will be made at the end of this chapter to suggest prescriptions for possible treatment. However, at this point it must be stressed that the above evidence of laggardness on the part of UK industries does not mean to say that all UK firms lag behind their foreign competitors. Far from it. Some UK firms are highly competitive but the overall national result (when compared to those of their foreign rivals) tends to undermine the effectiveness of their success. For example, some British firms have performed as well as or better than their foreign competitors. Rolls Royce (manufacturer of Rolls Royce Cars and aero engines (RB211 family)), Jaguar (producer of Jaguar Cars), Micro Cycles Ltd (manufacturer of a 'Bike in the Bag'), and Chubb Fire Security Ltd (producer of Pathfinder Airfield Crash Truck) are but a few examples.

However, when it comes to comparing the overall performance of British firms with those of their major competitors, as said earlier, it will be seen that even that small proportion of UK firms that are highly competitive in world markets will seem to be 'eclipsed' by the effect of poor performance of a much larger number of non-competitive UK firms, thus rendering the impact of successful firms on the British economy virtually insignificant. Hence it is generally argued that the UK lags behind its major rivals in international competition. As Rothwell *et al.* (1983) observe: 'Some British firms have been successful, but the failure of many companies to provide products that were technically superior to foreign machines has led to an overall decline of industry.' They argue that: 'Even when British products were relatively cheap they were still not bought because farmers perceived their lack of quality' and conclude, 'So it is clear that more British firms must make substantial efforts to compete in terms of both quality and price if they, and the economy as a whole, are to prosper.'

Conclusions

Our review shows that the UK lags behind its major competitors in almost every industry. Even in those industries where some companies (such as Rolls Royce) have been very successful, the failure of many UK firms to give customers products that are technically superior to rival product offerings has led to an overall decline of UK industries.

Two questions come to mind. Why has the UK failed to give customers/users what they want? Is Great Britain lacking in talents, creative ideas, inventiveness, or technological know-how? To the latter question the answer is negative as Table 1.13 clearly shows. If so, then the belief that Great Britain does not lack in inventiveness, new ideas, talents, or technological know-how appears to lead to the inevitable conclusion that what the UK lacks is the ability, to build what Pilditch (1978) has described as 'commercial success' as opposed to 'technological triumph'. He likens commercial success to the Boeing 747, 'which', he says, 'moved new ideas successfully into the economy' and technological triumph to 'Concorde', which he describes as a 'brilliant failure'.

It is suggested that to build 'commercial success' the UK has to put in effective practice a combination, of what Kotler and Fahey (1982) describe as 'creative imitation', improving existing products (where the Japanese talent mainly lay previously) and 'creative innovation' (providing new scientific breakthroughs), both of which are largely influenced by design, once again stressing the importance of design in international competitiveness.

It is worthy of note that what British customers need is not the

Table 1.13 Quotations illustrating UK's inventive ability

Author	*Quotation*
1. Akira Nagashima (1970)	The British are judged as being inventive and technically advanced.
2. James Pilditch (1978)	The British are supreme when it comes to making inventions. What they are less good at is innovating and there is an important difference.
3. M.J. Baker (1979c)	However, the major consequence of managerial incompetence and inefficient labour practices is not so much poor productivity *per se* but the frustration of inventive genius and innovative entrepreneurship which were the mainsprings of our industrial strength in the past.
4. David Mason (1980)	The US has a considerable lead in the field of microprocessors while the UK, despite its reputation for inventiveness and scientific competance, looks as though it will fall by the wayside.
5. David White (1980)	Britain has excellent ideas but lousy execution.
6. Robin Roy (1984)	The most radical innovations in cycle design and technology had tended in recent years to be developed abroad although deriving from earlier British design ideas which did not gain commercial success.
7. Aubrey Wilson (1984)	Certainly there is a history of brilliant innovative ideas and development, but other countries seem to have brought them to the market profitably.
8. Ken Gofton (1985)	British companies don't lack ideas but too often they are slow to exploit them.
9. Sir Bruce Williams (1985)	Britain had a good record for inventing new products but was less successful at exploiting them.
10. Ron Chilton (1985)	We have a remarkable array of talent.
11. Theodore Levitt (1985)	The interesting thing is that Britain has a very long history of originating knowledge, scientific knowledge.

Source: Ughanwa 1986

cheapest product (particularly in the mechanically-driven product sector) but one that will work reliably and if it fails, is capable of being repaired and maintained 'with simple tools' by operators or users themselves (i.e. a well-designed product).

To achieve this it is further proposed that the best possible alternative is for the UK 'to go the Japanese way'. This will involve the emulation of the Japanese new-product design and marketing strategies as well as management practices. For example, UK firms should attempt to improve their skills in marketing and market research, adjust to specific markets and customer needs, design products with ease of manufacture in mind, develop and manufacture high-quality and technically sophisticated products for world markets, provide prompt and efficient after-sales and delivery services, and be willing to spend, where need be, to achieve technical leadership.

As Sir Hugh Cortazzi (1985) puts it, 'A better approach is to isolate the qualities which are responsible for the crushing competitive performance of Japan; and enquire how British industry could adapt them for use in its own, very different conditions.'

In addition, British management should be prepared to change attitudes — attitudes to investment, competition, and labour management. Besides, management should avoid sticking to traditions, accept and implement promptly expert recommendations and be able to plan ahead, learn key foreign languages, recognize the fact that they have to go out and sell British products abroad, not concentrate on the home market, and last but not least, note that tastes and traditions differ from country to country.

Most importantly, 'There must of course be a clear recognition by both management and designers of *the leap-frogging nature of design competition.* It is essential, therefore, that the design team is constantly informed of what is going on elsewhere, what new materials and technologies are coming forward and particularly, as far as possible, the activities and potential of its competitors. ... In industry, as in the military world, time spent on reconnaissance is seldom wasted' (Nelson, 1984).

Above all, British graduate designers must be encouraged to stay in Britain and work for British industries. Available evidence suggests that most talented British graduate designers do not appear to have sufficient incentives to work in the UK. Instead, they prefer to work for continental firms from where, regrettably, they are sent back to Britain on assignments aimed, ironically, at competing against British firms.

Reading literature about the Japanese exemplary management and manufacturing practices (such as participative management style and management by quality (quality circles)) is not enough. British firms should attempt to send, with or without government support, efficient Britons (preferably selected across the board) to Japan, on short courses, to carry out on-the-spot study of the Japanese management, manufacturing, and marketing practices with the overall purpose of adapting them for use as they suit the British environment. It may be costly, but in the end, it will pay off. Further, UK firms have to find ways to reduce their cost disadvantage. To this end, it is suggested that thought be given to intensifying the setting up of plants in foreign countries, particularly where costs of labour and probably raw materials are low, provided the political economy is stable.

Also, UK firms should attempt to enter into collaborative arrangements with, say, major rivals, foreign users, etc. in such areas as new-product design and development, manufacturing, and the like. One hopes that this will not only help in reducing substantially the astronomical costs of new-product design, development, and manufacturing, but also in improving British technology.

Finally, both government and commercial banks must show their readiness to give positive support to UK industries (as is the case in Japan) to enable UK firms to finance major projects (even if risky), and be able to compete effectively in world markets. For example, if the British government or commercial bank had given financial support to the Cleveland Bridge Consortium, the Bosphorous Bridge in Turkey would not have been lost to a Japanese firm.

'Margaret Thatcher and Norman Tebbit criticised the Japanese for giving unnecessary aid to win the contract to build a new bridge over the Bosphorus, effectively outbidding the Cleveland Bridge Consortium which had been the favourite' (Tom Lester, 1985). This is not disputed. However, it is pertinent to note that financing major projects, such as the Bosphorus Bridge, is part of competitive strategy in world markets, which not only enhances the competitiveness of the firm concerned, but also promotes exports and thus, improves a country's economy. According to Tom Lester, 'It emerged only later that it was bureaucratic fumbling which, some participants allege, gave the Japanese their opportunity to step in.' He continues: 'If the Treasury and the Department of Trade and Industry had worked more closely and smoothly together, so that they were able to offer early on in the negotiations the terms which they were finally forced to concede (but too late), an order for 300,000 tons of steel and £45 million of work would have been won.'

In sum, it is fair to say, based on our review, that the actual causes of the rise and fall of many British industries are far beyond the reasons often put forward in explanation, namely, protectionist measures adopted by host governments, cheap foreign materials, labour, and so forth. At best, such reasons provide insight into some of the symptoms and causes; at worst, the root causes are yet to be found. Indeed, our detailed analysis strongly indicates that the root cause is the pursuit of short-term profits to the neglect of long-term gains and this appears to have a 'multiplier effect' on other factors. For instance, although a number of British firms expanded 'bountifully' through this short-term profit strategy during the nineteenth and twentieth centuries, they appeared to have been caught up in the web of what Levitt (1960) describes as 'a self-deceiving cycle of bountiful expansion (of the boom periods) and undetected decay', which, he says, characterizes 'every dead and dying "growth" industry'. The result is clear, 'if thinking is an intellectual response to a problem, then the absence of a problem leads to the absence of thinking'. This absence of thinking in response to the 'bountiful expansion' of the then boom periods appears to have resulted in a widespread complacency in many British industries, ironically at the time when most of its major competitors were upgrading their technology and modernizing their manufacturing processes. The result is that much of UK industry failed to capitalize on growth opportunities, particularly in the mechnical-engineering and chemical industries, and thus lost its dominant position to major foreign rivals (West Germany, Japan, the USA, etc.).

Even in those industries where some firms, (e.g. Rolls Royce and Jaguar) have been highly competitive in world markets, the failure of many British companies to offer customers and users goods that are technically superior to rival product offerings has led to an overall competitive decline of British industries in world trade. In this chapter, we have examined many of the factors accounting for the UK's weakening position in manufactured exports. In the following chapter, we shall attempt to distinguish between price and non-price factors to enable us to understand clearly how consumers assess value. It is hoped that the emerging results will enable innovators to differentiate their products aimed at satisfying the varying needs and wants of customers/users which, in turn, will make them more competitive, or at least as competitive, as their foreign counterparts.

Appendix

Why the UK lags behind in international competitiveness: a select tabulation

Author(s)	*Study*	*Reasons for lagging behind*
1. D. MacDougall (1951, 1952)	British and American exports	1. Price disadvantage 2. Inadequate commercial ties 3. Neglect of non-price factors
2. NEDO (1963)	Export trends	The slow growth of UK exports was due to: 1. Slow expansion of domestic output and productivity 2. Problems of design, quality, delivery and salesmanship possibly due to the UK's dependence on sheltered markets 3. Problems of costs and prices of UK exports 4. Geographical pattern of the UK's trade
3. Fielden Committee (1963)	Engineering design	Too many British products are being outclassed in: 1. Performance 2. Reliability 3. Sales appeal
4. S.J. Wells (1964)	British export performance: a comparative study	1. Price disadvantage 2. Lack of appropriate styling and finish 3. Inefficient salesmanship
5. NEDO (1965)	Imported manufactures	Major reasons for growth of import penetration in the UK: 1. Relative prices and costs

		2. A shortage of capacity in booms associated with an import penetration ratchet effect (e.g. chemicals, paper and board, steel) 3. Technical performance of imported products (e.g. mechanical engineering products, electronic capital goods, scientific instruments 4. A slow marketing response by UK manufacturers 5. The growing importance of international companies that rationalize their plant locations in different countries and that shop around the world for capital goods rather than first looking for them in the domestic market
6. NIESR (1966)	Export competitiveness: British experience in Eastern Europe	Why Eastern European buyers favoured other foreign products rather than the UK products: 1. High price for UK products 2. Design often not up-to-date 3. Slowness in providing quotations 4. High credit charges 5. Products less attractive than those available from other foreign suppliers 6. After-sales service often not very good 7. Not enough goodwill visits to interest buyers in new products 8. Public relations work often poor

Appendix continued

Author(s)	*Study*	*Reasons for lagging behind*
		9. Worse than other competitors in arranging counter-purchases and joint ventures
7. Steuer, Ball, and Eaton (1966)	The effect of waiting times on foreign orders for machine tools	Long lead time
8. Rothwell, R. *et al.* (1974)	Sappho updated — Project Sappho Phase II	1. Neglect of user needs and failure to interpret such needs into design 2. Inadequate market effort 3. Lack of R&D strength 4. Failure of management to assess or control the project satisfactorily during its development and launch 5. Failure to establish effective internal and external communication networks 6. Failure to integrate projects into overall strategy (such as linking new product designs to ease of manufacture and marketing)
9. Shankleman (1975)	Britain's post-war export performance	1. Weakness in marketing and market research 2. Failure to produce goods with the technological and design characteristics that are in demand.
10. CPRS (1975)	Future of the British car industry	1. Inadequate distribution networks

		2. Product unsuitability in terms of quality and the type of car produced (e.g. the lack of small cars after the oil crisis) 3. Slow delivery 4. High costs 5. Low productivity and under-investment 6. Over-manning 7. Poor labour relations 8. Slow work-pace and line speeds 9. Bad work practices and frequent stoppages.
11. Boston Consulting Group (1975)	Strategy alternatives for the British motor-cycle industry	1. Concern for short term profitability 2. Withdrawal from small-bike market in the face of Japanese competition 3. High cost of production 4. Failure to use best practice techniques in large-scale production 5. Inability to incorporate advanced design features in the product itself.
12. Panic (1975)	Why the UK's propensity to import is high	The UK economy was suffering from structural disequilibrium
13. The BETRO Report (1976)	Concentration on Key Markets	1. Inadequate sales coverage in export markets 2. Insufficient customer contacts in the export markets 3. Lack of in-depth market research in export markets 4. Late reply to correspondence 5. Infrequent visits by principals 6. Late deliveries 7. Higher prices

Appendix continued

Author(s)	*Study*	*Reasons for lagging behind*
14. N. Walker (1976)	Technical change and economic performance in the UK mechanical engineering industry: a review of the literature	1. Marketing and managerial practice tended to be inferior in the UK 2. Insufficient market research 3. Quality and sales effort of salesmen was poor 4. The feedback of information from salesmen to production managers about customer preferences tended to be ignored or was of poor quality
15. J.U.V. Edwardes (1976)	No substitute for good workmanship	UK products are: 1. Less reliable. 2. Less efficient. 3. Less value for money than their foreign counterparts. 4. Inadequate after-sales service. 5. Long lead time
16. I. Scheer (1976)	Design's key role in selling UK products	1. Use of poor quality materials 2. Reduction in company expenditure on product design
17. NEDO (1977) (Industrial Review to 1977)	Textile machinery	The UK textile industry fails to compete with EEC on: 1. Design 2. Packaging 3. Delivery 4. Service to customers
	Machine tools	1. Lower unit value of the UK exports compared to higher unit value of its imports 2. Machine tools not sophisticated 3. UK industry oriented more towards 'standard' products than its major competitors

	4. The UK industry's 'image' amongst its customers seems to be far from satisfactory in such matters as reliability of delivery
Pump and valves	A significant proportion of the UK industry lacks: 1. The organizational support 2. Resources necessary to match marketing, delivery, and after-sales performance of foreign competitors
Domestic electrical appliances	1. Higher prices 2. Unreliable deliveries 3. Inability to meet European standards
Electrical motors	1. Deficient in establishing overseas marketing and technical networks 2. Deficient in productivity level 3. Deficient in meeting delivery dates
Motor control gears	1. Less well advanced technically 2. Lack of innovation 3. Lags in technical specification 4. Deficient in establishing overseas marketing networks 5. Inability to meet delivery dates
Heavy electrical machinery	1. Inability to sustain technical capability comparable to those of its major competitors 2. Problem of delivery dates 3. Less able to compete effectively in severe price cutting, which often characterizes 'turnkey contracts' in a period of depressed demand
Organic chemical and plastic materials	1. Lower unit values of exports to other major industrial countries 2. Higher unit values of imports from major competing countries

Appendix continued

Author(s)	*Study*	*Reasons for lagging behind*
		3. Inability to compete successfully in the higher value groups of product offerings.
	Electronics	1. Electronic consumer goods have a reputation for poor quality 2. Have not in the past been designed with the European market in mind 3. UK does not have a large domestic market (compared with the US) in which it can exploit economies of scale 4. UK finds difficulties in competing with the aggressive marketing of IBM
	Textile	1. UK has the tendency to hold quality and design down to a price rather than up to consumer requirements for fit and certain features 2. Majority of wool textile imports from Italy are better designed and better priced 3. Designs and fashions for the UK market are often not in line with continental preferences
	Construction equipment	1. A decline in the unit value of exports relative to those of competitors 2. Less reliable 3. Inefficient in after-sales service
18. J. Pilditch (1978)	Why Britain cannot compete	1. Failure to give our home market what they want. 2. Britain is not being inventive enough

		3. Engineering firms, whether of industrial goods or consumer products, generally expect technical staff to develop new products that manufacturing makes and salesmen sell 4. In saturated markets when competitors fight for the same customers, the competition sets standards — of performance, price and delivery. Few UK companies study their competitors in such depth 5. Many development programmes remain too abstract too long (instead of turning words on paper (i.e. design) into tangible objects (products))
19. J. Piper (1978)	Making the most of West Germany	Many would-be UK exporters: 1. Have failed to recognize the market needs 2. Have not adapted their products to suit German standards and tastes, which are often different from the British and have, as a result, got their fingers badly burned. 3. In the past British manufacturers have adopted a 'take it or leave it' attitude — which spells instant disaster in the German market
20. R. Rothwell and W. Zegveld (1978)	Small and medium-sized manufacturing firms: their role and problems in innovation	1. Lack of in-house management expertise 2. Lack of technical expertise 3. Lack of production capacity 4. Lack of market power 5. Scarcity of finance for funding costly (and often risky) development projects
21. Corfield Report (1979)	Product design	1. Inappropriateness of British products to world market requirements

Appendix continued

Author(s)	*Study*	*Reasons for lagging behind*
		2. In some cases it appears that we are making the wrong products 3. In many instances it seems that we are not doing well enough what we attempt 4. Imperfect distribution which may prevent the excellent from being as accessible as the mediocre 5. Lack of discrimination and knowledge on the part of the purchaser, so that the excellent is not perceived as being more desirable than the mediocre 6. The price of acquisition, which may lead to the inferior being chosen on grounds of economy
22. Keith Lucas Report (1980)	Design education at secondary school level	1. Sticking to tradition of according much higher priority to the pure sciences and mathematics than the applied or practical arts 2. Few bright pupils continued with 'practical/aesthetic' subjects beyond the age of 13 3. Lack of exploitation of opportunity 4. All too often the spark of originality and the urge to make have been neglected and submerged in the requirements of the traditional education system
23. P. Ambridge (1980)	Metro: open up or shut down for BL	1. Unwillingness of engineers to involve industrial designers at the early stage of new product development 2. Lack of aesthetic appeals in the UK cars 3. Unwillingness of management to adhere to specialists' advice

24. A.W. Watt (1980)	Managing design function in the UK textile industry or 'I'm no expert, but I know what I like.'	1. Neglect of design and design management 2. No real career structure for design staff 3. The actual management of design was carried out by committees that comprised functional heads, none of whom usually had training in design or design appreciation
25. Mason (1980)	Why Britain's chips lag behind	1. Inability to develop and market new products 2. Inadequate spending on R&D 3. Lack of sales push or government procurement stimulus 4. Late entry into the market (e.g. one publicity poster epitomizes the British inability to commercialize thus: 'Ferranti — first into the future and last into production')
26. S.T. Parkinson (1980)	New product development and international trading success (a comparison between West Germany and the UK)	1. British companies have continued to base their marketing strategy on price competitiveness 2. Comparative lack of technical sophistication in their products 3. Singularly less able to identify user needs 4. Singularly less able to develop new products in line with these needs profitably than their foreign counterparts
27. E.H.A. Zeid (1981)	Marketing and export success	1. Selling orientation approach to export markets 2. UK firms pay greater attention to developing the marketing efforts to the home markets rather than the export market 3. Undertaking inadequate market research 4. Laggardness in innovation 5. Poor design and performance 6. Bad delivery 7. Inefficient after-sales service

Appendix continued

Author(s)	*Study*	*Reasons for lagging behind*
28. R. Rothwell (1981)	Non-price factors in export competitiveness of agricultural engineering products	1. Lack of technical sophistication 2. Lack of design excellence 3. Heavy reliance on exports of low technology, traditional goods (textiles) mainly to Empire Countries during a period when West Germany and others were rapidly developing chemical, machine building, and other 'modern' technologies 4. Lack of sufficient growth in the UK industry's production capacity 5. Inability to meet both home and export demand 6. 100% tax allowance on machinery purchases (introduced in 1971) contributed to the dramatic increase in imports to the UK in the 1970s of high priced — highly sophisticated — foreign-made machinery 7. Lack of systematic market research or detailed market information 8. The stance of many UK firms towards product development is reactive rather than positive: they develop machinery in the light of what competitors have done, or on the basis of a feed-back from the farmer rather than as a result of their own forward-looking development and marketing plans 9. Few firms, especially smaller firms, which compose the bulk of the industry, contain the management expertise to achieve satisfactory long-term planning 10. Many UK firms lack technical resources to enable them to draw ahead of their continental rivals even if they had plans to do so

11. Continental firms nearly always have a 'Head of Development' who possesses a technical qualification, usually an engineering degree, and often a doctorate unlike the UK firms
12. The percentage of graduate engineers associated with research, design, and development is higher in the continental firms than in the UK
13. There is generally a higher level of technically qualified managers in the continental firms, than in the UK
14. Generally speaking, continental firms place greater emphasis on seeking more efficient manufacturing procedures, and the use of automatic and semi-automatic features, than the UK firms
15. Many UK firms place small emphasis on production engineering and on linking design closely to manufacturing. They simply do not possess the professional competence to enable them to achieve this
16. Lack of professionalism on the part of many UK managers; they lack the ability properly to plan ahead and formulate explicit and detailed development and marketing strategies and to link design and production closely in order to achieve maximum production efficiency
17. Close and continuous interaction between product development and design personnel, and those involved in production planning and design, was seen to be a significant factor in the greater success of continental manufacturers over their UK counterparts
18. British manufacturers use inferior materials resulting in a good deal of excessive wear on moving parts

Appendix continued

Author(s)	*Study*	*Reasons for lagging behind*
29. P. Popham (1982)	Spaghetti Eastern	1. Unwillingness to spend heavily on R&D 2. Inability to offer unique attraction in cars 3. Lack of aesthetic appeals 4. Inability to study specific markets and satisfy needs in line with those markets
30. Rothwell *et al.* (1983)	The role of design and innovation in the prosperity of industrial companies	1. Poor design 2. Failure to innovate 3. Lack of engineering and design skills 4. The tendency in Britain to design something and then to hand the design to the production side and let them make it. In the US something is designed and then costed and then redesigned to lower the costs, and this might be done several times before the design is finally adopted 5. The British industry hesitates over anything new, looking for similar sort of design already on the market for encouragement 6. The Italian industry has the reputation for flair and imagination while the British industry languishes beneath a reputation for dull and dated designs
31. Lord Nelson of Stafford (1984)	Export competitiveness: a factor of management and/or design	1. UK is too slow in shaking off the effects of long years of easy markets and full employment 2. UK's failure to match the development spend of her competitors 3. Unreliability 4. Non-competitive costs 5. Lack of skilled managers

32. G.F. Turnbull (1984)	Management for profit by product design	1. Too often, in the UK, it was a case of a product looking for a market, whereas what was needed was exactly the opposite 2. Lack of flexibility in the management's approach to an effective product policy 3. Lack of proper analysis of customer needs in the chosen business sector 4. All too often those in the engineering management were the wrong people: they are frequently overqualified and have insufficient drive and determination 5. Obsession in management for ratios
33. I. Wicks (1984)	Organizational structure	1. Market research, marketing, and selling are terms which are frequently mixed and muddled 2. The inability of general management to relate to ideas that often fall outside their immediate knowledge areas (suggests that UK management are not outward-looking) 3. The inability to recognize that R&D on the whole project, not just the engineering aspects, is critical 4. Problem of the methodology of training customers to accept new techniques
34. A. Sutherland (1984)	Getting the right product at the right time	1. More willing to adopt a defensive posture 2. Slip in quality 3. Bad timing 4. Lack of understanding and agreeing upon business objectives for growth and recognising their consequences, by the company as a whole

Appendix continued

Author(s)	*Study*	*Reasons for lagging behind*
35. M.A. Hall (1984)	A passive or active role for development?	1. Development department is often isolated from the outside world and is denied or denies itself contact with the market place 2. Inadequate inculcation of the 'right' values at Universities and even at pre-O-level schooling, to would-be managers
36. J.V. Osola (1984)	Money	1. Companies far too often develop new or improved products but fail to develop simultaneously the manufacturing methods by which those products are to be made 2. Insufficient consideration given to auditing of products
37. B.T. Dastur (1984)	Management of quality: an Indian view	1. Inadequate expenditure on R&D 2. Wild-cat strikes
38. B.G. Dale and R.J. Mortiboys (1984)	Quality circles in the UK	1. Lack of controlled research into quality circle activities 2. Trade Union opposition to Quality Circle activities 3. Quality Circles were found to have little or no advantage 4. Lack of support from management
39. CNAA (1984)	Managing design: an initiative in management education	1. The inability of much of industry in the UK to use design effectively 2. The needs of industry are not fully met by graduates of UK courses in design 3. Managers in the UK generally do not appreciate the role of design 4. Virtual absence of the management of design from the curricula of business and management courses in public sector institutions of higher education

		5. Lack of insight into the relationship between effective design on the one hand and industrial performance on the other
40. Sir Hugh Cortazzi (1984)	Why 'eyes and ears' are necessary	1. Inability to adjust UK products to specific market needs 2. Not competitive on price 3. Not competitive on delivery 4. Inability to study certain foreign markets in sufficient depth 5. Unwillingness of most British businessmen to learn foreign languages
41. D. Channon (1985)	Managing in tomorrow's world	1. Whereas Japan designs products for a world market, the British design for a home market 2. Lack of competitor intelligence 3. Banks become less closely involved with the organisations with which they are working 4. Rise in unemployment
42. F. Griffiths (1985)	Is the customer sometimes right?	1. Concentration of the bulk of trade on Empire and Commonwealth markets 2. Slip in prices 3. Slip in delivery performance 4. Slip in quality 5. Problem of attitudes: — attitudes to investment — attitudes towards labour management — attitudes to competition
43. CPDM[1] (1985)	Survey of current practices UK exporters to Europe	1. A significant number of UK exporters ship their products to Europe following systems and sell on delivery terms which have no basis other than 'tradition'

[1]CPDM (Centre for Physical Distribution Management) is one of the specialist organizations within the British Institute of Management

Appendix continued

Author(s)	*Study*	*Reasons for lagging behind*
		2. A significant number of UK exporters sell 'ex works' as a matter of policy 3. Very few UK exporters, in small and medium range of companies, appear to sell in Europe on a delivered basis 4. Payments for all charges after FOB are left in most cases to the importer or importing agent 5. Senior management in small- to middle-sized companies do not seem to concern themselves with distribution in Europe 6. British exporters who traditionally have been trained in Commonwealth and former Empire exporting markets have traded 'only in sterling'. There is fear of incurring an exchange-rate loss by quoting the currency of the customer 7. British exporters appear to have an inherent fear of European languages 8. An historic difficulty for British exporters in accurately calculating delivered costs
44. T. Bushell (1985)	IM's lesson for industry	One of the biggest problems facing British industry is its failure to recognize that it had to go out and sell its products. Company's reliance on 'old boy network' no longer works

Source: Ughanwa 1986

Chapter two

Price and value

An economist is a man who knows the price of everything and the value of nothing (Oscar Wilde)

Introduction

In the previous chapter we reviewed a cross-section of British industries that once dominated the domestic and many foreign markets but have now become a spent force both at home and overseas. From the case histories and an analysis of over forty specific studies of competitiveness, it became clear that while the precise nature of this concept is intuitively simple, its operational definition is complex and comprises a multiplicity of factors whose importance will vary from situation to situation. However, to assist analysis we categorized the factors into two broad groups — those depending upon price and those that were design and design-related, which we classified as non-price factors.

Contrary to popular belief, derived from a naive understanding of economic theory, price is not the determinant factor in most purchase decisions. Indeed price is only really relevant as a basis for enabling one to discriminate between the *value* of different products in terms of their utility to prospective users. Only when products or services are identical in every respect in the user's judgement will the latter be justified in stating unequivocally that the lowest price represents the best value. Accordingly, this chapter is concerned largely with a consideration of how users evaluate products and services as 'bundles' of attributes — both tangible/objective like speed, output, etc., and less tangible or subjective like quality, after-sales service, etc., in coming to a purchase decision.

The chapter begins with an overview of the theoretical basis of price and non-price competition from which it is concluded that in making a purchase decision, the buyer takes account of both sets of factors. A composite model of buyer behaviour developed by one of the authors (Baker, 1975) is then proposed as a useful framework for analysing purchase decisions and, thereby, developing appropriate marketing strategies.

In order to test the emphasis that firms put on the price and non-price factors, a survey was undertaken of commercially successful firms (Queen's Award winners) and the chapter concludes with a review of the key findings.

The nature of price theory

Until the end of the nineteenth century, economists upheld *price* as the most influential selling factor for any competitive trade, and more importantly, for the success or failure of any new product. However, the industrial revolution brought with it many changes. For example, new systems and techniques were introduced. Transport and communication improved. International trade was enhanced. Many monopolies and cartels were broken up by law, and supply tended to concentrate in the hands of a few sellers (oligopoly). Taken together, these changes resulted in a very different competitive climate from that which had spawned the theory of perfect competition, in which choice decisions are based solely on price considerations.

The pure theory of trade depends on the assumption of perfect competition which states that:

(1) Many independent firms produce a homogeneous product (hence an individual firm cannot influence price).
(2) There are no legal, financial, technical, or other obstacles to entry into the industry.
(3) There are many independent buyers, none of whom can influence price.
(4) Each firm has full knowledge of cost and demand data for both the present and future (economic risk and uncertainty are non-existent).

These four assumptions suggest that perfect competition is associated with conditions of stability, such as constant tastes/preferences, constant factor supplies, unchanging technology, etc. They are also associated with conditions of unsatisfied demand and low discretionary purchasing power under which the greatest good of the greatest number will be achieved through product standardization and pursuit of scale economies.

It should be remembered, however, that the concept of perfect competition is just that — a concept that defines one end of a spectrum of competitive states, the other end of which is anchored by the concept of pure monopoly. As noted, where there is an endemic supply deficiency then the theory of pure competition is correct in pointing out that profits (for the industry) will be maximized by concentration on product standardization and price as the measure of efficiency in production and distribution. This is not to say that some users would not pay more for a supply than the going market price, or that differentiated products are not sold at higher prices. (The theorist would point out that a differentiated product is not homogeneous and so is not competing in the same market!) That said, 'perfect' or price competition offered the optimum theoretical solution to the fundamental economic problem of maximizing satisfaction from the consumption of scarce resources (it also determines the proportion of different products that will be produced) in the early nineteenth century.

Unfortunately, too few students of economics progress beyond the definition of the nature of perfect competition to understand that efficiency will be improved by a degree of concentration when economies of scale come into play, which reduce costs and enable the producer to eliminate less efficient and smaller competitors by using price as a competitive weapon. Ultimately, one might reach the other end of the competitive spectrum and become a monopolist but, because such power might be used against the consumer interest (a reduction in overall satisfaction), most countries have legislated against it. (There are, of course, diseconomies of scale but usually these are insufficient to enable a new entrant to challenge the monopolist on price, given their massive share of the market.)

As a consequence of the anti-monopoly legislation of the nineteenth and twentieth centuries, producers had to find a new basis for competition. While they were forbidden to become monopolists on an industry-wide basis, there was no reason why they should not seek to become monopolists in respect of the demand of a segment of the total demand for the industry's output. To this end, producers sought to differentiate their output in the mind of consumers, most notably in the markets for classical homogeneous products like soap and baking powder through the use of branding and advertising. For more complex products with greater opportunities for physical differentiation (e.g. consumer durables and industrial goods) such objective changes were preferred. By the early 1930s these changes began to be reflected in the articulation of a new theory of *imperfect* competition.

Robinson (1932) was the first to examine critically some of the assumptions of perfect competition. She recognized that the changes brought about by the industrial revolution were a reality and found that the assumption of product homogeneity (availability of products of similar description) was no longer valid. Put another way, she found that differentiated products offered more unique selling propositions than price in competition. This led to her abandonment of the product homogeneity proposition in preference to the theory of monopolistic competition.

In 1933, Chamberlain conducted a study that confirmed Robinson's finding that the basis for the assumption of a product homogeneity in the twentieth century no longer existed (Chamberlain, 1962). In addition, he developed the theory of monopolistic competition (including the concept of product differentiation). This was an attempt to remedy the anomalies of the theories of perfect competition and monopoly, for instance, the idea that the output of all producers was homogeneous, that perfect competition was linked with a non-dynamic or stable economy where tastes, preferences, technology, etc., were presumed constant. The theory, therefore, sought to introduce additional competing factors (non-price factors) to complement existing price factors. These included product attributes, such as performance in operation, reliability, flexibility, safety in use, after-sales service, delivery, etc., most of which are a function of design.

According to Chamberlain (1957), 'the management of individual firms will seek in their control of their market relationships not only to choose price most advantageously but with it to combine the best choice of the product itself in its various qualitative aspects'. He argues that 'the admission of the product as a variable not only adds to the picture an alternative area in which competition may in fact be quite active; it does much more than this; it supplies a powerful new force working against price competition.'

Although Chamberlain (1957) did not make direct reference to 'design' in the discussions of his theory, he does indirectly refer to it when he describes a product as a 'bundle of utilities' that makes one firm win more customers than another, or a customer prefer one supplier to another. This is strongly implied in his subsequent statement, which indicates that a general class of product is differentiated if any significant basis exists for distinguishing the goods of one seller from those of another. Such basis may be real or fancied, so long as it is of any importance whatever to buyers and leads to a preference for one variety of the product over another. Where such differentiation exists, even though it be slight, buyers will be paired with sellers, not at random (as reminiscent of perfect competition) but according to their preferences. This suggests that customers tend to look for 'added value', which apparently makes them prefer one product to another. This appears to be in line with the view expressed by Baily (1978) who argued that although 'price may be the buyer's special concern, price alone means little. It is what you get for the price that matters (i.e. value for money).'

The role of price and non-price factors in competition

Price and non-price factors are in their own rights genuinely competitive, each distinct from the other, and each capable of occurring either along with or without the other.

Abbott (1955) argues that 'when quality (non-price) competition occurs alone (as under conditions of perfect price rigidity), it reveals one form of competition in isolation. Thus quality competition, though competitive is only incompletely so. Price competition', he continues, 'is likewise a form of competition that can occur either along with quality competition or without it. It too is an example of partial competition, for if quality alternatives are lacking a genuinely competitive element is omitted.' It is his contention that 'no product can be considered satisfactory or acceptable in quality unless it stands the test of comparison with other quality alternatives. Even in the 'purely competitive' markets someone, somehow, must undertake the job of quality comparison in order to determine that the competing products are in fact identical.'

If this argument is right, then it is reasonable to propose that the basic point about price and non-price factors in competition is that when products are homogeneous, price is the only rational way in which to discriminate between alternative suppliers. To avoid this lack of control over demand, suppliers seek to differentiate their output so that it is seen as separate from competitive offerings and direct price comparisons are not valid. This supports Shankleman's (1975) study, which shows that 'as internationally traded goods become increasingly differentiated, the demand for them becomes less responsive to price'.

Rothwell *et al.* (1983) contend that 'if two products of similar quality are available, and the price of one product is less than the other, the lower priced product will be preferred by all rational purchasers. Similarly, if two products of similar price and different quality are available, all rational consumers will prefer the better quality product.' This corroborates Lancaster and White's (1977) finding, which states that 'the industrial buyer (consumer) is regarded as a rational being whose purchase decisions rest upon such criteria as : price, quality, delivery, etc.'

On the other hand, McConnel (1970), Gabor (1979), and Berkman and Gibson. (1981) all conclude that price is an indication of quality. McConnel, for instance, shows how consumers exposed to three brands of beer – containing exactly the same brew – perceived the highest-priced brand as the highest quality. The medium-priced brand was only seen as marginally better than the lowest priced. In their study, Berkman and Gibson (1981) confirm that 'price was perceived to be an indication of quality in high ticket items such as stereo sets, and tennis rackets'.

Gabor (1979) argues that 'only seldom, if ever, is better quality not coupled with price above that of lower quality'. He contends that where new products are 'functionally similar or functionally identical', price becomes the determining factor. This implies that where new products are functionally dissimilar or non-identical, then non-price factors will be the major discriminating element. Gabor (1979) points out that 'the reason why housewives use price as a quality indicator is that by and large, their experience suggests that "you get what you pay for"'. This has been confirmed by Labouchere (1984) in his article captioned 'You get what you pay for', which argues that 'a customer will buy an item only if he considers that the benefits from doing so are likely to exceed the costs incurred'.

Since consumers often go for products with better value for money, it follows that they will tend to perceive, and probably choose, products with 'a higher price as better when no differences exist with those carrying a lower price and even when the higher-priced items are objectively inferior'. For example, a higher-priced product may result from the higher costs of labour, raw materials/components, etc., but this may not necessarily indicate that the product is of a superior quality.

One can reason that consumers tend to interpret and respond differentially to both price and non-price factors when taking purchase decisions. Baker (1983) describes this pattern of behavioural response as 'conscious selective perception', which enables a potential customer 'in a multiple choice situation' to decide whether 'to buy, reject or defer' this or that product. He argues that 'while consumers are unable to distinguish between unbranded products (largely influenced by price — a homogeneous situation), they have no such difficulty when brand names are given (a heterogeneous situation often characterised by non-price factors)'. This appears to demonstrate the role of price and non-price factors in competition in terms of homogeneous and heterogeneous situations (i.e. situations when product offerings are of similar description and when they are not, respectively). As Onah

(1977) points out, 'there are so many similar products available to the consumer that they lead to confusion, and make it difficult for the consumer to use his purchasing power judgement'. This confusion tends to increase in a competitive and volatile market as more new products are introduced. Such confusion may lead to uncertainty, which in turn, may delay purchase, invoke outright rejection, cause deferment or the switch of brand loyalty, or may even introduce cognitive dissonance.

This argument raises the following question. What can innovators do to monitor and interpret this buying behavioural phenomenon inherent in consumers to be able to differentiate their products, satisfy the varied needs and wants of consumers, and be competitive? To address this problem, one of the authors (Baker 1975; 1983) has developed a composite model of the purchase-decision process that seeks to combine both objective (price-related) and subjective (non-price-related) factors and also to allow for the mediating effect of the prospective purchaser's own experience (selective perception). For a review of this model, see Figure 2.1.

Baker's composite model of buying behaviour

The elements of the model may be defined as follows:

Figure 2.1 A composite model of buying behaviour

$$P = f[SP, (PC, EC, (T_A - T_D), (E_A - E_D), BR)]$$

Where

P = purchase

f = a function (unspecified) of

SP = selective perception

PC = precipitating circumstances

EC = enabling conditions

T_A = technological advantage

T_D = technological disadvantage

E_A = economic advantage

E_D = economic disadvantage

BR = behavioural response

Source: Baker (1983) *Market Development*, Penguin, London, p. 61

Selective perception (SP)

A customer is often exposed to a staggering variety of items when he goes to the shop to buy, say, a tooth-brush. The chances are that he may go away with just that one item, given that there is no other item to arouse his desire to buy (it is assumed that the customer has money to buy more). Such items are often differentiated by stimulus factors such as colour, size, shape, brand name, etc. SP enables the potential customer to cope with such a situation, or narrow the choices by selecting one or two items among the many items available, depending on his need, interest, value, cognitive set, or what Baker describes as 'preparatory set' (i.e. the idea that customers tend to perceive products in terms of their expectations).

Knowing the customers' needs, wants, interests, values, and their expectations of new products in terms of value for money, are but a few factors that will enable innovators to distinguish their products so that customers can perceive and select them as better than those of their competitors.

Precipitating circumstances (PC)

These are those circumstances or cues that prompt a customer to consider, for example, a replacement of an existing washing machine, a purchase of a new stereo set, a switch from one brand to another, say from Coca Cola to Pepsi or from Kitkat to Penguin biscuits. Dissatisfaction with the existing product, changes in taste or preferences, etc., could cause such circumstances. In the case of a washing machine, for instance, the customer may be dissatisfied with its functional performance, style, colour, appearance, size, safety, or increasing operating costs due to wear and tear.

These are the distinguishing factors to which innovators should address themselves to ensure that the right features are incorporated in new products.

In addition, economic factors such as shortage of supplies, non-competitiveness, or a gap in the market can cause innovators to take positive action to differentiate their products. Thus PC coincides with the interest stage in the buying decision process as it includes those incidents or elements that move the firm from a possibly passive awareness to an active consideration. Thus, breakdown of plant and equipment, a shortage of fabricated or raw materials, components, or sub-assemblies, loss of market share due to price and/or quality differentials, or opportunity to enter new markets are but a few of the factors that might precipitate active consideration of an innovation. All of these factors actively encourage product differentiation.

Enabling conditions (EC)

To purchase, a customer's interest must be activated or triggered off by one or a combination of product attributes such as, price and/or non-price factors (design, quality, reliability, performance in operation, etc.). EC, therefore, appears to encompass all those elements that offer customers/users better value for money. EC are a *sine qua non* of consideration — in their absence evaluation of a new product or process will cease.

Since EC are factors without which consideration customers can neither evaluate nor buy a product, it is proposed that these are factors to which innovators/firms should pay particular attention during new product development, both in terms of product differentiation and effective competition.

Objective factors $(T_A - T_D)$, $(E_A - E_D)$

Rationally, consumers tend to consider the net economic and technical outcome of almost every purchase they make. That is, they tend to do cost/benefit analysis as to the relative advantage or disadvantage of what they buy. This is largely so, not only to ensure value for money, but also to reduce post-purchase anxiety.

Firms and innovators also weigh up the pros and cons of courses of action they take when developing new products. For example, Reid (1984) found that in an attempt to reduce the frequency of accident repairs when developing their 'Fiesta' model, for customers, Ford hypothesized that 'high repair costs reduce an owner's brand loyalty and intent to repurchase'. Based on this proposition, the management decided that 'it is economically viable to invest 20p in the product cost in order to save the customer £1 during the first four years of ownership'. Reid added: 'In this way, it was possible to incorporate maintainability features and designs and justify that they were economically viable, not only to the customer, but that added manufacturing costs could be justified on the basis of long-term repeat sales intent.' This formula has been so successful that it has been applied to the 'Escort' and 'Sierra' models. According to Reid (1984), 'if we want a satisfied customer who will purchase again, we must not alienate him by producing a design which is easy to manufacture and fulfil its design purpose, only to find it is almost impossible to service or economically non-viable to repair.'

It is argued that if innovators know what economic and technical factors their customers consider before purchase, as well as those characteristics that attract repeat buys, then they will be able to incorporate them in new products and thereby differentiate them from the rival products.

Behavioural response (BR)

This is the last activity in this 'sequential process model'. It is the product of all the series of processes already discussed, which implies 'action' taken.

BR only assumes importance when an 'objective' techno-economic analysis still leaves more than one alternative from which to choose. But in most competitive markets, there is often little to choose objectively between alternative offerings, and the buyer will have to make deliberate recourse to subjective value judgements to assist in distinguishing between the various items available.

To the customer, the 'action' may mean buy, reject, or defer the purchase of the product; to innovators, it can imply proceed with development, suspend action, or abort development. In this instance, it is important that the cost/benefit of each course of action is 'weighed up' properly to ensure effective results.

The introduction of this model at this juncture is particularly important because it provides the tool with which new products can be differentiated. For example, if firms know:

What creates product awareness in the market place?
What arouses or triggers off the consumers' desire to purchase?
What factors do consumers consider or evaluate before purchase?
Why do customers/users prefer one product to another, etc?

then they will be able to use these factors to differentiate their products in order to satisfy the customer's needs.

Second, firms are expected, in a dynamic environment, constantly to differentiate their products in order to meet the increasing and varying tastes and preferences of consumers, to maintain or increase market share, and to be competitive. Again, to distinguish such products it is essential first, to know the consumer buying behavioural pattern (what he buys, how he buys, why he buys, etc.). All these are incorporated in the model.

It is suggested, therefore, that the model will not only enable innovators to organize their own thinking towards their overall approach to innovation, but also will enable them to differentiate their products in terms of price and non-price factors.

It is hoped that this background discussion of the nature of price theory and the role of price and non-price factors in competition will provide a base upon which price and non-price factors can now be examined in the light of how they apply to the business world.

Price and non-price factors in practice

The price factor

The price factor comprises sale price, list price, net price after trade-in allowance, leasing, and so on.

Consider Korean and UK clothing goods. Buying experience shows that Korean sportswear, for example, is relatively cheaper than its British-made counterparts. The price advantage can be explained by the lower production costs in Korea. The cost of hiring labour, for instance, is much lower in Korea than in Britain. Raw material is another explanatory factor. Whereas a large proportion of raw material is grown locally in Korea, most of the raw materials used in the UK manufacture are imported. With the combined comparative advantage of labour, raw materials, and of course, efficient machinery, the Korean manufacturer can afford to hold down his price to the disadvantage of the UK producer, particularly when both products are competing in the same market.

To withdraw or 'get out' of the market as a result of the developing countries' cost advantage appears to be a painful decision to take by advanced countries with a strong industrial base, such as Great Britain, which led the industrial revolution and once controlled nearly 50 per cent of the world trade.

Thus, to 'get on' or compete effectively, the UK manufacturer must find an alternative strategy to fight off keep price competition from developing countries such as Korea. Sir Fredrick Catherwood suggested a way of resolving this problem. When addressing an Overseas Trade Board Conference in November 1978 he said, 'To avoid the fierce competition of Third-World goods, British manufacturers should concentrate on the manufacture of higher 'added value' products by developing and marketing high technology products' (Catherwood, 1978). The alternative strategy as suggested in Catherwood's statement can be found in 'concentration on high technology', which stimulates the rapid introduction of new products (due to product obsolescence), which are largely differentiated by the extent of 'added value' they contain.

This reasoning supports Freeman (1978) who argued: 'To improve the competitiveness of British manufactured goods, the emphasis on low price, achieved by high labour productivity and low costs (such as the Korean case), should be redirected towards improving product design and technical quality features, both of which make demand less sensitive to the ordinary forces of competition.' Other examples abound.

In October 1978, the British Institute of Management warned in their report that 'unless the industrialised countries, particularly Great Britain, concentrate on the development of high-technology products, they will suffer from the low wage competition of developing nations, (Korea, Taiwan, etc.) — British companies should specialise in the manufacture of sophisticated high value products' (BIM, 1978). Barratt (1984) sounded a similar warning to purchasing and material managers when he argued that 'unless the UK purchasing and material managers change their priority and practices fast, and abandon the obsession with price, the UK will accelerate its decline into the status of a fourth-rate economic entity'.

It has to be remembered that manufacturers depend largely on buyers for product specifications and requirements, and that if the UK purchasing and material or component managers emphasize 'price' as their major requirement, then UK firms may continue to concentrate on price as a major selling and distinguishing factor for their products. As a consequence of this, British firms may continue to lag behind their foreign counterparts, as other factors abound to compete with price. As Zeid (1981) points out the price factor 'is just one element of the mix determining competitiveness ... and is only relevant in the light of other elements such as reliability, after sales service, and delivery'.

Non-price factors

A product can also be differentiated by non-price factors, which may be divided into two broad categories — economic and technical.

The economic dimension of non-price competition encompasses: delivery dates, after-sales service, availability of parts, education and training facilities for users, depreciation, etc. For example, a well-designed harvesting machine reduces the frequency and cost of after-sale service because of infrequent breakdown of the machine. The rate of replacement of parts is also minimized for the same reason, which in turn reduces the overall maintenance cost. Even when the replacement is frequent, probably because of the nature of operation, a well-designed component tends to fit well into the overall machine hardware and thus makes both replacement and maintenance easier and quicker.

The technical aspect of non-price factors is associated with: performance in operation, reliability, durability, flexibility, ease of use, ease of manufacture, safety in use, aesthetic appeals (attractive appearance, style, colour etc.), ergonomics, and graphics, all of which are functions of design. Technically, therefore, a good design is expected, among other things, to perform well, to be reliable, to be safe to use, and to appeal to a wider segment of the market than a low-sale-price product. This suggests a wider acceptance advantage that design appears to have over price.

As will be seen later in this chapter, it is evident that design is increasingly becoming the primary focus for buying decisions rather than price. Consequently, most products are being distinguished from each other through the process of design, because it appears to satisfy the increasing variety of customer/user requirements more than price.

In his study, Alderson (1965) found that 'behind the acceptance of differentiation are differences in tastes, desires, income, location of buyer, and the uses of commodities'. He argued that these differences were often influenced by 'certain characteristics of the product itself, patented features, trade marks, trade names, peculiarities of the package or container, singularity in quality, design, colour, style'.

Alderson's findings suggest that product differentiation is brought about by the often varying needs and wants of consumers and that these requirements can be satisfied by the product mix — presentation, quality, colour, style, etc., all of which are strongly associated with design. Design, for instance, creates more uses, through improvements in existing products, to satisfy various customer tastes and desires. Alderson's (1965) study also confirmed earlier work by Smith (1956) who found that a 'lack of homogeneity on the demand side may be based upon different customs, desires for variety, or desire for exclusiveness or may arise from basic differences in the user needs'.

Udell (1968) was one of the pioneers to research into the relative importance of non-price factors compared to price. He found that non-price factors, according to the groupings — sales efforts (advertising and promotion), and product efforts (R & D, of which design is a part) were ranked first and second, respectively, while 'price' was ranked third. Again, this stresses the growing importance of non-price factors, especially design over price.

Wentz, Eyrich, and Stevenson (1973) confirmed Udell's (1968) findings when they suggested that 'prior to the twentieth century price was the main instrument of competition and the primary weapon for the destruction of competing firms. Today the product plays the role.' In other words, product design has taken the lead as the major distinguishing factor that tends to make customers (existing and potential) prefer one competitor's product to another when compared to price. Other similar findings exist.

In his study, Thompson (1962) found that 'the two most important factors in marketing are: the product and the ultimate consumer (people) ... the obvious (firm's) objective is to get these two in perfect harmony ... if this situation does not exist which of the two major elements is the easiest to change: product or people?' He attempted to provide an answer by arguing: 'You can change products: it is a comparatively simple matter of decision and cost. You can't change people — but you can influence them ... but seldom if ever cheaply. It is far easier — and thus far more economical — to find out what people want and to supply it than it is to influence them to want what you make.'

Since the firm cannot change people, but can change its product offering, and since the only way to change a product in order to compete effectively in a volatile environment is to increase the added value (value for money) in the product, it is reasonable to suggest that the best alternative for meeting such a crucial requirement is through the process of design. *Design shapes the product.* By improving upon the existing product through redesign, a firm should attempt to incorporate various needs and wants (based on customer specifications), which are absent in the rival's product.

Perhaps, one of the significant differences between 'price' and 'design' seems to be that while price is controllable, design is not. Oligopolists and cartels, for example, can collude to reduce or increase the price of a commodity. In the design process, the situation is different. Given that design is an incremental change, from the existing product to a future product, which it is, let us assume for one moment that market research reveals that in the year 2000, a 'one-man' car will be in great demand. How this car will be designed, developed, and presented to the car market by each competing firm within the motor-car industry, cannot easily be forecast by one's competitors. It is entirely within the portfolio of the designer or design team of the competing firm.

Thus the secret of design appears to lie behind the subjective and creative ability of the designer to search for, discover, and create new products based on customer requirements. By doing so, design not only differentiates the competing products, but also stimulates and satisfies demands, and promotes international competitiveness or technical progressiveness as shown in Figure 2.2.

Figure 2.2 suggests that the secret behind design lies in the subjective and creative ability of the designer or innovator to search for new ideas, to discover and create new products with value for money. In turn, this stimulates and satisfies demand (different needs and wants of customers/users). Consequent upon this, customers tend to switch loyalty to the firm that presents products with added values and this will give the firm a competitive advantage in the international market. For example, Rothwell (1983) found in his study that the British farmers switched their loyalty, and bought foreign-built agricultural machines in preference to the British-made, on the ground of added value (value for money).

Unlike price, therefore, which can be manipulated by the oligopolists, good design dictates the price of a product. According to McTavish (1974), 'If a given product can be established as different from other products of the same general description, and on the assumption that demand for differentiated product exists,

Figure 2.2 The process of achieving international competitiveness through design. *Source:* Ughanwa 1986

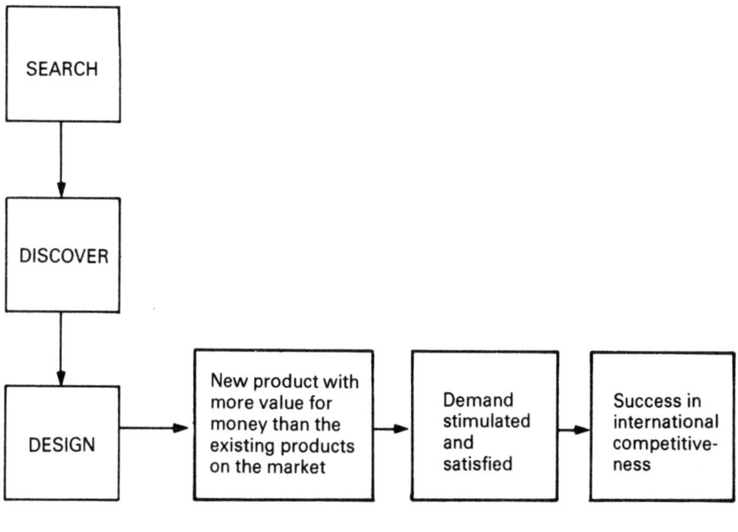

then a degree of price control accrues to the manufacturer in that he possesses at least a temporary monopoly of the distinguishing element.' This implies that until the manufacturer is dispossessed of the monopoly of the differentiating element (added value) that has enabled him to control the price, he remains the market leader — an evidence of success in international competitiveness.

This argument corroborates the NEDO Report (1977) that stated that 'Where oligopolist market structure prevails, however, the size of the margin is controlled by the goals of the dominant firms in the industry or by collusive agreement. Consequently, a firm wishing to grow faster than its competitors and thereby increase its share of the market has to find methods other than price competition to influence the growth of future demand.'

In order to alter the product qualities or to technically differentiate one competitor's product from another, the existing product has to be improved upon through the process of design (that is, by redesigning or reinnovating the existing product, or by designing a totally new product based on customer/user specifications). This will enable the firm to incorporate in its products those values that are wanting in the rivals' products, and are capable of distinguishing the competing products.

Another marked difference between design and price is that there are certain values in a product that design can produce but price cannot. Aesthetic appeal (attractive appearance, style, colour, etc.) for example, is increasingly becoming important in new-product design. Once it is not created in a product, it becomes absent, and no element of price (low/high) can bring it about. Only design can do it. According to Chuck Jordan, Director of Design at General Motors, 'The style of a vehicle, its appearance and its emotional appeal make all the difference. The primary reason for selecting one car over another in the future will be *appearance*, and companies that survive the 1990s will be the ones to give their designers the freedom to step out ahead' (Jordan, 1984).

Figure 2.3 suggests how a good design can influence price and achieve international competitiveness and thereby render dependence on product homogeneity, which is essentially influenced by price, less effective. It is assumed that:

the firm's product can no longer compete on the basis of price;
the firm wishes to continue to sell the existing product in the existing market (market penetration strategy);
there is no change in the environment.

Figure 2.3 Good design can control price and achieve international competitiveness. *Source:* Ughanwa 1986

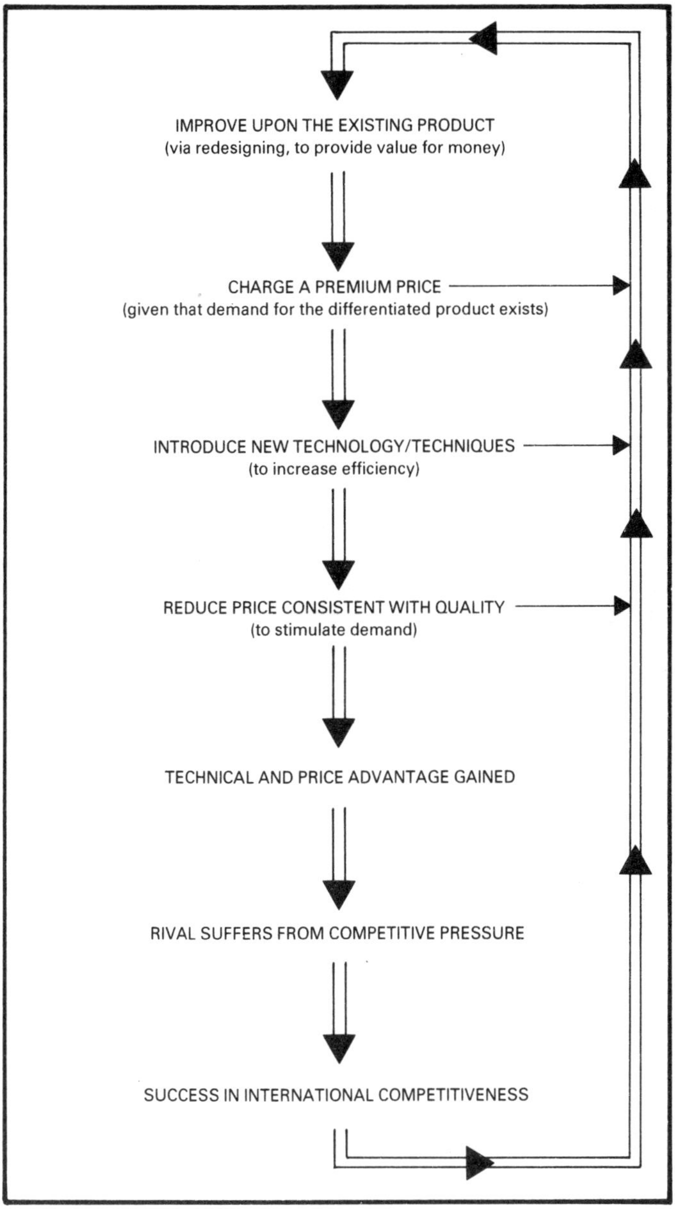

Given these conditions, then the options suggested by the model (Fig. 2.3) may be described as follows.

Improve upon existing product

Since the firm's existing product can no longer compete on price, and since the firm wishes to continue the same business (selling the existing product) in the same market, one can argue that the best available option, is to improve upon, or redesign the existing product to differentiate it from competing products, by incorporating certain added values that are absent in the rivals' products, e.g. Japanese motor cycles as described in Chapter 1.

Able to distinguish its products from those of the competitors, the firm may:

charge a premium, if it is the first to introduce them or has no major challenger in the market, and there is the customers' willingness to buy, or

reduce price, consistent with quality to stimulate demand.

Charge a premium price

Given that the firm is first to market with a well-designed or differentiated product (there is no major competitor) and evidence of customers' willingness to purchase, the firm can 'skim the cream' off the market by charging a premium (additional charge on top of the going price). As a result of the high margin, the firm may be able to increase its cash flow, which can be used for expansion or improvement upon the existing product.

Introduce new technology techniques to increase efficiency

A firm that enjoys an increase in sales is expected, *inter alia*, to look for new technology or techniques to increase efficiency (productivity) and to produce a similar product at an even lower cost (cost advantage) than its competitors.

Thus in the motor cycle industry, Japanese efficiency increased as a result of improvements in production techniques. The overall result of the Japanese strategy was that the productivity improvements resulting from their growth and scale were sufficient both to allow rapidly advancing rates of pay for their employees, and to bring down the real costs and prices of their products in the market place. Comparing the British and the Japanese motor-cycle industries, it could be seen that UK long-term commercial success in fact depended on achieving sales volume at least equal to those of the Japanese and employing equally sophisticated low-cost production methods.

To accomplish sales volume at least equal to those of the Japanese, the British industry would have had to achieve some economies of scale, particularly technical, and to do this, there would have to be a significant improvement in the UK's production techniques to match those of the Japanese. Unfortunately, they were unable to do so and because of the economies of scale that accompanied this mass production, the Japanese were able to use this as a springboard for a highly competitive export drive. The British found it impossible to match the low Japanese price levels on small bikes profitably in the short term. They therefore responded to the Japanese challenge by withdrawing from the smaller bike segments that were being contested.

Reduce price consistent with quality to stimulate demand

In effect this is the strategy followed by the Japanese motor-cycle (and car) manufacturers in that they used the cost advantages created by their manufacturing efficiency both to penetrate overseas markets and win share. In doing so, they also helped to stimulate primary demand.

Technical and price advantage gained

By following the preceding step-by-step strategy, the firm should gain a combined technical/cost and price advantage as a result of improvement (redesign) option, which amounts to 'killing two birds with one stone'. The Japanese motor-cycle industry, for instance, appears to have gained technical cost and price advantage over its competitors, particularly its British counterparts, and thus is now the world leader in the motor-cycle market — an evidence of how good design (a non-price factor) can differentiate a product, enhance perceived value, which is tantamount to a price *decrease*, and achieve international competitiveness.

Rival suffers from competitive pressure

From the foregoing analysis, it can be seen that a firm that enjoys economies of scale as a result of increased efficiency can afford to sell at a reduced price and thus bring competitive pressure to bear on financially 'handicapped' rivals (competitors with insufficient financial resources to introduce and sustain new technology or techniques, because of the heavy capital investment involved). In order to combat effectively the low-cost advantage enjoyed by the developing countries, it is generally accepted that one must innovate by developing a more efficient means of production or a substitute product with more desirable characteristics. In both instances, the capital investment required will usually place such innovation beyond the resources of developing countries (competitors) and so enable the advanced country to maintain a competitive edge.

Consequently, some rivals may withdraw from the market because they can no longer compete. Others may resort to non-price factors, such as, design, but may be too late to catch up as their competitors may have gone steps ahead.

This step-by-step approach will enable a firm to achieve success in international competition. However, the world does not stand still and the firm must repeat the process in order to maintain and sustain international competitiveness.

Non-price competition and its effect on trade

Analysing 'Non-price competition and its effect on trade', Root (1978) commercial: 'Most of the manufactured goods that enter international trade are sold in either oligopolistically or monopolistically competitive markets. In the latter variety of market, each good is produced by a large number of producers, but each producer has succeeded in differentiating the product from competitive products in some way. This differentiation is achieved by the technique of non-price competition, which is the hallmark of monopolistic competition. Although price competition is close and competitor's prices can never be disregarded, it is non-price competition in quality, style, and services that makes up the aggressive front of monopolistic competition.'

He went on to cite instances: 'American consumers encounter non-price competition at every turn. Television, radio, and other advertising media daily assault our eyes and ears with clever appeals to buy this or that product. *Price appeals are relatively rare*, and they are usually drowned out by the host of selling points centred on non-price factors.... *The effect of non-price competition on international trade is expansive.* Non-price competition stimulates demand by acquainting buyers with goods, by transforming latent demand into active demand, by introducing new products and new uses for old products, by heightening the availability of products, and in other ways. ... The dynamism of non-price competition in creating new demand contrasts with the passivity of price competition — accepting demand but doing nothing to change it' (Root, 1978).

Root's assertions summarize the discussion so far about price and non-price factors, and particularly, the confirmation of the growing significance of design in international competitiveness. First, it highlights the major advantages of design and its attributes in international competitiveness (in the end, the ultimate objective of a progressive firm is to compete internationally for otherwise it would have to accept a limit to its growth). Second, it portrays price as a less appealing factor when a customer wants to decide whether or not to purchase this or that product.

According to Barratt (1984), 'If we are to give any consideration priority in our dealing with suppliers, we believe it should be *quality not price.* If quality is understood as 'conformance to specification every time then price will look after itself'. This is yet further evidence of how good design can control price and strongly supports a survey by Rothwell *et al.* (1983) who quoted Duncan Carmichael as saying: 'Today, he who meets the specifications and builds the best product for the lowest cost gets the order.' Rothwell *et al.* (1983) also quoted Peter Horvath of Milford Clothing as summing up his perception of design thus: 'Without design and designing a product accepted by the market and filling a gap in the market, I can't see a future for any one.' Finally, Rothwell (1983) concluded, 'Design matters more than price.'

That said, price remains an important element in the purchase decision. If nothing else, it provides the benchmark against which one evaluates the benefits/satisfactions that will flow from purchase and so enables a comparative assessment between competing, albeit differentiated, offerings. Clearly, 'price' and 'design' are two interrelated and integrated components of something broader, that is, 'value for money'. As a result, each factor should be given as much attention as the other, so that every new product design should reflect its price and in turn, every price should be consistent with the quality of the product.

It will be observed that in most parts of this chapter and, indeed, elsewhere in the book that 'design' is used interchangeably with 'non-price factors'. This may be explained by the fact that other aspects, such as delivery, after-sales service, advertising, etc., even though they are distinct, appear to be closely related to design, and are complementary to successful presentation or introduction of a newly designed product in the marketplace. For example, delivery dates, after-sales service, and so on, are all part of a new-product design package, which can detract from success if not timely and properly executed.

Thus, this review of price and non-price factors appears to suggest that although product design has always been important, the greater awareness of design was created in the marketplace for the following reasons:

(1) as an alternative approach to price competition;

(2) as a method of differentiating competing products;
(3) as a means of providing a wider variety of choices to satisfy the increasing needs and wants of consumers;
(4) as a key factor to ensure that customers/users obtain maximum value for money.

These conclusions strongly support Watt's (1980) findings, which showed that the purpose of 'design' is 'to give the product added value, to differentiate it, from other similar products and, in the final analysis, to sell it'. The question remains, 'What is the appropriate balance between price and non-price factors?' In an attempt to answer this question, one of the authors (Ughanwa 1986) carried out a survey of 'the role of new product design in international competitiveness'.

In sum, the overriding conclusion is that although design (non-price) factors appear to play a more influential role in the sale of new products in international markets when compared to price, design factors (e.g. performance in operation, reliability, etc.) alone cannot maximize the sale of a product without compromising with price. In other words, both price and design (non-price) factors are interrelated and integrated components of something broader, that is *value for money*, suggesting the greater need for a 'balanced design', a design in which price and non-price factors may be adjusted so every new product should reflect its price and every price should be consistent with the quality (design) of the product.

In the following chapter, we shall examine the reasons for the increasing penetration of foreign manufactured imports into the UK.

An empirical survey

If one is to draw sound conclusions on the role of price and non-price factors in competition, it would seem sensible to establish the degree of importance placed upon both factors by commercially successful firms. One group that would seem to meet the 'commercial success' criterion is the winners of the Queen's Award for Technology or Export Achievement. It would seem reasonable to claim that those firms that have received this award must have survived the past recession(s) and competed successfully in international markets on either price or non-price factors.

Accordingly, a sample of 138 Queen's Award winners for technological and export achievement was 'judgementally' selected from a pool of 2,092 winners of the Queen's Award for the period 1966-85, with all the four main regions of the UK (England,

Scotland, Wales, and Northern Ireland) represented. Based upon an extensive review of the literature, a questionnaire was developed and, after pilot testing, administered to the sample companies.

Of the 138 firms contacted, 119 (86 per cent) responded, of which ninety-four (79 per cent) replies were usable. Sixty-three per cent of these usable questionnaires were completed by company directors and the remainder by senior managers.

Factors influencing the sale of products in international markets

Table 2.1 presents the results of the question that asked: 'To what extent has each of the following factors influenced the sale of your products against those of your competitors in international markets? For ease of analysis and in the light of the many variables involved, Table 2.1 is categorized into three groups of influencers;

1. Group 1 influencers – where 75–100 per cent of the respondents rated a factor as influential, with less than 10 per cent stating it was not.
2. Group 2 influencers – where 50–74 per cent of the respondents perceived a factor as influential, but less than 20 per cent oppose this view.
3. Group 3 influencers – where 0–49 per cent of the respondents considered a factor as influential with greater than 20 per cent rating it as not influential.

Group 1 influencers

Among the factors in this group are:

(1) Performance in operation. Of the ninety-one firms who rated this factor, 93 per cent said it was influential, compared to 5 per cent who perceived it as not influential.
(2) Reliability. 86 per cent (77 out of 89) of the respondents considered this factor as influential, as against 3 per cent who said it was not influential.
(3) Sale price. Although 'sale price' was third in order of importance (mean = 1.56), it was the only factor that was considered by 100 per cent (92 out of 92) of the sample, with only two firms (2 per cent) rating it as not influential, in marked contrast, 82 per cent of the sample considered it to be influential.
(4) Efficient delivery. Of the eight-four firms who rated this factor, seventy-one (84 per cent) said it was influential, three firms (3 per cent) felt it was not.

Table 2.1 To what extent has each of the following factors influenced the sale of your products against those of your competitors in international markets?

Factors	*Very influential* 1 %	2 %	3 %	*Not influential at all* 4 %	5 %	*Base* %	*n*	*N*	$\bar{x}$
Performance in operation	71[a]	22	1	3	2	100	91	92	1.42
Reliability	64	22	10	2	1	100	89	92	1.53
Sale price	54	28	18	2	—	100	92	92	1.56
Efficient delivery	46	38	12	2	1	100	84	92	1.73
Quality of after-sales service	41	33	12	11	4	100	85	92	2.02
Technical sophistication	35	36	23	3	2	100	86	92	2.02
Durability	24	36	23	12	6	100	84	92	2.04
Easy to use	38	32	20	8	2	100	85	92	2.05
Safe to use	46	17	24	10	4	100	83	92	2.08
Easy to maintain	23	31	28	5	14	100	80	92	2.56
Parts availability	23	23	27	14	12	100	73	92	2.68
Attractive appearance/shape	17	21	30	22	10	100	81	92	2.86
Flexibility and adaptability in use	13	28	31	15	13	100	78	92	2.87
Advertising and promotion	11	19	39	17	14	100	84	92	3.04
Operator comfort	16	19	24	16	24	100	79	92	3.12
Style/fashion	16	16	20	16	31	100	80	92	3.30

Notes: a. To be read: 71% of the responding firms indicated that performance in operation was very influential in the sale of their products against those of their competitors in international markets rating 1 on a 5-point scale (1 = Very influential to 5 = Not influential at all).
n = the number of respondents who rated each factor
N = the number of firms in the sample who responded to this question
$\bar{x}$ = the mean number attached to each factor
NB: the lower the mean, the more influential the factor.

It is interesting to observe that these four factors belong to one group. Almost without exception, every product is expected to perform well in operation — be it car, washing machine, agricultural implement, fountain pen, or tin-opener. Similarly, these days, no customer wants to buy a washing machine, only to have to call for a maintenance engineer the next month. Thus, a product is expected to be reliable. In other words, firms must live up to the guarantee they offer to customers/users.

Furthermore, a reliable product or a product that performs well in operation is expected to be available for purchase at the right time, in the right place; so it must be promptly delivered to the shop, dealer, distributors, etc., particularly in overseas markets.

Finally, what a customer pays for is not only the physical product but a package of utilities – performance in operation, reliability, efficient delivery, etc. This means that the price of a product must be consistent with the benefits it offers to the customer and vice versa. It is, therefore, not surprising that these four elements belong to the top group of factors influencing the sale of the Queen's award winners' products against those of their competitors in international markets. This implies that innovators should give special attention to performance in operation, reliability, sale price, and efficient delivery when developing new products to enhance international competitiveness.

Group 2 influencers

(1) Quality of after-sales service. 92 per cent (85 out of 92) of the sample responded. Seventy-four per cent (63 out of 85) rated it as influential, 15 per cent did not.
(2) Technical sophistication. Of the eighty-six firms who rated this factor, 71 per cent (61 out of 86) said it was influential, compared to 5 per cent who were opposed to this view.
(3) Durability. Rated by eighty-four firms, of which 60 per cent (50 out of 84) stated it was an influential sale factor, compared to 18 per cent who believed to the contrary.
(4) Easy to use. Of the eighty-five firms who rated this factor, 70 per cent believed it was influential, 10 per cent believed to the contrary.
(5) Safe to use. Rated by eight-three firms. Sixty-three per cent (52 out of 83) perceived it as an influential sale factor, as against 14 per cent who did not consider it as influential.
(6) Easy to maintain. Eight-seven per cent (80 out of 92) of the sample said it was an influential element in the sale of their products in international markets, 19 per cent opposed this view.

The analysis of Group 2 influencers reveals three major elements that appear to influence the sale of the Queen's award winners' products against those of their competitors in international markets. They include:

(a) Services (e.g. quality of after-sales service);
(b) Technical quality (e.g. technical sophistication and durability);
(c) Ergonomic features (e.g. easy to use, safe to use, and easy to maintain).

Again these are interesting findings in the sense that factors in this group, like those in Group 1, appear to interrelate. For example, a customer who buys, say, a sophisticated combine harvester is invariably buying after-sales service at the same time. Even without breaking down, there is still the need for a regular maintenance service of the machine to enhance its performance in operation, as well as durability.

Similarly, a product, even when sophisticated, is expected to be easy to use, safe to use and easy to maintain by the operator.

In sum, the aforementioned findings, suggest that firms should give consideration to services, technical quality, and ergonomic aspects when designing and developing new products.

Group 3 influencers

Group 3 is comprised of the following:

(1) Parts availability. Of the seventy-three firms who responded, 46 per cent (34 out of 73) considered it as influential; in contrast, 2 per cent (19 out of 73) did not see it as an influential factor.
(2) Attractive appearance/shape. Rated by 38 per cent (31 out of 81) of the respondents as influential, compared to 32 per cent (26 out of 81) who said it was not influential.
(3) Flexibility and adaptability in use. Rated by 85 per cent (78 out of 92) of the sample. However, only 41 per cent of the respondents considered it as influential, whilst 28 per cent stated it was not.
(4) Advertising and promotion. Of the eighty-four respondents, 31 per cent (26 out of 84) said it was not influential compared to 30 per cent who believed it was an influential sale factor.
(5) Operator comfort. Although 86 per cent (79 out of 92) of the sample rated this factor, 40 per cent (32 out of 79) felt it was not an influential factor, *vis-á-vis* 35 per cent (28 out of 79) of the respondents who believed to the contrary.
(6) Style/fashion. Eighty firms rated this factor. Forty-seven per cent (38 out of 80) of the respondents said it was not influential, in marked contrast, only 32 per cent (26 out of 80) considered it as an influential factor.

From this analysis, two major influencers of the sale of new products in international markets are drawn to innovators' attention. They are: technical factors, (parts availability, flexibility, and adaptability in use), and aesthetic factors (attractive appearance/shape). If a product, say a car is introduced onto the market without making parts available for repairs and routine maintenance, the car may be useless to the user if it breaks down. Similarly, a product that is flexible in use or that can be adapted to suit different environments may gain a wider international market acceptance than a product that does not possess these qualities. Examples of flexible/adaptable products include: mains and battery radios, shavers, TV sets that can be used in Europe as well as in Africa, and so on.

Further, a product is expected to be attractive to look at and live with. Such attraction can be in the form of appearance, shape, colour, style, etc. In realization of this important role of aesthetic features in a product, manufacturers now produce bulldozers, earth-moving equipment, and the like, in glittering colours, even though they know such products will soil in use.

From Group 3, the following have been identified as the least influential factors in terms of relative importance (i.e. those with mean scores greater than 3): advertising and promotion (mean = 3.04), operator comfort (mean = 3.12), and style/fashion (mean = 3.30). The lower rating given to these factors reflects both upon their higher mean ranks when compared to other factors (see Table 2.1), as well as the 'reverse' trends (more respondents said that the factors are not influential than said they are) exhibited by the rating itself. However, the lower rating given to these factors may be explained by a number of factors.

First, a number of these variables interrelate and overlap. For example, style/fashion relates to attractive appearance/shape. Thus, those who rated appearance/shape may regard this as being subsumed by style/fashion. Second, many of the overall sample are winners of the Queen's award for technological achievement (forty-nine of the awards were for technology and forty-five for exports). Since technological innovation tends to emphasize engineering (as will be argued in Chapter 7), and since engineers appear to be more concerned with functional performance of a product than aesthetic features such as style, fashion, etc., (as discussed in Chapter 5) then it is unsurprising that style/fashion received lower rating. In fact, in discussions with companies producing products like household textiles and carpets, where surface design is a key factor, many senior managers rated style/fashion number one!

As for advertising and promotion, Britain appears to export more than its rivals to non-OECD markets (largely composed of developing countries) as will be seen in Chapter 3. It is therefore assumed that a large proportion of the UK's manufactured products are exported to developing countries where success in advertising and promotion may be limited by such factors as disposable income, in the sense that some target audience may not be able to afford television and radio sets, regular newspapers/magazines, or be able to attend exhibitions where much of the advertising and promotion are carried out, hence less emphasis was placed on advertising and promotion by respondents. This appears to be in line with Zeid's finding, which shows that developing markets (Africa, Asia, the Middle East, South America, the Caribbean, etc.) 'have inefficient infrastructure (i.e. inefficient channels of distribution, mass media, state of technology, and low levels of income) and that such deficiencies may make the needs of customers in developing markets quite different from those of customers in developed markets . . .' (Zeid, 1981).

However, a close look at Table 2.1 p. 84 indicates that the rating of advertising and promotion appears to be normally distributed, strongly suggesting that in spite of the lower rating, advertising and promotion still seem to play an important role in the sale of new products in the marketplace.

Finally, the lower rating given to operator comfort may be explained by the fact that a product that is easy to use, safe to use, and easy to maintain (all of which were rated highly) by the operator, may be perceived as implicitly providing comfort to the operator. Further, the concept may be seen as irrelevant for products that are consumed rather than 'used'.

Conclusion

With the insight from the aforementioned findings, the following conclusions can be drawn:

1. Commercially successful firms tend to attach great importance to design factors (such as performance in operation, reliability, etc.), design-related factors (such as efficient delivery, quality of after-sales service, etc.), and design-influenced factors (such as easy to use, safe to use, etc.), in the sale of their products against those of their competitors in international markets. If so, then this strongly confirms our hypothesis that commercially successful firms are reputed for the use of design, design-related, and design-influenced factors to achieve international competitiveness.
2. Technical qualities contribute significantly to the sale of new products in international markets. This is attested to by the great importance attached to performance in operation, reli-

ability, technical sophistication, durability, etc. (see Table 2.1). Therefore, our hypothesis that technical quality plays a significant role in determining the competitiveness of commercially progressive firms is confirmed.

3. Ergonomic and aesthetic features are very necessary ingredients in the success of manufactured exports.
4. Price appears to play a lesser role in the sale of new products in international markets when compared to design factors. This appears to strengthen our argument that although price is important, good design tends to control price (see Figure 2.3). This is based on the premise that if you get the design of a new product right, it is almost certain that its price will be right, given that there is no 'judgemental error'* in pricing; hence Gabor (1979) argues that price is often used as an indicator of quality, as alluded to earlier.

   Further, this conclusion strongly corroborates the conclusion of earlier studies. According to NEDO (1977) and EDC (1968), 'The engineering design, performance, reliability of delivery, and after-sales service are usually more important in meeting foreign competition than price.' Similarly, Baker (1979b) maintains that 'price is but one dimension of the purchasing decision (or the rejection of a proposition/deferment of judgement) and is only relevant in the context of other parameters such as peformance, reliability and after-sales service.'
5. The overall findings appear to suggest that given the choice, firms will rationally select 'performance in operation' and 'reliability' (composite factors) before considering other factors or before doing cost/benefit analysis as to what other factors to consider (e.g. price, delivery, after-sales service, safe to use, etc.). If the foregoing argument carries conviction, then the practicality and applicability of 'Baker's Composite Model of Buying Behaviour' (see Figure 2.1) seems to be strongly supported (Baker 1975). According to many buying behavioural models, technology, performance, and the economics of cost/benefit of a purchase are at the very heart of the Marshallian and 'rational' schools of buying behaviour. All of these factors are implicit in the $(T_A - T_D)$, $(E_A - E_D)$ aspect of Baker's model. These technological and economic advantages and disadvantages may be explained in part, by the fact that people do weigh up the

* According to J.V.Osola (Consultant), 'Too often a new product is underpriced because the assessment of the value to the buyer is faulty' (Osola 1984).

pros and cons of courses of action and more so because if one is going to use a model such as this, then one should first come to grips with the situation specifics, that is, one should specify as fully as possible what these advantages and disadvantages are, weight these if necessary, and only then can one come up with an overall judgement as to how the new product measures up against the competitive product offerings. If so, then this finding appears to negate the criticism made by Foxall (1980) that Baker's model (alongside such other buying behavioural models as Nicosia, Howard and Sheth, and Engel, Blackwell, and Kollat) is 'inadequate' in that it is perceived as 'pre-scientific' in the sense that it is 'untestable'. The preceding evidence shows that this model is practical and testable. The testability of this model will be demonstrated in the subsequent analyses.

6. Finally, the fact that some of the factors clustered in terms of 'frequency' suggests, first, that these factors interrelate. Second, it shows how difficult it is, for one factor alone to influence the sale of a product. This implies that it is a combination of these factors that actually influenced the sale of one product against the other.

Having established the nature of the factors that influence the nature of competitiveness, we turn in the next chapter to look at the impact that the interplay between them has had on the UK market in recent years.

Chapter three

Taking stock

Introduction

Thus far we have examined a number of case histories of declining competitiveness in a cross-section of industries in which Britain formerly was acknowledged as a world leader, reinforced by the evidence from over forty studies of particular instances of deteriorating performance. Analysis of the reasons underlying this deterioration indicates that 'performance' or success in the market-place, is a consequence of a combination of both price and non-price factors.

Recognition of this is contained in the composite model of buyer behaviour, which proposes that a purchase decision is the outcome of a process in which an individual or organization recognizes an unsatisfied need, searches for a means of satisfying it, evaluates carefully the objective performance and cost-benefit characteristics of possible solutions and takes a decision. However, in doing so the model indicates that the whole process is moderated by the selective perception of the decision-maker, which means that the information selected for consideration and the criteria used to evaluate it are mediated by the beliefs, attitudes, and past experience of the decision-maker. In this sense all the data reviewed is interpreted subjectively from the decision-maker's perspective.

It is for this reason that we argued that a price is really only an indicator of a much more complex concept of value. If the decision-maker's evaluation leads to the conclusion that one alternative is superior to all others then his behavioural response (BR in the model) is simple — he adopts it. We thus can have a position in which a disinterested third party may perceive competition between homogeneous products, which economists would classify as perfect substitutes (e.g. baked beans, detergents, lubricating oil, steel sheet, etc.), but the propsective buyer perceives one offering as superior for subjective reasons (they taste better, wash cleaner, afford better protection, give fewer rejects when formed). (One author has direct experience of the latter two situations where tests proved the preferred product was identical to the non-preferred product in technical terms but the way in which the user handled them varied, giving rise to the so-called difference).

Our analysis of the components of value suggested a simple distinction between price and non-price factors and the conclusion that many non-price factors are design-related. These were discussed in some detail. In this chapter we develop the argument further by looking at patterns of international trade between the UK and other countries. The chapter itself contains four sections. In the first we take a broad look at UK exports and imports of a wide range of manufactured goods that are traded internationally and then compare these figures with those of major competitors. Our objective here is to try and establish the reasons why the UK imports 'dear' and exports 'cheap'.

In the second section we look at the consequences of this pattern of trade (it's hardly a policy!) and its costs in terms of revenue and jobs. In the process it will become clear that UK customers have no intrinsic preference for foreign products and often would prefer to 'Buy British'. Enlightened self-interest and the need to remain efficient and cost-effective in their own activities lead to preference for imported products and we examine a number of specific industries (machine tools, textiles, and agricultural machinery) in the third section.

The fourth section leads to the inevitable conclusion that the reason the UK imports dear and exports cheap is that in most cases, its products are technically inferior (less reliable, less durable, offer less performance in operation, are ergonomically unsatisfactory and aesthetically less appealing) when compared to rival foreign offerings. No matter how chauvinist the buyer (selective perception), the objective evidence is too great to be ignored.

Import dear — export cheap

Available evidence suggests that the UK imports higher value products, on which rival producers have the advantage of charging a premium, but exports lower value equivalents, which may be sold at a cheaper price to overseas markets; hence it is hypothesized that the UK imports 'dear' and exports 'cheap' manufactured products. For example, Rothwell (1979) found that 'the UK has a marked tendency to 'export cheap' and 'import dear' in mechanical engineering product'. This strongly supports Baker's (1979c) contention that 'Britain is rapidly becoming a country which exports cheap unsophisticated products to developing countries, while depending upon expensive sophisticated imports to maintain its own production efficiency.'

Similarly, NEDO (1977) maintains that 'there are signs that the UK, by comparison with France and particularly West Germany, tends to export relatively low-valued products particularly within engineering sectors'. Again, this suggests that the value of foreign imports into the UK is higher than that of the equivalent UK-manufactured exports.

The general purpose of this section, therefore, is to find out why the UK imports higher value (expensive) foreign products but exports lower value (cheap) ones to overseas markets. Specifically, this section is designed to determine the relationship between the rapidly increasing import penetration into the UK and the rapidly eroding shares of the UK's world manufactured exports. In turn, this will involve an examination of various techniques, such as value per ton, unit value, patent activities, and R & D expenditure, often used as rough indicators of trading success. Customer's points of view as to why they prefer foreign to home-made products will be reviewed, and finally the section summarized.

Values per ton

Table 3.1 has been compiled by NEDO (1977) to compare the average value per ton of West German and French manufactured exports to the equivalent UK exports.

Columns 1 and 2 (Table 3.1) represent the ratio of the average value per ton of West German and French exports to the corresponding figure of the UK exports. The figures are expressed as index numbers with UK unit values set equal to 100. In other words, any figure in excess of 100 in columns 1 and 2 means that the average value per ton of the West German or French exports is higher than that of the UK in the relevant product category, and vice versa.

As Table 3.1 (column 1) illustrates, the West German exports have higher unit value per ton than those of the UK in twenty-nine (82.8 per cent) of thirty-five product sectors surveyed by the sector working parties (SWP), especially in the mechanical engineering sector, meaning that the UK has higher value per ton of exports than West Germany in only six product groups. Similarly, the average value per ton of French manufactured exports is higher

Table 3.1 International comparisons of unit value 1974

Rank: Sector working party	*Unit values of German and French Exports (of UK unit value = 100)*		*Ratio of import unit value to export unit values*		
	West Germany	*France*	*UK*	*West Germany*	*France*
	(1)	(2)	(3)	(4)	(5)
1 Construction equipment/ mining machinery	276	244	2.11	0.83	0.96
2 Office machinery	255	—	1.62	1.46	—
3 Telecommunications	222	167	1.39	1.03	0.79
4 Machine tools	199	153	1.53	0.74	1.04
5 Food and drink packaging equipment	192	123	1.84	0.89	1.13
6 Mechanical handling equipment	188	121	1.07	0.64	1.23
7 Pharmaceuticals	180	100	1.60	1.13	3.75
8 Heavy electrical machinery/ industrial equipment	176	123	1.63	0.83	1.25
9 Pumps and valves	170	108	1.55	0.73	1.10
10 Brewing and malting	168	109	0.82	0.59	0.79
11 Automation and instrumentation	147	152	1.73	1.26	1.15
12 Electronic consumer goods	146	133	1.21	0.81	1.14
13 Petrochemicals/speciality chemicals	145	125	1.47	0.72	1.13
14 Biscuits	142	102	1.14	0.64	1.03
15 Bearings	139	110	1.28	0.74	0.98
16 Process plant fabrications	139	92	1.90	0.89	1.20
17 Industrial trucks	138	98	1.10	0.67	1.09
18 Industrial engines	134	120	0.91	0.67	0.93
19 Textile machinery	134	101	1.11	0.66	1.05
20 Man-made fibres	130	127	1.12	0.85	0.96
21 Meat and meat products	130	94	1.17	1.03	1.99
22 Constructional steel	126	114	1.27	0.72	0.81
23 Clothing	122	127	0.61	0.59	0.57
24 Iron and steel	122	109	0.91	0.87	0.94
25 Rubber	117	106	1.14	0.85	0.92
26 Domestic electrical equipment	110	117	0.96	0.76	0.73
27 Printing machinery	110	92	1.16	0.91	1.23
28 Ferrous foundries/ drop forgings	108	70	1.57	0.74	1.89
29 Wool textiles	103	86	0.37	0.83	0.61
30 Hosiery & knitwear	100	143	0.93	1.13	0.62
31 Electronic components	92	96	1.06	1.43	1.06
32 Milk & milk products	89	112	1.85	1.97	1.88
33 Paper and board	84	82	0.48	0.63	0.71

34 Space heating and ventilating machinery	72	63	0.98	0.83	0.95
35 Radio, radar and electronic goods	43	87	1.00	1.40	0.44

Source: NEDO, (1977) 'International price competitiveness, non-price factors and export performance', NEDO, London, p. 28.

than those of the UK in twenty-three (67.6 per cent) of the thirty-four product groups investigated (see Table 3.1, column 2), suggesting that the UK is more competitive than France in terms of value per ton of exports in only eleven of the thirty-four sectors.

According to the NEDO Report (1977), 'The UK share of world trade fell in all but three of the 35 product groups over the period 1970 to 1974. 'In contrast', the report continues, 'Germany increased its share in 31 of them over the same period and France increased its share in 27 of the 35 sectors.' It would seem that both the West German- and French-manufactured exports are technically superior to those of their UK counterparts.

However, the use of the 'value per ton' technique as an indicator of high-quality goods may be criticized on the grounds that the differences between the lower value per ton of UK exports and the higher value per ton of both the West German and French exports may be explained by factors other than quality differentials.

First, the statistical data in Table 3.1 appear to be aggregated. For instance, the products are grouped, such as, construction equipment and mining machinery; radio, radar, and electronic goods; etc. As a result, it is difficult to pinpoint any singular factor as being the cause of the relatively higher or lower value per ton of exports.

Second, Britain may have lower prices for higher value goods similar to those of France and West Germany due to economies of scale (that is without necessarily reducing the product quality or the value content of the manufactured exports). For example, much Japanese electronic equipment, say, stereo sets, imported into the UK is in most cases cheaper, yet has much higher value than most of the equivalent UK exports.

Third, the UK's cheap exports could result from possible weakness in marketing and salesmanship. For instance, NEDO (1977) states that 'The BETRO report documented the (UK's) weakness of salesmanship at great length.'

Last but not least, Osola maintains that there is a considerable feeling that British firms are 'usually far too timid' at pricing. He argues that 'Too often a new product is underpriced because the assessment of value to the buyer is faulty.'

However, as Table 3.1 strongly suggests, the differences in unit values in many product groups are 'so large and so widespread' that one can reasonably suppose that they are likely to represent differences in design or product quality between British manufactured exports and those of the French or West German, particularly in the mechanically engineered product sector. One can then conclude that since the higher unit value per ton of manufactured exports appears to signify high quality which, in turn, is a measure of export success, it follows that the UK's poor performance in these thirty-five product groups is as a result of its inability to compete with its West German and French rivals on technical quality, product superiority, and uniqueness, all of which are largely influenced by design.

Unit value

Unit value is another technique used by NEDO to compare the UK's trade performance with those of its major competitors (West Germany and France) between 1970 and 1974. Unit value ratio is derived by dividing a country's unit value of imports by the unit value of its exports, expressed in formula thus:

$$\frac{\text{unit value of imports}}{\text{unit value of exports}}$$

Table 3.1 (columns 3, 4, and 5) indicates that any ratio greater than 1.00 suggests that 'the country imports products of a higher unit value than the product it exports in the group'. As Table 3.1 (column 3) demonstrates, of the thirty-five product sectors, the UK has ratios in excess of 1.00 in twenty-five (71.4 per cent), compared to only nine (25.7 per cent) for West Germany, and eighteen (51.4 per cent) for France, during the same period. Once again, the UK appears to lag behind its major rivals, especially in the mechanical-engineering sector. According to NEDO (1977) 'Of the 12 mechanical engineering sectors listed, the UK imported goods with a higher unit value than it exported in 11 (91.6 per cent). In the case of West Germany only one product group fell into this category.'

Simple arithmetic shows that the West German construction equipment/mining machinery (Table 3.1, column 1, row 1) imported into the UK between 1970 and 1974, was about two and a half times more expensive per ton and thus by implication, more competitive than equivalent products exported to West Germany by the UK during the corresponding period. On the other hand, the French exports to Britain were about twice as costly as similar products (per ton) imported into France from the UK, for the same period.

This analysis strongly supports Rothwell's (1976) finding, which shows that 'in the great majority of cases, West German products are more expensive than their UK equivalents: significantly, they are more competitive'. Taking 'unit value' as a rough indicator of technical sophistication, Rothwell argues that the higher unit value of the West German mechanical-engineering products, including textile machinery, 'would appear to indicate a positive relationship between this (i.e. technical sophistication) and export competitiveness'. If so, then Britain is clearly less competitive than West Germany and other major competitors in design, technical excellence, and higher value for money, particularly in the mechanical-engineering industry. This may help to explain why the UK imports 'dear' and exports 'cheap' manufactured products.

The fork-lift trucks industry

Rothwell (1979a) drew up Table 3.2 to compare the trade performance between the UK and West Germany in the fork-lift trucks industry, in terms of imports expressed as a percentage of exports and the ratio of value per ton of exports to the value per ton of imports to all countries between 1970 and 1973. He found that 'while the value per ton of UK exports was consistently lower than that of imports, the value per ton of West German exports was, in contrast, consistently and substantially, higher than that of

Table 3.2 Fork-lift trucks: imports, exports, and value per ton, UK and West Germany

Imports as a percentage of exports to all countries 1970–73 (value)				
	1970	1971	1972	1973
United Kingdom	33.5	26.1	30.6	41.1
West Germany	52.8	45.7	33.1	37.3
Value per ton of exports to all countries 1970–73				
	1970	1971	1972	1973
United Kingdom	0.88	0.82	0.96	0.87
West Germany	1.37	1.45	1.51	1.51

Source: Senker, P., *et al.*, 'Forklift trucks: a study of a sector of the UK engineering industry', Science Policy Research Unit, University of Sussex, 1977.

imports; a positive relationship between high value per ton and export success thus appears to be indicated.' He also found that 'During this period (1970–1973), UK imports rose much faster than exports: West Germany's performance was substantially better, with exports rising faster than imports.'

This finding suggests that West German fork-lift trucks are technically superior to those of its UK counterpart, hence they have higher values. It also suggests that whereas the West German users seem to be satisfied with the overall performance and reliability of home-built fork-lift trucks (hence fewer imports), their UK counterparts are not, hence they tend to prefer foreign-built fork-lift trucks with higher value-satisfying characteristics than domestically produced ones.

This pattern may be summarized by the following import/export matrix (see Figure 3.1).

Figure 3.1 Import/export matrix

Import/Export	UK	WEST GERMANY	
Import	Expensive	Cheap	L
Export	Cheap	Expensive	H
	L	H	

Note: H = High value (expensive), L = Low value (cheap)

It can be argued that since higher value (expensive) products are more competitive, then there seems to be a link between expensive products and export competitiveness. It is assumed that all expensive products have higher added value and thus are more competitive. According to Gabor (1979) 'only seldom, if ever, is better quality (higher value) not coupled with a price above that of the lower quality (lower value) item.'

The electrical power-tools industry

Drawing on Walker's (1976) study, Rothwell (1979) also compared the UK's value per ton of manufactured exports with those of West Germany, Austria, and Switzerland in the electrical power-tools industry, as shown in Table 3.3. Once again, the UK is seen to lag behind its major competitors.

Table 3.3 indicates that the UK's percentage share of the world export trade is almost halved in 1972/3 — from 16.5 per cent in 1963/4 to 8.3 per cent in the 1972/3. On the other hand, the

Table 3.3 Electrical portable power tools: world export trade share and value per unit weight of exports

	% Share of world export trade 1963/4	1968/9	1972/3
United Kingdom	16.50	13.08	8.39
West Germany	32.68	36.13	39.19
Austria/Switzerland	9.26	14.14	18.19
	Value ($000) per ton of exports		
United Kingdom	4.12	3.38	4.37
West Germany	7.56	7.95	11.66
Switzerland	7.40	8.99	13.50

Source: Walker, (1976), 'International trade in portable power tools', Science & Policy Research Unit, University of Sussex, Brighton.

value per ton of the UK's exports is the lowest when compared to those of its rivals for the same period.

As Walker (1976) puts it: 'Even after making allowance for shifts in exchange rates and different factor costs, UK value to weight ratios lag behind those of our main continental competitors by a substantial margin. If one believes that there is a direct link between value to weight ratios and technical sophistication, this would lead to the conclusion that UK exports are in general relatively unsophisticated ... It can hardly be a coincidence that the two countries with the highest value to weight ratios in electrical portable power tools trade – West Germany and Switzerland – have also peformed the best in world markets.'

The agricultural-machinery industry

Comparing bilateral trade between West Germany and the UK in the agricultural machinery sector, (see Table 3.4), Rothwell (1981) points out that 'between 1964 and 1974 the UK's trade balance with West Germany declined drastically'. Table 3.4, for instance, shows that West Germany has consistently performed better than the UK in export trade. In 1964, the UK imported agricultural machines worth about £3.8 million from West Germany, but exported only £1.5 million worth of equivalent products to West Germany (less than half the West German exports to the UK). Similarly, in 1974, the UK exported about £3.9 million worth of agricultural machines to West Germany; in contrast, it imported about five times this figure from West Germany during the same period, indicating that many UK users preferred West German-built to British-made agricultural machines.

Table 3.4 UK trade with West Germany in agricultural machinery (1964 and 1974)

	1964 *Value (£000)*	*units*	*1974* *Value (£000)*	*units*	*Unit value 1974 (£1,000)*
UK exports to W Germany	1,594	—	2,764	3,903	0.71
UK imports from W Germany	3,825	—	20,764	19,591	1.06
Ratio, UK exports WG/UK imports from W Germany	0.24	—	0.13	—	UV ratio/0.67

Source: Rothwell, (1981), 'Non-price factors in the export competitiveness of agricultural engineering products', *Research Policy,* vol. 10, p. 280.

A comparison of unit-value ratios (Table 3.4) also shows that the West German unit value was much higher (1.06) than that of the UK (0.71), suggesting that the West German agricultural machines were of superior quality when compared with their UK counterparts. This, in turn, helps to explain why the UK imported more West German agricultural machines than it exported to West Germany between 1964 and 1974. As revealed in Rothwell's (1981a) finding, 'the contention that technical quality or sophistication is an extremely important ingredient in determining competitiveness of agricultural machinery' is strongly supported by the 'opinions of both machinery users and producers in the UK'. This is consistent with an earlier study carried out by the Department of Trade and Industry (1978b), which found that, 'almost all dealers questioned on the success of imported (agricultural) machines agreed that ... foreign manufacturers were innovatory, their machines being more in line with contemporary farming requirements.' Clearly, West German agricultural machines are perceived as being more modern than their British rivals.

UK vs West Germany in OECD and Non-OECD markets

Rothwell also compares UK and West German exports of agricultural machines to OECD and non-OECD countries in 1971 and 1975 as shown in Table 3.5.

As Table 3.5 indicates, whilst the UK exported 41 per cent of its agricultural machines to non-OECD countries in 1971, the West German export to non-OECD nations was only 15.7 per cent. Again, in 1975, the UK appears to have performed better than West Germany, having exported 44 per cent of its agricultural machines to non-OECD countries compared to 25.6 per cent for West Germany for the corresponding period. Thus, as far as

Table 3.5 UK and West German exports of agricultural machines to OECD[a] and non-OECD[b] countries, 1971 and 1975

Exports	*$ × 10³ OECD*	*$10³ non-OECD*	*$ × 10³ Total*	*Increase to OECD value*	%	*Increase to non-OECD value*	%	*Increase in total exports*	%	*% of exports to non-OECD*
UK										
1971	39,036	27,158	66,194	18,011	46	17,832	66	35,843	54	41
1975	57,047	44,999	102,037							44
W Germany										
1971	138,191	25,748	163,939	217,457	157	96,739	376	314,196	191	15.7
1975	355,648	122,487	478,135							25.6

Note: The percentage of UK and W. German exports of all mechanical engineering products to non-OECD countries in 1971 and 1975 were:

	UK	*West Germany*
1971	43.9	25.5
1975	45.6	38.0

Note: a. OECD (Organization for Economic Co-operation and Development) was set up under a convention signed in Paris on 14 December, 1960. As per 1971 statistics, the members of OECD were: Austria, Belgium, Canada, Denmark, Finland, France, West Germany, Iceland, Ireland, Italy, Japan, Luxembourg, the Netherlands, Norway, Portugal, Spain, Sweden, Switzerland, Turkey, the UK, and the USA.
b. Non-OECD countries mean non members of OECD, especially developing countries.

Source: Rothwell (1981) 'Non-price factors in the export competitiveness of agricultural engineering products', *Research Policy*, vol. 10, p. 281.

exports of agricultural machinery to non-OECD countries are concerned, West Germany clearly lags behind Britain.

However, when it comes to exports of agricultural machines to OECD countries, West Germany performs better than the UK. For example, for every one agricultural machine shipped for sale to OECD countries by the UK in 1971, West Germany exported about four equivalent machines.

In 1975, the situation appears to have worsened as exports increased to a ratio of about 6:1, that is, for every one agricultural machine exported to OECD markets by the UK, its West German rival exported six times that number during the same period (see Table 3.5). Why did the UK perform better than West Germany in non-OECD markets but lag behind its West German competitors in OECD markets? Previous studies have made several attempts to answer this question.

Rothwell (1981a) argues that the 'high percentage of non-OECD exports can be interpreted as a sign of technological backwardness: i.e. it represents the exports of relatively unsophisticated goods to less developed countries.' He maintains that 'Previous work at SPRU has shown that for historical reasons, i.e. the existence of captive Empire markets — many UK firms have strong traditions of exporting to Empire markets where demand has been largely for less sophisticated goods.' (For details, see Walker (1980) 'Britain's industrial performance 1850–1950: a failure to adjust', in K. Pavritt, (ed.) *Technical innovation and British economic performance*, Macmillan, London.) He asserts that 'This has placed them at a disadvantage in more advanced OECD markets. Thus, a high percentage of exports to non-OECD countries is not, normally, the result of an explicit marketing policy, but rather represents a historical pattern of exporting goods with a relatively low technical content.' This 'low technical content' of British products has been well documented in two separate studies conducted by Nagashima 1970 (see Tables 3.6 and 3.7).

Tables 3.6 and 3.7 speak for themselves. Table 3.6, for example, appears to indicate that all but one (automobiles) are products of 'low technical content' that first come to the mind of both Japanese and US businessmen when they see 'Made In UK' labels. Compare these with technically sophisticated West German products such as machinery and tools (which includes agricultural machinery), medical equipment, cameras, and the like. It can hardly be a coincidence that both American and Japanese businessmen perceive 'Made In UK' as products of low technical content.

In 1975, Nagashima carried out a similar survey, which virtually

Table 3.6 Please list the product names which come first to your mind when you see the following 'Made In' names

'Made In'	*Products associated with concepts*	
	Japanese businessmen	*US businessmen*
West Germany	1. Automobiles 2. Precision machinery 3. Razors 4. Medical & pharmaceutical products 5. Machinery & tools 6. Cameras & optical goods	1. Automobiles 2. Machinery & tools 3. Electrical products 4. Cameras & optical products 5. Toys and sporting goods
England (UK)	1. Cloth 2. Scotch 3. Automobiles 4. Textiles & woollens 5. Confections	1. Automobiles 2. Woollens & textiles 3. China & silverware 4. Whisky 5. Leather & leather goods

Source: Adapted from Nagashima (1970) 'A comparison of Japanese and US attitudes toward foreign products', *Journal of Marketing*, vol. 34, January, p. 73.

Table 3.7 Please list the product names which come first to your mind when you see the following 'Made In' labels

'Made In'	*1967*	*'Made In'*	*1975*
Germany	1. Automobiles 2. Precision machinery 3. Razors 4. Medical & pharma-ceutical products 5. Machinery & tools 6. Cameras & optical goods	Germany	1. Automobiles 2. Razors 3. Precision machinery (watches) 4. Medical & pharma-ceutical products 5. Machinery & tools 6. Cameras & optical goods
England (UK)	1. Cloth 2. Scotch 3. Automobiles 4. Textiles & woollens 5. Confections	England (UK)	1. Textiles 2. Whisky 3. Black tea 4. Automobiles 5. Lighters

Source: Adapted from Nagashima (1977) 'A comparative "Made In" products image survey among Japanese businessmen', *Journal of Marketing*, vol. 43 July, p. 98.

confirmed his 1977 study (see Table 3.7).

Baker (1979c), in his study, contends that the reason many UK firms export mainly to overseas markets, notably the developing countries is, amongst other things, to get rid of declining products that are obsolescent in the context of competition in the home market. Because the UK is not up to date in modern technology, in the agricultural machinery sector, its existing machines are seen to be rendered obsolescent by comparison with rival machines with higher technological content, such as those from West Germany. It also suggests that the UK produces 'cheap' agricultural machines, hence they are targeted at Third-World countries where disposable income is usually very low. If such an argument carries conviction, then the UK's larger export share of the non-OECD market appears to be the result of sales of low-value items, not as a consequence of technical sophistication or design excellence as reflected in the West German agricultural machinery. This shows itself in Table 3.5, which indicates that the percentage value of the UK exports to both OECD and non-OECD countries increased by only 46 per cent and 66 per cent respectively between 1971 and 1975, compared to 157 per cent and 376 per cent for West Germany for the corresponding periods. This indicates that the UK's manufactured exports to OECD markets have deteriorated in value added by 111 per cent, ostensibly indicating that the UK's agricultural machines are inferior to those of West Germany. Similarly, the percentage increase in value of West German agricultural machines exported to non-OECD markets its notably higher than that of the UK by 310 per cent between 1971 and 1975 (see Table 3.5).

However, a striking result given in Table 3.5 also reveals that West Germany is rapidly making in-roads into non-OECD markets that have historically been dominated by the UK. For example, in 1971, the UK's export share of all mechanical engineering products to non-OECD countries was 43.9 per cent compared to 25.5 per cent for West Germany (a difference of 18.4 per cent); but by 1975, West Germany was only 7.6 per cent short of catching up with the UK in non-OECD markets. As Rothwell (1981a) describes it, 'with the recession in Western Europe, and general stagnation in OECD markets, West German exporters have, while maintaining their share of OECD markets, increasingly turned their attention to markets which have long been the preserve of UK machinery producers.' This suggests that the UK is seemingly lagging in two fronts – in OECD markets, West German manufacturers lead their UK counterparts; and in non-OECD countries, West Germany appears to be rapidly catching up with the UK in the export share of agricultural machinery.

Rothwell concludes, 'If the West German agricultural machinery industry is as successful in these markets (i.e. non-OECD) as it has been in the markets of the West (i.e. OECD), then the future facing much of the UK industry would seem to be, to say the least, rather bleak' (Rothwell, 1981).

Next, (see Table 3.8) Rothwell compares the UK and West Germany in terms of total engineering exports to OECD countries, total value of all engineering exports, and unit value of exports between 1971 and 1975.

Table 3.8 shows that the West German shares of OECD engineering exports were more than double those of the UK in 1971 and 1975. Moreover, whereas the UK export shares decreased from 10.9 per cent in 1971 to 9.2 per cent in 1975, the West German shares marginally increased from 22 per cent in 1971 to 22.09 per cent in 1975.

In terms of total value of engineering exports, it can be seen that the West German export shares are again more than double those of the UK. In 1971, for example, the value of West German engineering manufactured exports was $12.4 billion compared to $6.2 billion for the UK. Similarly, in 1975, the total value of West German engineering exports was $29.2 billion compared to $12.3 billion for the UK.

It is also noticeable that the unit value of West German engineering exports is much higher than that of the UK in spite of the strong German Deutschmark, which could have given Britain some export advantage (see Table 3.8). It seems reasonable to assume, with regard to the unit value, that both countries operate over the same range of engineering products with no marked specializations to ensure that like is compared with like and to prevent possible 'unit value' differences that may arise as a result of, say, West Germany specializing in harvester machines while the UK specializes in ploughs, in the agricultural sector.

The fact that the total value of West German engineering

Table 3.8 UK and West German engineering export market shares, 1971 and 1975[a]

	Year	*UK*	*W Germany*
Share of OECD engineering exports	1971	10.9%	22.0%
	1975	9.2%	22.09%
Value of engineering exports	1971	$6.2 bn	$12.4 bn
	1975	$12.3 bn	$29.2 bn
Unit value	1975	$5,710	$8,160
Index of effective exchange rates (1971=100)		78	123

Source: Rothwell (1981) 'Non-price factors in the export competitiveness of agricultural engineering products', *Research Policy*, vol. 10, p. 266.

Note: a. Based on the product groups.

exports exceeded that of the UK by a staggering 137 per cent in 1975, despite its higher unit price plus the strong Deutschmark, appears to suggest that West German engineering products are not only more technically sophisticated and better designed, but also more competitive than those of the UK. There seems therefore to be a relationship between high unit value and international competitiveness. This above argument lends support to Rothwell's (1981a) study which states that 'in engineering goods, price is not the major determinant of competitiveness', suggesting that low unit price does not make a product more competitive it is the value it contains. As Rothwell puts it, 'it can be seen that high unit value is associated with greater competitiveness'. He concludes, 'It seems reasonable to assume from these data that if the customer is willing — as he clearly is — to pay significantly more for an imported (West German) engineering product, then this is a reflection of that product's greater technical sophistication and/or design excellence.'

Thus it is evident from this review that the UK lags behind its West German counterparts in technical sophistication, design excellence, and value for money in nearly all international markets.

Patent activity

Patent activity can be defined as a process by which an exclusive right is conferred on an inventor or holder to protect his or her new invention from being copied by competitors for a period of time. This in itself gives the patentee the right to prosecute anyone who copies such an invention within the period specified in the patent agreement. Using patent data as a proxy measure for technical change, Rothwell (1976a) examines the exports of five countries to the US relative to the number of patents issued to them by the US Patents Office in textile technology between 1970 and 1973, as shown in Figure 3.2. His findings reveal that 'there is a good positive relationship between patent activity or inventiveness and export success, once again indicating a positive link between technical change and export competitiveness in the textile machinery area'. This seems to support the hypothesis that 'trading success is dependent on technical leadership'. For example, Figure 3.2 shows that West Germany registered the highest number of patents (400) between 1970 and 1973. It also had the highest export share in textile machinery when compared to those of its competitors, during the corresponding period. Furthermore, Rothwell points out that 'many of the Swiss patents ... were in the weaving machinery area, where the Swiss industry has gained a

Figure 3.2 Exports to USA and patents registered in USA

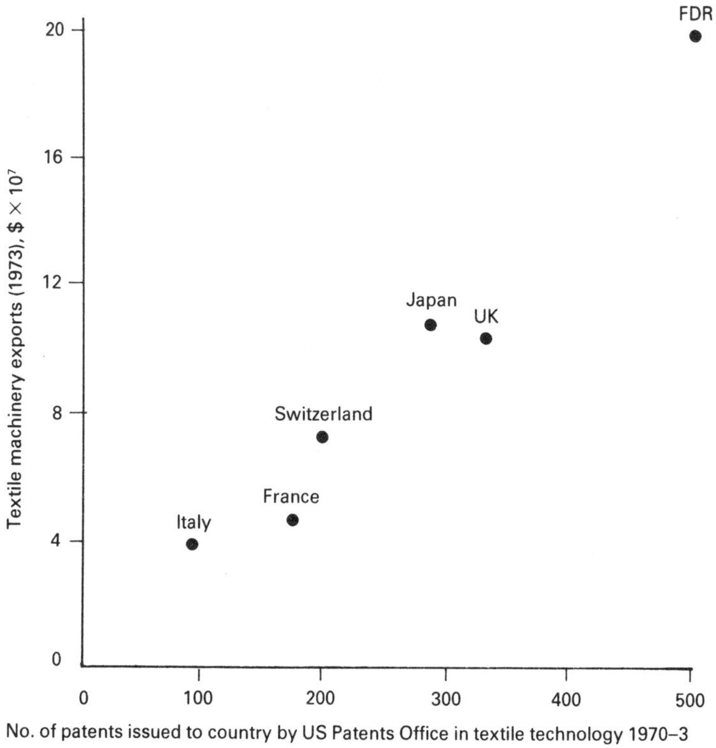

Source: Rothwell (1976a) 'The role of technical change in international competitiveness: the case of textile machinery industry', *Management Decision* 15(6), p. 544

dominant trading position internationally through a strong technical lead.'

Surprisingly, Figure 3.2 indicates that although the UK registered more patents than Japan, in marked contrast, Japan exported more textile machines to the US than the UK. This may be put down to Rothwell's (1979) argument that 'not every patent registered is necessarily transformed into a new or modified piece of equipment', hence Britain registered more patents in, but exported less to, the US than Japan.

This is not disputed; however, this may be seen as a sign of weakness on the part of the UK for not transforming all registered patents into marketable or commercial products, since one can assume that the UK, at least for the period under investigation, was neither lacking in manpower, materials, nor technical know-how to make it happen. This seems to agree with Johne's (1982) finding that states that 'British innovative activities have been less than effective when compared with those of our competitors in translating R & D effort into competitive products.'

Another explanation of the patents/exports difference between the UK and Japan may be related to the Japanese technical (design) excellence incorporated in its textile machines. This again appears to link technical sophistication to export success or international competitiveness, thus, lending weight to the contention that trading success depends to a large extent on technical leadership or technical superiority, as referred to earlier.

To confirm this relationship between trading success and technical leadership, Rothwell (1979) examined the case of circular knitting machines and found that 'In both 1967–8 and 1972–3 UK machines had a 20 per cent higher value/tonne than German built machines ... and bilateral trade with Germany improved from 0.9:1 in 1967–68 to 1.03:1 in 1972–73.' Moreover, he found that between 1963 and 1970, the UK exports grew faster than imports in 'hosiery and other knitting machines' where the UK knitting-machine industry was highly innovative, particularly during the latter half of this period.

The fact that exports grew significantly higher in the 'hosiery and other knitting machines' where the UK firms were highly innovative tends to suggest that a relationship does exist between export success and innovativeness, thus, adding to the validity of the argument that export competitiveness is positively correlated with technical superiority, which is largely influenced by design.

According to Walker (1976) 'Comparison between the number of patents filed by a major US/UK company (25 patents registered between 1966 and 1975) and its major West German competitor (55 patents registered during the same period), adds a little more flesh to the thesis that a relationship exists between economic performance and inventiveness, whether they be symptoms or causes.'

Patent data: portable power tools

Next, Rothwell (1979) examined the patent data derived from the portable power tools sector as shown in Table 3.9. As can be seen, the percentage share of the UK's world export trade in portable power tools (electric and non-electric) has consistently declined. For example, the UK's export share of electric portable power

Table 3.9 Patent activity in portable power tools and export shares

	US patents as a % of total surveyed 1935–60	*UK % of world export trade 1963/4*	*UK patents as a % of total surveyed 1961–75*	*UK % of world export trade 1972/3*
Electric	33.3	16.5	12.8	8.4
Non-electric	23.1	8.5[a]	19.4	7.3

Source: Rothwell (1979) 'The relationship between technical change and economic performance in mechanical engineering: some evidence', in M.J. Baker (ed.) *Industrial innovation, technology, policy, diffusion,* Macmillan Press, London, p. 42.

Note: a. Estimated.

tools has substantially fallen, from 16.5 per cent in 1963/4 to 8.5 per cent in 1972/4. Similarly, its share of non-electric power tools has also fallen by 1.2 per cent, from 8.5 per cent to 7.3 per cent during the corresponding period. It can also be seen in Table 3.9 that the percentage of patents filed in the UK between 1935 and 1975 has decreased by 20.5 per cent for electric, and 3.7 per cent for non-electric power tools.

As Rothwell (1979) puts it, 'A very detailed study of patents registered in the UK since 1935 suggests that British portable power tool companies have lagged behind their US and continental rivals as far as invention is concerned, and that there might be a link between this apparent failure to keep abreast of state of the art and declining UK trade shares.' To explore this relationship further, Rothwell plotted the shares of world-manufacturing exports against the share of all foreign patents (for five countries) registered in the US for the period 1899 to 1975, as shown in Figure 3.3.

Figure 3.3 suggests that in 1899, the UK was clearly ahead of its competitors, registering about 40 per cent of its patents in the US and capturing about 40 per cent of the world's manufactured exports.

In 1913, the UK still maintained its lead in manufacturing exports, but appeared to have lost ground in patent activities (13 per cent) between 1899 and 1913. However, by 1937 the UK seemed to have lost ground on both fronts (manufacturing exports and patent activities), having been beaten to second place by West Germany.

By 1965, the UK still trailed behind West Germany in both patent activities and manufacturing exports appeared to be even more marked when it dropped to the third place behind France in its share of exports but still maintained a marginal lead over

Figure 3.3 Patent share of manufacturing exports vs share of patents registered in the US in 1899, 1913, 1937, 1965 and 1975

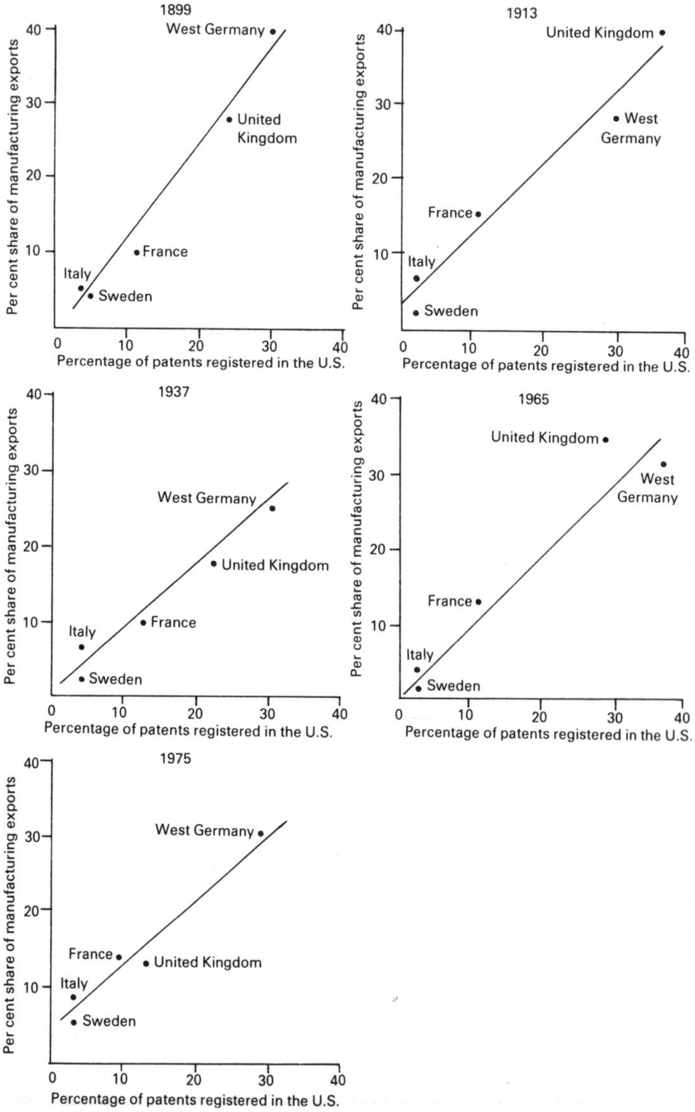

Source: Rothwell and Zegveld (1981) 'Industrial innovation and public policy', Frances Pinter (Publishers) Ltd, London, p. 31

France in patent activities (similar to what happened in 1913 between the UK and West Germany). The causes of these variations of 'up and down' trends are not clear. Nevertheless, Rothwell (1981a) has put down the major cause (particularly in 1913 when the UK's share of manufacturing exports exceeded that of West Germany, but fell short of the West German share of patent activities) to the UK's historical connections with the then 'Empire' states where it exported technology with 'low content' and large quantities of textile materials. In 1913, about 40 per cent of UK exports were in textiles, i.e. in an area that involves little technical activity.

Rothwell (1981a) asserted: 'Indeed; it is probable that this heavy reliance by the UK on exports of low technology, traditional goods (textiles), mainly to Empire countries, during a period when Germany and others were rapidly developing chemical, machine building and other 'modern' technologies, is the root cause of British relative technological backwardness in many classes of engineering goods today.' This appears to confirm an earlier study conducted by MacFarlane (1977) who found that 'many factors contributed to this ... but briefly, this position was largely due to the existence of the Empire, which provided what was in effect an extension of the home market.'

It could, therefore, be argued that during the nineteenth, and first half of the twentieth centuries when the UK's exports were booming, British manufacturers appeared to be complacent in the belief, to borrow from Levitt: (1960)

> '... that growth is assured by an expanding and more affluent population.
>
> '... that there is no competitive substitute for the industry's major product ...'

However, Levitt (1960) has strongly argued that 'there is no such thing as a growth industry ... There are only companies organised and operated to create and capitalise on growth opportunities.' He maintains that 'An expanding market keeps the manufacturer from having to think very hard or imaginatively' and asserts: 'If thinking is an intellectual response to a problem, then the absence of a problem leads to the absence of thinking.' It is Levitt's contention that 'If your product has an automatically expanding market, then you will not give much thought to how to expand it.' This view is supported by Baker (1982) when he argues that 'a dominant position (in the market) can result in complacency and loss of touch with the consumer'.

The aforementioned factors seemed to characterize British manufacturers' behaviour during the boom period, 1899 to 1913 (see Fig. 3.3). They held dominant positions in most of the world markets, particularly Third-World markets, but gradually became complacent and appeared to have lost touch with the exact customer needs and wants, hence the switch of loyalty to competitors' brands, which appeared to have resulted in the loss of many markets by the UK.

Regrettably, these markets were lost, not necessarily because Britain lacked intellectuals or imaginative ability, but because it seemed to have assumed that the growth of its industries had been assured and that the industries' products had no rivals and/or substitutes. In the words of Pilditch (1978) 'Britain does not lack the skills or power to win markets but it uses them in the wrong way to solve the wrong problems.'

Consequently, UK manufacturers appear not to have capitalized the growth opportunities that occurred during these boom periods, including growth potentials in the chemical and machine-building industries, probably undetected by British innovators or detected but under utilized or ignored, hence, the UK seemed to lag behind its major competitors in these sectors. It would seem that UK manufacturers were caught up in the web of what Levitt (1960) describes as 'a self-deceiving cycle of bountiful expansion (of boom periods) and undetected decay', which he says, characterizes 'every dead and dying "growth" industry'.

Finally, Rothwell (1981a) points out that a thorough examination of patent applications filed in the UK by about twenty well-known UK and European producers of agricultural machines produced two major results:

(i) It confirmed the 'evolutionary' nature of technical change in this area. The bulk of patented changes (generally more than 95 per cent and often 100 per cent) were technically relatively small improvements to existing machines; a small number (generally less than 5 per cent) represented fairly major structural change to an existing machine; a still smaller number represented a major design breakthrough or major technical step forward (about 1 per cent or less). The leading continental European firms produced a higher percentage of major structural and design breakthrough changes than their UK counterparts.

(ii) It indicated that foreign manufacturers adopted a more systematic approach to machinery development, in that their patents appeared to be more closely related to each other: there was a more logical progression of interrelated changes. Once one problem was solved they seemed to proceed onto the next most important or pressing problem in a logical sequence. It also showed that foreign patents were more detailed and appeared to reflect a deeper understanding of basic engineering principles.

In both these findings, the UK has been found to lag behind its major European competitors in terms of sophisticated design and engineering; in both results, patent activities appear to be associated with technical progress and thus, international competitiveness.

Are patents a useful indicator?

The use of patent statistics as an 'aggregate measure' of a nation's innovative activities has been criticized by a number of researchers.

Although Johne (1982) admits that 'patent applications filed in a country do provide some guide to overall trends', as illustrated in Table 3.10, he has criticizied the use of patent figures as an aggregate measure of 'national innovative activities' on four counts:

Table 3.10 Patent applications filed in 1976

		No.	%
USA	Residents	65,050	64
	Non-residents	37,294	36
	Total	102,344	100
UK	Residents	21,797	40
	Non-residents	32,764	60
	Total	54,561	100
France	Residents	11,471	29
	Non-residents	28,419	71
	Total	39,890	100
Japan	Residents	135,762	84
	Non-residents	25,254	16
	Total	161,016	100
West Germany	Residents	31,065	50
	Non-residents	30,640	50
	Total	61,705	100

Source: Johne (1982) 'Innovation, organisation and marketing of new industrial products' PhD thesis, University of Strathclyde, vol. 1, p. 15.

(1) Whilst patent offices contain detailed records, registered patents of inventions are unlikely to accurately reflect differences in innovative activity, i.e. the development of inventions into commercial products. Examples can be found in Figure 3.2 (p. 107) where the UK registered more patents, but exported less textile machines to the US than Japan. Similarly, in Figure 3.3 p. 110, the UK registered more patents than France in the US, in contrast, the French exports of the equivalent products to the US exceeded those of the UK.

(2) Some patents cover major discoveries, some are almost trivial.

(3) Patent law varies in different countries and the propensity to patent also varies in different industries. For example, Levitt argues that some industries deliberately prefer to adopt an innovation policy of 'watchful waiting' or 'The Used Apple Policy'. TI-Raleigh is a typical example.

(4) The purpose of the patent can be widely different — some are taken to block competitors and some are simply filed for later use. This confirms Rothwell's (1976a) study (an attempt to elicit information concerning patent practice in the UK textile machinery), which reveals that 'in six instances patents were registered even though they were not deemed to be absolutely necessary: it is presumably seen to be better safe than sorry'. In another three cases he discovers that 'a secondary technical solution was forced on the innovator because of patents held by competitors'.

Johne's fourth criticism is particularly important to this study. First, although any patent registered in 1976 (Table 3.10) either for the purpose of blocking rivals or for later use, has helped to increase the number of patent applications filed by a country, in practice it has contributed little to the technical progressiveness (conversion of an invention into a commercial success) of the country concerned.

It is also worthy of note that Table 3.10 indicates that the UK has more non-residential patent applications (60 per cent) than residents (40 per cent) when compared to Japan with 84 per cent for residents and only 16 per cent for non-residents; or the USA with residents (64 per cent) and non-residents (36 per cent); or fifty-fifty residents/non-residents registration for West Germany in percentage terms. In numerical terms though, West German patents for residents exceeded those of non-residents. This appears to suggest that the UK relies more on foreign inventions or 'imported innovation' than its major competitors. This has been confirmed by Rothwell (1981b) and Pavritt (1983). For example, Rothwell points out that 'most of the significant innovations come from abroad'.

Similarly, Pavritt's (1983) finding shows that 45 per cent of innovations of foreign origin (about 10.6 per cent of the sample) were introduced by foreign multinationals. He reveals that of 118 innovations (about 5.6 per cent of the sample) that were original innovations made in the UK by foreign multinationals, 75 per cent were from USA, 14 per cent from Holland and 11 per cent from West Germany.

In their criticism, Rothwell and Zegveld (1982) maintain that the use of patent activity 'as a proxy measure for technical sophistication and national trade share' has three main flaws:

(1) Not all patented inventions carry the same weight in terms of their potential commercial impact (radical versus incremental innovations).
(2) Nor are all patents subsequently commercialised.
(3) Nor do the aggregate data allow for national industrial specialisation.

They however, concede to the fact that 'rigorous use' of patent data does indicate their 'validity' (i.e. that the use of patent statistics often shows the general trend of a nation's trade performance in world markets), thus lending support to Johne's (1982) findings as discussed previously p. 113–14.

Thus, from this review, it is reasonable to argue that patent activity, in spite of all criticisms, is indicative of technical activities going on in a country, whether or not it evolves from residents or non-residents, and thus gives a rough indication of technical progressiveness. It also suggests that a link exists between patent activity and international competitiveness.

Research and development expenditure

It has been argued that technical progressiveness and/or increase in productivity can be measured by the extent to which a nation expends its financial resources on Research and Development (R & D). According to Johne (1982) 'National R & D effort can be measured in two main ways:

(1) In terms of the amount spent on R & D within a country –

a measure of technological innovation input, and
(2) In terms of the number of patent applications made by a country — a measure of technological innovation output.

Clearly this assumes a link exists between R & D expenditure and technical progress.

On the other hand, Rothwell and Zegveld (1982) suppose that, 'a relationship might exist between national R & D expenditure and GNP, or between R & D expenditure and growth in GNP'. This hypothesis derives from their argument that 'Most technical change that contributes to economic growth might be expected to arise as a result of organised R & D effort.' Consequently, they proceeds to compare trends in R & D expenditure as a percentage of gross domestic product (GDP) in some OECD countries on the one hand, and the 'global economic trends' (an indicator of economic performance of advanced market economies), on the other (see Tables 3.11 and 3.12).

By comparing Table 3.11 with Table 3.12, Rothwell and Zegveld have demonstrated that 'it is those countries that have spent proportionally less of their national R & D resources on defence that have enjoyed the greatest GNP and productivity growth rate'. Table 3.11, for example, shows that Japan spent least on defence — just 0.01 per cent. In turn, Japan's growth rate in GDP was the highest for the period 1960 to 1985 (see Table 3.12) when compared with that of its major competitors during the corresponding period. In marked contrast, the UK's GDP was the lowest when compared with that of its foreign counterparts, for the same period.

One can argue that it is those countries that spend a large proportion of the national R & D budget on non-defence-related applications that are likely to be more competitive in the world trade of manufactured goods than those that spend more on defence. This is in line with Johne's (1982) study, which argues that 'the relative success of Japan and West Germany in international trade in manufactured products may be connected with the high share of R & D aimed at non-defence applications.'

Drawing on the work of Rothwell and Zegveld (1981) Johne (1982) illustrates the aforementioned reasoning with a modified version of Table 3.11. He splits each country's national expenditure on R & D between 'defence' and 'other' applications, as shown in Table 3.13. As can be seen, in 1975, the UK allocated 30 per cent of its national expenditure budget to 'defence' and 70 per cent to 'other' applications. In sharp contrast, the UK's major competitors budgeted more than 70 per cent of total R & D

Table 3.11 Trends in expenditure on R & D as a percentage of GDP in selected countries, total, defence, and other

	1963[a]	*1967*	*1971*	*1975*
FRANCE				
Total	1.60	2.20	1.90	1.80
Defence	0.43	0.55	0.33	0.35
Other	1.17	1.65	1.57	1.45
GERMANY				
Total	1.40	1.70	2.10	2.10
Defence	0.14	0.21	0.16	0.14
Other	1.26	1.49	1.94	1.96
JAPAN				
Total	1.30	1.30	1.60	1.70
Defence	0.01	0.02	—	0.01
Other	1.29	1.28	—	1.69
UNITED KINGDOM				
Total	2.30	2.30	2.10[b]	2.10
Defence	0.79	0.61	0.53	0.62
Other	1.51	1.69	(1.57)	1.48
UNITED STATES				
Total	2.90	2.90	2.60	2.30
Defence	1.37	1.10	0.80	0.64
Other	1.53	1.80	1.80	1.66
NETHERLANDS				
Total	1.90	2.20	2.00	1.90
Defence	—	—	0.04[c]	0.03
Other	—	—	(1.96)	1.87
SWEDEN				
Total	1.30	1.30	1.50	1.80
Defence	0.40	0.43	0.23[b]	—
Other	0.90	0.87	(1.27)	—
CANADA				
Total	1.00	1.20	1.20	1.00
Defence	0.09	0.09	0.06	0.04
Other	0.91	1.11	1.14	0.96
ITALY				
Total	0.60	0.70	0.90	0.90
Defence	0.01	0.02	0.02	0.02
Other	0.59	0.68	0.88	0.88

Source: Rothwell and Zegveld (1981) *Industrial innovation and public policy*, Frances Pinter (Publishers) Ltd, London, p. 27.

Note: a. Germany, Netherlands, Sweden, United Kingdom.
b. 1972.
c. 1970.

Table 3.12 Global economic trends

	1960–65	*1965–70*	*1970–75*	*1975–80*	*1980–85*
GDP:					
World	5.0	5.5	3.8	3.9	4.5
Developed countries[a]	5.1	5.5	3.3	3.3	(4)
of which:					
Canada	5.7	4.8	5.0	3.9	7
United States	4.6	3.1	2.4	2.7	4.1
Japan	10.1	11.6	5.5	5.0	5.9
France	5.8	5.4	4.0	2.8	4.4
Germany	5.0	4.5	2.0	3.4	3.7
Italy	5.1	6.0	2.5	3.4	4.8
United Kingdom	3.1	2.5	2.1	1.4	3.4
TRADE:					
World	6.8	9.2	5.7	5.6	5.8
INFLATION:					
World[b]	4.0	5.0	10.0	11.4	7.5
CURRENT ACCOUNT:					
OECD (end of period – $ bill	3.8	6.7	−0.3	−24.4	−90.0

Source: Rothwell and Zegveld (1981) *Industrial innovation and public policy*, Frances Pinter (Publishers) Ltd, London, p. 28.

Note: a. 13 major industrialized OECD countries.
b. Excluding centrally planned economies.

expenditure for 'other' applications and less than 30 per cent for 'defence'.

This may help to explain why the UK lags behind its main foreign rivals, since those countries that budget more for 'other' and less for 'defence' applications appear to perform better than the UK in the world manufactured export trade. As Rothwell and Zegveld (1982) point out, 'the opportunity cost to economic growth of defence (non-economic) R & D might be high'. They argue that, 'Although there have been a number of notable spin-offs from defence-related R & D, this would appear to be a rather inefficient way to achieve commercial ends.' This seems to be in line with the NEDO (1981) study, which contends that the effect of such spin-offs tends to be marginal when compared to spin-offs from non-defence related R & D.

Mason (1980) cites an example of defence spin-off when he reveals that 'it was US military spending (on R & D) that helped electronic companies develop new advanced products'. Johne (1982) acknowledges this when he states that 'it is sometimes claimed that government funded R & D aimed at meeting defence

Table 3.13 National expenditure on R & D split between defence and other applications

		1963 %	*1971* %	*1975* %
United Kingdom	Defence	34	25	30
	Other	66	75	70
USA	Defence	47	31	28
	Other	53	69	72
France	Defence	27	17	18
	Other	73	83	81
Japan	Defence	0	0	0
	Other	100	100	100
Western Germany	Defence	1	1	1
	Other	99	99	99

Source: Johne (1982) 'Innovation, organisation and marketing of new industrial products', PhD Thesis, University of Strathclyde, p. 22.

needs can foster technological innovation in the commercial sector'. Whilst there is no intention to refute these arguments, one may be tempted to ask: How often does this defence spin-off occur compared to spin-offs from non-defence-related R & D? How much did Japan spend on defence to excel its major competitors in electronics and/or other major high-technology industries?

Finally, using NEDO data, Johne (1982) has illustrated that a low GDP tends to reduce the spending propensity of a nation on R & D. He argues that although 'in proportion to GDP, the UK R & D spend has been in line with that of our major trading competitors, at around 2 per cent' (see Figure 3.4), in reality, however, 'the absolute level spent on R & D in the UK (as shown in Figure 3.5) is considerably smaller than that of our major competitors because of our lower GNP'. This shows that the UK lags behind its main foreign rivals in R & D spending, particularly in non-defence related applications, which, in turn, tends to retard the UK's competitiveness in world-manufactured products. This has been confirmed by Bond (1985).

Reporting on a study by the London-based Technical Change Centre, described as 'a far-reaching examination of the way in which R & D is structured in both public-funded and private sectors', Bond (1985) found that 'UK food manufacturers are not devoting sufficient resources to research and development'. She maintained that, 'In many cases, R & D has been cut back, jeopardising the industry's competitiveness in European markets. ... The vast majority of privately funded R & D takes place in large corporations, and it is in this sector that cost-cutting has led

Figure 3.4 Total R&D expenditure as a proportion of national GDP

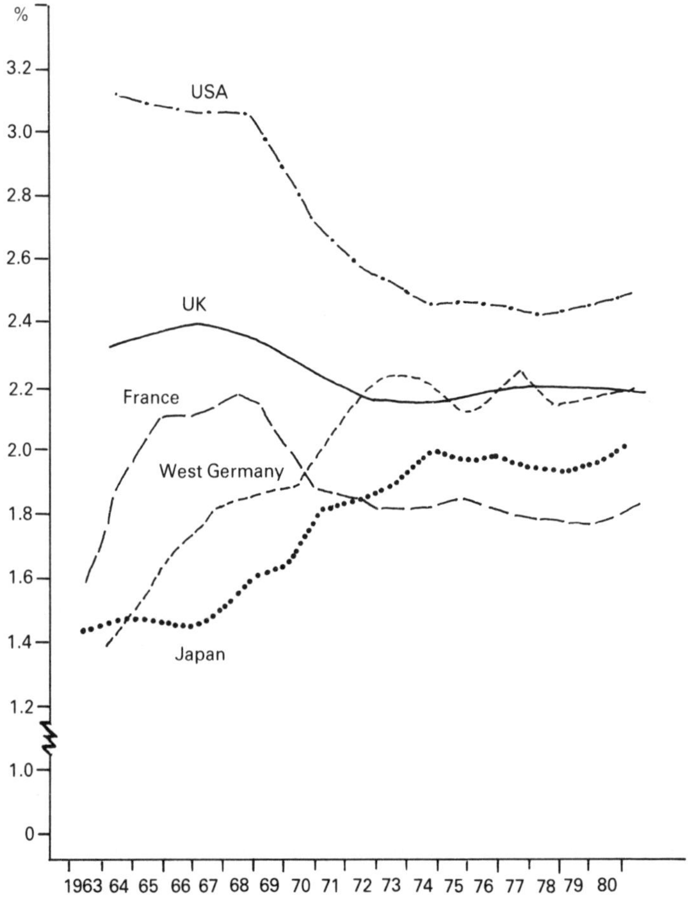

Source: NEDO (1981) 'Industrial performance: R&D and innovation', NEDO, London, p. 12

to shrinkage in the research field ... Large manufacturers have become unwilling to devote time and money to the type of smaller projects — perhaps in the dairy or delicatessen field — with which foreign competitors are forging ahead.' She then concluded: 'The whole structure of the UK food industry is set against innovation, rather than encouraging it.'

These examples indicate that the UK lags behind its continental counterparts because of an apparent lack of adequate spending on

Figure 3.5 National gross expenditure on R&D

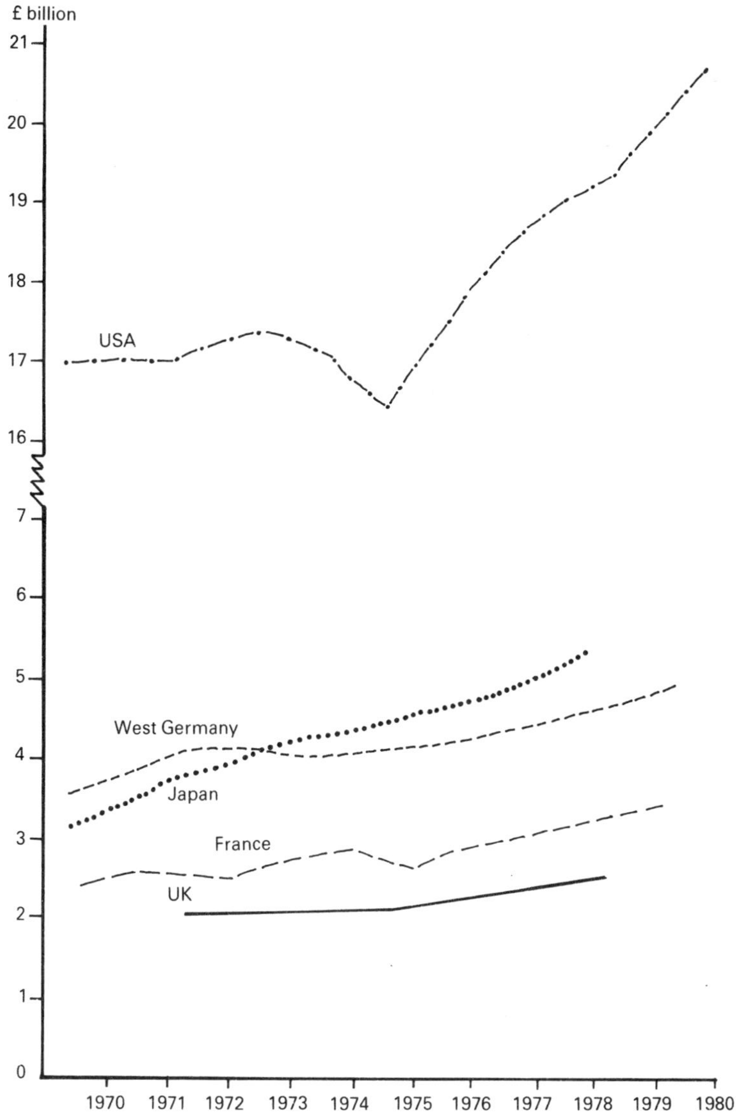

Source: NEDO (1981) 'Industrial performance: R & D innovation', NEDO, London, p. 13

research and development. This has been highlighted by Schott (1981) when she pointed out that in 1975, both the UK and USA spent about 50 per cent of the R & D budget on defence alone. In marked contrast, Japan and West Germany spent over 50 per cent of the R & D resources on basic research, which gives birth to new inventions that are then translated into manufacturable and marketable new products (innovation) (see Table 3.14).

However, the use of R & D expenditure as a measure of technical progress or international competitiveness has been criticized by a number of researchers.

Rothwell and Zegveld (1982) point out that individuals could achieve technical progress with little R & D expenditure; and that users' experience rather than R & D effort, had resulted in 'many improvements in plants and equipment'. This confirms the earlier works done by Peplow (1960), Meadows (1969), Utterback (1971), and von Hippel (1977). They all found in their respective studies that the great majority of product and/or process innovation works had been initiated by customers/users rather than by R & D effort.

Johne (1982) argues that the use of R & D expenditure as an indicator of technical progress is fraught with difficulties. He reasons that although 'national R & D statistics are commonly used by policy makers and academics ... they do not embrace

Table 3.14 Deployment of government-financed expenditure on R & D in selected countries, 1975

	UK	*Canada*	*US*	*France*	*West Germany*	*Japan*
	%	%	%	%	%	%
1. General advance of knowledge (basic research)	21.4	25.2	3.9	25.3	51.0	55.8
2. Agriculture and health	6.5	25.4	13.7	8.4	5.2	16.6
3. Defence	48.9	6.1	49.8	29.5	11.1	2.2
4. Civilian industry	26.8	29.5	21.3	25.8	22.3	20.0
5. Quality of life	4.7	14.7	11.2	10.1	10.3	5.9
6. Other	0	0	0	0.4	0	0
Total	100	100	100	100	100	100

Source: Schott (1981) 'Industrial innovation in the UK, Canada and the U.S.', British North American Committee, London, p. 56.

innovative activities outside formal R & D departments'. He cites earlier works by Mansfield (1972) and Stead (1976) both of whom found that formal R & D effort only accounts for about 50 per cent of the company's total innovation expenditure — 'the rest being represented by tooling, manufacturing and marketing expenditures', thus suggesting that the use of R & D expenditure as an 'absolute' measure of a country's innovative activities and thus, technical progress, may present inaccurate results. However, Johne admits that 'Despite the (above) deficiencies inherent in national R & D statistics they are able to show aggregate trends.'

In her study, Schott (1981) contends that virtually all R & D activities targeted at commercial innovation are carried out by manufacturing firms with or without financial support from government, suggesting that reliance upon the 'national gross expenditure on R & D' as a tool for measuring a country's technical progress may not present the true picture of a nation's overall competitiveness in manufactured export trade.

Finally, Pilditch (1978) has criticized the UK industries for concentrating on R & D (i.e. inventiveness) at the expense of innovation. He argues that 'too great a reliance on R & D or whoever the inventors are, makes us fail to recognise, respect or harness other skills at least as important to successful innovation'. Differentiating invention (R & D) from innovation, he contends that whilst Britain excels most of its major competitors in invention, it lags behind them in innovation. He asserts: 'If invention were the same as innovation, Britain should be well off, that is, if we accept Nobel Prizes as a sign of inventiveness. For a population a quarter the size of the US, the British achievement in science, is remarkable.' However, 'the truth is that while Britain wins Nobel Prizes it loses markets. So, if we say we urgently need more innovation we must ensure that we mean the whole process (by which ideas are translated into economy) not just more R & D;' the reason being 'the academic aura that surrounds so much R & D actually reduces the chance of innovation'.

Wilson (1984) appears to support this view when he argues that 'innovatory inspiration alone (presumably R & D effort) may not provide the key to commercial, competitive performance'. Wilson then cites two classical examples where Britain was inventive but not innovative: 'De Haviland initiated the commercial jet business, but America's Boeing reaped the commercial harvest. Fleming found penicillin: the US cashed in.' This suggests that excellent ideas (born out of R & D effort) alone may not produce fruitful results without proper execution (i.e. without being translated into commercial products).

Combining all of these criticisms, one can argue that although a nation may not, in the short term, get a technological output that is proportional to the technological input (R & D resources) just as a customer/user would expect to get back what he/she pays for in value, it is nevertheless expected that adequate spending on R & D (be it in the public or private sector) should reflect a substantial increase in the overall productivity of a nation, given that R & D effort and other related activities, such as manufacturing and marketing, are well co-ordinated; and that labour and capital are held constant. If this carries conviction, then a sufficient spending on R & D (discovery of new concepts) is more likely to improve the innovative activities of a nation, which in turn, enhances the country's technical progress or international competitiveness in manufactured products.

This reasoning is important because it tends to be consistent with the findings of previous studies, by Brown and Conrad (1967) and Minasian (1969). Brown and Conrad studied the link between R & D spending and increase in productivity in some US companies and found that expenditure on R & D had some relationship with increase in productivity. A similar result was established by Minasian, in his study of the link between cumulated R & D expenditure and value added, capital, and labour in seventeen firms in the US chemical industry. He found that there was a close correlation between the cumulative expenditure of R & D and the firms' value added, given that the inputs of both capital and labour remained unchanged.

One can conclude, based on this review, that the UK lags behind in world trade because it tends to spend less on R & D than its major competitors. For example, Kotler and Fahey (1982) argue that 'Innovative leadership requires that US companies invest substantial sums in R & D.' Thus for the UK to regain its leadership or compete effectively, it must expend adequate resources on R & D with less emphasis on defence-related areas.

Taking stock of import penetration

Rothwell *et al.* (1983) reveal that 'In 1982, import penetration by foreign manufactured goods took a record 28.7 per cent of Britain's home market (in 1981, the figure was 25.5 per cent).'

Table 3.15 shows some of the product areas worst hit by the rapidly increasing import penetration.

Using two separate graphs (Figures 3.6 and 3.7), Rothwell *et al.* (1983) demonstrate the UK's import and export trends between 1950 and 1981. Figure 3.6, for example, shows that in 1950, the

Table 3.15 Product areas worst affected by import penetration in the UK

1. Music centres	84%	imported
2. Sports equipment	74%	imported
3. Scissors	80%	imported
4. Black & white TVs	84%	imported
5. Luggage	81%	imported
6. Freezers	70%	imported
7. Electric cookers	43%	imported

Source: Rothwell (1983) 'The effects of design and innovation on the National economy', *Design and Economy*, The Design Council, p. 6.

UK's export share of the world market was 25 per cent, but by 1981, its share of the world manufactured exports had dropped to 9 per cent. Conversely, Figure 3.7 appears to show a rapid increase in import penetration into the UK from about 6 per cent in 1955 to a 'worrying' level of over 25 per cent in 1981 – almost the equivalent of its export achievement in 1955 (as shown in Figure 3.6).

Comparing Figures 3.6 and 3.7 one can say that the UK has lost its export markets much more rapidly to its foreign competitors between 1950 and 1982, than it gained. Rothwell *et al.* (1983) attribute the lost export markets to the gains made by West Germany and Japan between 1950 and 1981, during which time the West German share of the world-manufactured exports increased from 7 per cent to 18 per cent; and Japan, from 3 per cent to 18 per cent.

These findings are reinforced by Nelson's (1984) analysis, which has demonstrated, based on the Bank of England's statistics (Figure 3.8), that the import penetration of foreign manufactured goods into the UK, 'in terms of volume, has doubled since 1972, having risen from 20 per cent in 1972 to 40 per cent in 1982. 'Many of these imports', he regrets 'are in just those areas of advanced technology and high added value in which we ourselves should be excelling'. Why is the UK backward in these areas of advanced technology and high added value?

One possible answer to this question appears to lie in the UK's apparent lack of technical sophistication and design excellence that has enabled foreign competitors to overtake it in the world share of manufactured exports. For example, Table 3.16 illustrates how rapidly the UK has lost its share of the world export market to its major competitors. It suggests that the percentage share of the UK's world manufactured goods has been halved from about 16 per cent in 1960 to 8 per cent in 1982. At the same time, it shows that the UK's rivals have either consistently increased their shares

Figure 3.6 Export shares: world trade of manufacturers (1950–81)

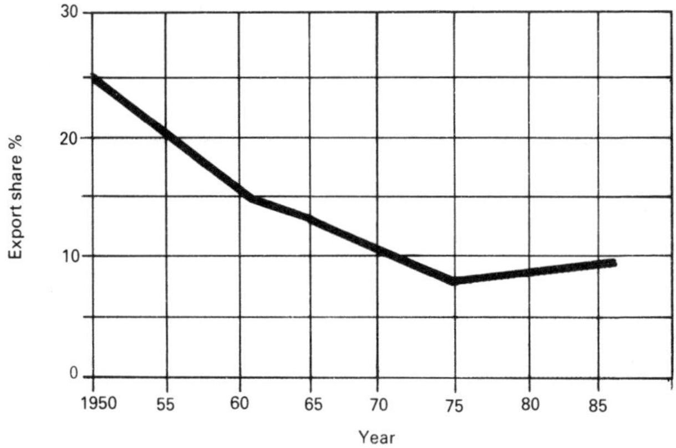

Source: Rothwell *et al.* (1983) *Design and Economy*, The Design Council, London p. 5.

Figure 3.7 UK manufacturing import penetration % (1955–87)

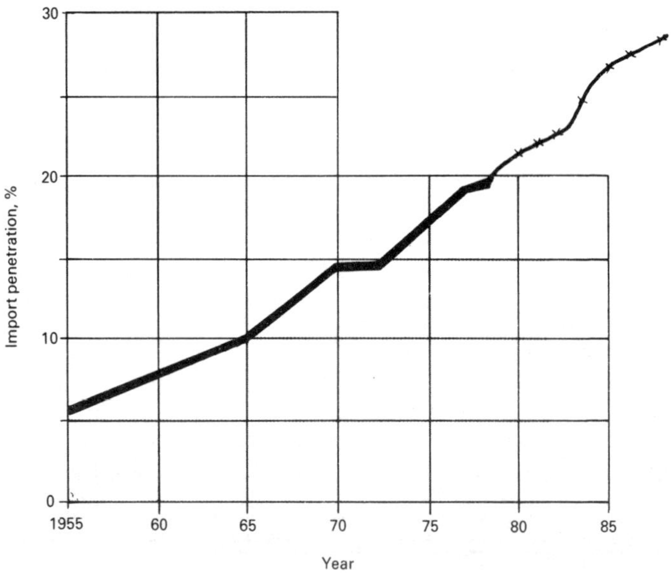

Source: Rothwell *et al.* (1983) *Design and Economy*, The Design Council, London, p. 6.

Figure 3.8 UK import penetration (in terms of volume)

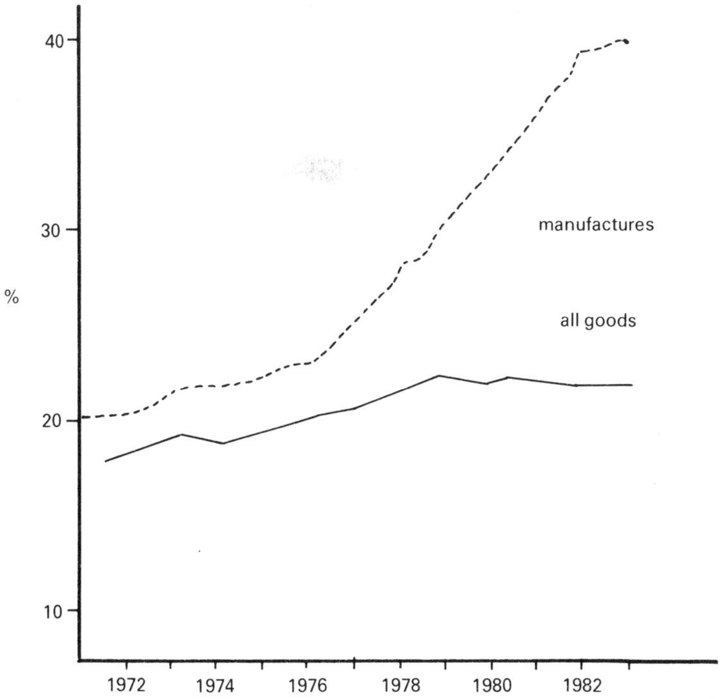

Source: Nelson (1984) 'Export competitiveness: a factor of management and/or design', IEE Proceedings vol. 131, no. 8 November, p. 627

Table 3.16 Percentage share of world export in manufactures

	1960	*1970*	*1975*	*1980*	*1982*
UK	16.3	10.6	9.1	9.7	8.2
Japan	6.2	11.7	13.6	14.9	17.5
W. Germany	19.3	19.9	20.3	19.9	19.0
France	9.7	8.8	10.2	10.0	8.9
Italy	5.2	7.2	7.4	7.9	9.2
USA	21.6	18.6	17.7	17.1	17.8

Source: Nelson (1984) 'Export competitiveness: a factor of management and/or design', IEE Proceedings, vol. 131, no. 8, November, p. 627.

of the world export trade, such as Japan and Italy (particularly Japan, which almost tripled its share), or steadily maintained their export shares, as reflected by the West German and French trade figures.

A comparison between the UK's export share of 8.2 per cent in 1982 (Table 3.16) and its import penetration figure of 28.7 per cent as noted earlier, or 40 per cent in terms of volume, as shown in Figure 3.8, for the corresponding period, leads to the conclusion that the UK significantly lags behind its major competitors in the world export trade.

While Figure 3.9 (The Marketing Team, 1985) shows that Britain performs better than the USA in terms of 'total UK exports to the US' which includes oil exports, what happens if the North Sea oil declines or runs out? According to Lord Nelson (1984) 'The country's balance of payments would indeed now be in a critical

Figure 3.9 US and UK trade (£bn)

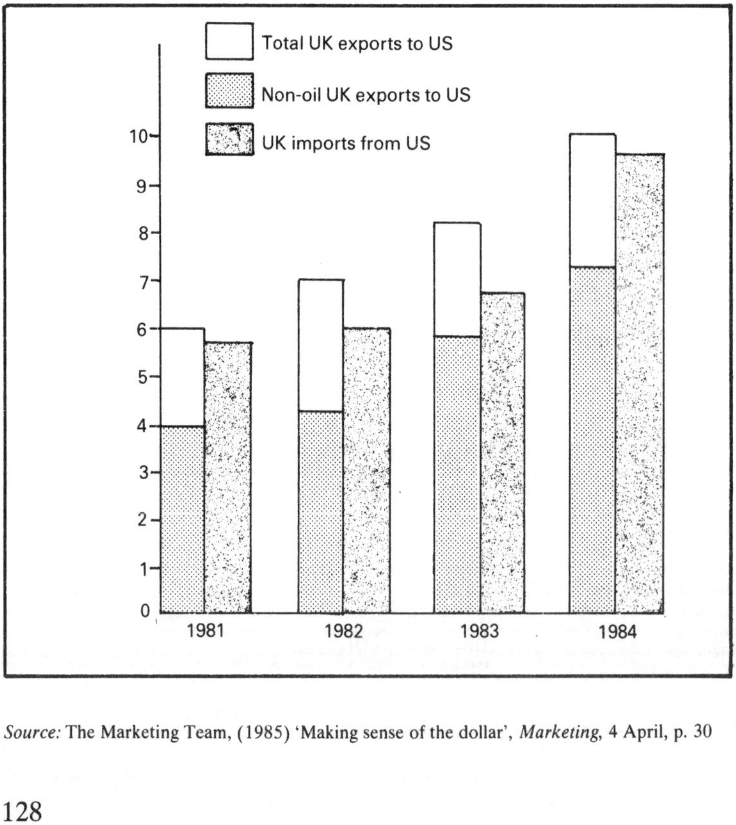

Source: The Marketing Team, (1985) 'Making sense of the dollar', *Marketing*, 4 April, p. 30

position if it were not for North Sea oil and gas, and it is clear that it will become critical in the future if this situation is not rectified before the North Sea production begins to fall.'

A case for import substitution

Given the UK's natural resource base, it is evident that we must participate in international trade if for no other reason than because the theory of comparative advantage demonstrates that by doing so we will enhance our overall standard of living. However, much of the previous evidence has indicated that the UK appears to be importing many high value-added products (machinery, consumer electronics, cars, etc.) and exporting lower value-added versions of the same products. Clearly, skilled manpower is a resource when we appear (in light of unemployment statistics) to have an excess supply and the question that must be asked is why don't we use this more effectively in adding value to imported raw materials rather than let our competitors do it for us?

One explanation implicit in much of the preceding review and analysis is that Britain has developed a national attitude that it must import to survive and export to pay for these imports. Traditionally, this implied foodstuffs and raw materials from developing countries in exchange for basic manufacture and appropriate technology. Unfortunately, as other countries have industrialized and our own population has become more affluent, British manufacturers appear to have lost sight of the fact that success in exporting is invariably founded upon dominance of the domestic market. Instead of seeking to determine why British customers buy foreign and then taking these competitors on where they should have the greatest natural advantage (the domestic market) they continue to behave as if exports are automatically more worthwhile than domestic sales. This mindset probably owes much to ignorance of the size and nature of import penetration and the effect this is having on our economy.

Pilditch (1978) demonstrates this alarming increase in the cost of imports with a graph (see Figure 3.10) which shows that in 1956 the total cost of importing foreign manufactured products into the UK was £234 million; but by 1976 the cost has risen to about £8.8 billion — a staggering increase of £8.6 billion or 3,683 per cent in twenty years (i.e. an average increase of 184 per cent per annum). His survey indicates that the 'Total imports of finished manufactured goods cost us £8853 m in 1976 — more than the National Health Service (£6,169 m), and more than all education (£7,000m).' He asserts: 'As a matter of fact, these imports — largely of goods we ought to be making ourselves — cost every household in Britain the equivalent of more than £9 a week.'

Illustrating with consumer electronics (Figure 3.11), Pilditch (1978) shows that in 1970, the total value of consumer electronic goods imported into the UK was £34.2 million; but by 1976 imports have increased to £155.8 million, an increase of £121.6 million or 355 per cent (an average annual increase of 51 per cent). It is his contention that these import figures mean that 'Britain is losing revenue it can ill afford . . . as well as losing jobs.'

Using the UK passenger-car market as another example, Pilditch points out that 'In 1976, we imported 537,785 passenger cars, and if we take the European average of one man in the car industry making eight cars in a year, it means that imports of passenger cars into the UK last year cost us some 67,000 jobs in car factories alone. Add to that all the people that could be busy in supplier companies — in Lucas and Dunlop, for example — and we get a hint of the scale of unemployment being caused by our failure to give our home market what it wants.' He concludes 'UK

Figure 3.10 How the cost of imports is rising

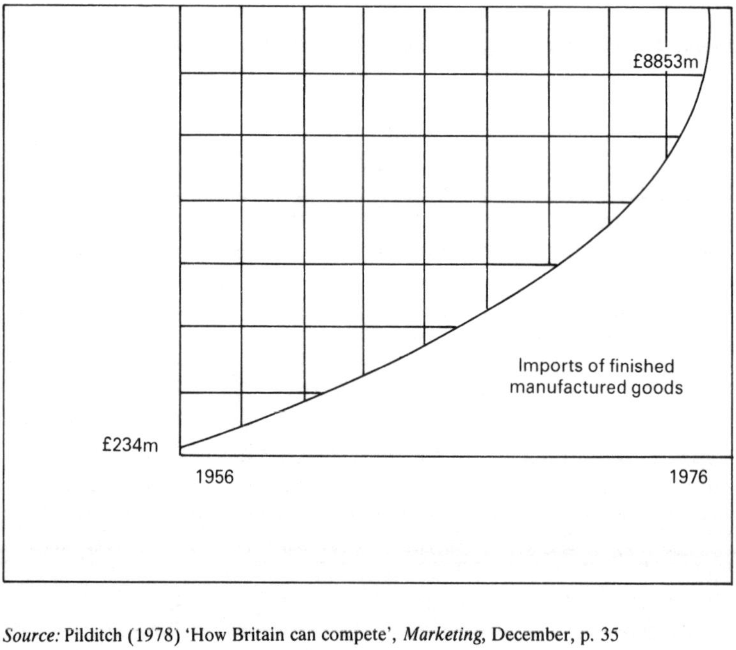

Source: Pilditch (1978) 'How Britain can compete', *Marketing*, December, p. 35

Figure 3.11 The UK consumer electronics market

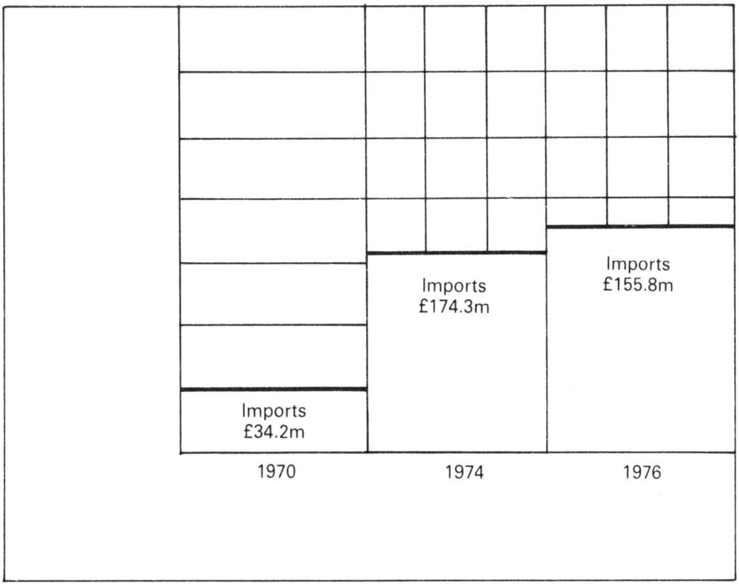

Source: Pilditch (1978) 'How Britain can compete', *Marketing*, December, p. 35

industry must provide products people in this country want to buy and can buy in preference to any other (they want to buy Land Rovers, but cannot, so production is a factor, too)' (Pilditch 1978).

Figures 3.10 and 3.11 give a rough indication of how much Britain loses in terms of revenue and jobs for importing more foreign manufactured goods than it exports. Blame is therefore, heavily shifted to the UK manufacturers for failing to design and produce unique and technically superior products for domestic consumption, which in turn has given rise to the rapid increase in import penetration by more technically sophisticated and superior foreign products into the UK. Perhaps the customers' point of view as to why they prefer foreign to home-made products will help to reinforce this conclusion, and it is to this we turn in the next section.

Satisfying customer needs

While much has been said about the use of such techniques as value per ton, unit value, patent activities, and R & D expenditure as measures of export success or international competitiveness, scant consideration has been given to customers/users who buy the manufactured product itself, and whose demand and preparedness to pay a premium for high-quality or sophisticated products, in preference to low-quality items, appears to have stimulated the increase in imports in the first place.

Although manufacturers are capable of assessing the importance of technical quality or design excellence in competition, it is the customer or user, out there in the marketplace, who appears to be even better qualified to evaluate these products according to the technical and economic values they contain. As Rothwell (1979) puts it, 'probably the person best qualified to determine the importance of technical sophistication is the user.' He maintains that, 'It is he, after all, who must make the decision whether or not to pay a premium for the more advanced models, or to purchase less-sophisticated models relatively cheaply, a decision on which the future of a company might depend.'

For one thing, a product that is technically sophisticated or superbly designed in the eyes of the manufacturer may not be perceived as technically superior or unique by the customer or user when compared to other rival product offerings. It is assumed that all technically progressive firms seek to manufacture what seems, in their eyes, to be the best they can afford to make, as no producer likes to offer products inferior to those of its competitors, particularly in a highly competitive market.

Textile machine users' viewpoints

In a series of studies, Rothwell (1976a, 1979, 1981a, 1981b), explores the viewpoints of UK customers/users as to why they prefer foreign-built to home-made textile machinery, machine tools, and agricultural machinery. Table 3.17 shows the answers given by 107 UK textile companies (users) to the question 'If, during the period 1970–1976, you have purchased machinery from abroad in preference to UK built models, could you please state your reasons for doing so?'

It is interesting to find that 32 per cent of the sample preferred foreign-made machines on the grounds of superiority in both performance and design. Next, 13 per cent said they preferred foreign machines to home-made ones, for being more technically advanced in design. A further 11 per cent of the sample bought foreign because there was 'no suitable UK alternatives', the word 'unsuitable' was interpreted by users to mean 'machinery of poorer quality', which is probably another way of saying that foreign-

Table 3.17 Reasons for UK textile companies purchasing foreign-built machinery, 1970–76[a]

	No.[b]	%
Not available in UK	39	27
No suitable UK alternative	15	11
Superior overall performance and design of foreign machines (more reliable, more productive, greater operational efficiency)	45	32
Foreign machinery technically more advanced in design	19	13
Better service provided by foreign suppliers (i.e. spares and after-sales service)	8	5.5
Foreign firms offer more reliable delivery dates	6	4.0
Foreign-built machinery cheaper	6	4.0
Foreign manufacturers more aware of specific requirements	5	3.5
	143	100

Source: Rothwell (1976a) 'The role of technical change in international competitiveness: the case of textile machinery industry', *Management Decision*, vol. 15, no. 6, p. 545.

Note: a. 89 per cent of the 107 companies completing the questionnaire bought foreign machinery during the period 1970–76.
b. A number of respondents gave more than one reason for buying foreign.

made textile machines were 'generally better engineered or worked better' than their UK counterparts. According to Rothwell (1979), 'This means that in those cases where UK machinery was available, 62 per cent of the reasons for buying foreign related to the performance or technical 'quality' of the machinery.' This appears to be supportive of the fact that only 4 per cent of the population (Table 3.17) bought foreign on the basis of 'cheapness' — evidence that the great majority of textile-machine users are not prepared to buy cheap home-made machines (an apparent reflection of poor design or quality) at the expense of superior, technically advanced foreign-built textile machines. This is particularly important because it tends to lend weight to the hypothesis that technical sophistication or technical/design excellence is closely associated with high-export performance or international competitiveness.

Machine-tool users' viewpoints

To confirm the proposition that 'technical (design) quality is a key determinant of customer preference of product offerings',

Rothwell (1979) questions the UK users of machine tools as to why they were 'increasingly opting for foreign-built machines'. In Table 3.18, Rothwell shows that the great majority (30 per cent) of the UK machine tool users have bought foreign machines on the basis of technical superiority. Once again a negligible percentage (5 per cent) purchased foreign machine tools on grounds of price advantage, suggesting that what UK machine-tool users really need is not the cheapest machine but one that is of excellent design, technically advanced and reliable.

Similarly, Rothwell (1979) notes that an investigation conducted by NEDO (1965) into the primary reason for importing foreign mechanical products (machine tools inclusive) as perceived by purchasers themselves, shows that, 'imported machines had performance and design characteristics which could not be matched by UK suppliers', — a clear evidence that the UK lags behind its foreign rivals in excellent design and technical sophistication.

Table 3.18 Why UK machine-tool users were increasingly opting for foreign-built machines

Reasons for opting to buy foreign	%
Price advantage	5
Better after-sales service	5
Willingness of foreign producers to meet special requirements	8
Prospects of reciprocal trading arrangements	11
Machine specification not available in the UK	21
Technical superiority	30

Source: Adapted from Rothwell (1979), 'The relationship between technical change and economic performance in mechanical engineering: some evidence', in M.J. Baker (ed.), *Industrial innovation, technology, policy, diffusion*, Macmillan Press, London, p. 45.

Agricultural-machine-users' viewpoints

Rothwell then surveyed the agricultural machinery industry with the objective of soliciting the views of 150 farmers as to the reasons for purchasing both UK and foreign-made agricultural machines between 1972 and 1977, as shown in Table 3.19. As can be seen, about 55 per cent of the sample bought foreign machines on the basis of performance characteristics, that is, better overall functional performance, superior and more advanced overall design, more reliable and more efficient. In marked contrast, only 16 per cent bought British on grounds of performance factors, suggesting that the UK agricultural machines are less reliable, inefficient, and

Table 3.19 Reasons given by 150 UK farmers for buying foreign-built and British-built agricultural machinery between 1972 and 1977

	Foreign machinery		*British machinery*	
	Number	*Percentage*	*Number*	*Percentage*
Performance factors: (better overall performance, superior overall design, more reliable, more efficient, more advanced design)	146	54.9	26	16.0
Price factors: (machinery cheaper, better value for money, i.e. equivalent performance at lower costs)	25	9.4	46	28.4
Convenience factors: (convenience of local manufacturers, local dealers, local source of spare parts)	—	—	52	32.1
Better service provided	31	11.6	13	8.0
Availability (off the shelf)	21	8.0	6	3.7
No UK equivalent	31	11.6	—	—
Other factors	12	4.5	19	11.8
	266[a]	100.0	162[a]	100.0

Source: Rothwell (1981a), 'Non-price factors in export competitiveness of agricultural engineering products', *Research Policy*, vol. 10, p. 274.
Note: a. A number of farmers gave more than one reason for their purchasing decision.

offer poor performance in operation when compared with the rival product offerings.

It is interesting to note that although 32.1 per cent bought British for convenience factors (i.e. the convenience of local producers and agents with possible local availability of spare parts), only 3.7 per cent purchased British-made on the basis of availability of the machine itself, compared to 8 per cent who bought foreign for the same reason. The implication is that foreign sales could have been even higher if their distribution network were improved.

The reasons for buying agricultural machinery on the basis of price factors were 'foreign' (9.4 per cent) and 'British' (28.4 per cent); once again indicating that buyers of foreign machines are not prepared to sacrifice high quality, i.e. superior and more advanced overall design, better performance, reliability, etc., for low price.

All in all, the great majority of UK farmers bought foreign-made agricultural machines on grounds of technical sophistication and design excellence confirmed by Rothwell's (1981b) study, which states that 'most UK farmers felt that, generally speaking, foreign-built machinery was generally more sophisticated and more robust than UK-built machinery'. He adds that 'Foreign machinery was generally felt to be better built and designed, requiring a good deal less weight (and thus less energy consumption) to obtain the required degree of strength and robustness'. He also points out that 'Some subjects complained that the UK manufacturers used inferior materials in building their machines which often resulted in a great deal of wear and tear, particularly for moving parts.'

Interestingly enough, Rothwell finds that those who bought British did so either because they wanted 'conventional' unsophisticated machines or on grounds of patriotism or brand loyalty — another evidence that UK customers buy home-made agricultural machines, not on the basis of technical sophistication, design excellence, reliability, or use of superior materials, but simply on grounds of brand loyalty and chauvinism.

It has to be emphasized that British customers have always indicated their willingness to buy British if only British products are better than, or even as good as, the foreign product offerings. For example, Baker (1979c) found that '*Almost without exception, leading British companies stated that they would be more than willing to buy from UK sources if these (i.e. British-made machines) could meet their requirement.* 'However' he commented, 'as they could not (i.e. satisfy the home consumer requirements), it was necessary to import high price, sophisticated foreign machinery in order to maintain their own competitiveness.' Similarly, Rothwell (1981a) found that 'British farmers demand high quality, innovative machines but UK manufacturers have failed to meet their needs.'

This review of customers' viewpoints is vital for a number of resons. First, it has strengthened the argument that foreign-made machines (textile machinery, power tools, agricultural machinery etc.) are more technically sophisticated and better designed than home-built machinery. Second, it tends to support the contention that it is those manufacturers that supply sophisticated and high-quality products who have been more successful in the world trade than those with low quality and unsophisticated products, as reflected in UK machinery. Third, it appears to confirm the validity of the use of such techniques as value per ton, unit value, etc., as measures of export success or international competitiveness. Finally, it has identified various other reasons for the non-competitiveness of many UK industries in world trade.

We set out in this chapter to find out why the UK 'exports

cheap' and 'imports dear' manufactured products, with the main purpose of establishing the relationship between the rapidly increasing import penetration into the UK and the rapidly eroding shares of the UK's world manufactured exports. Consequently, four techniques (value per ton, unit value, patent data, and R & D expenditure), which have been widely used as indicators of technical progress or international competitiveness were examined. The following section will present the conclusions drawn from the overall review.

Conclusions

A review of both 'value per ton' and 'unit value techniques' shows that foreign imported manufactured products have a higher value per ton as well as higher unit value than the equivalent UK manufactured exports. Furthermore, foreign imports are not only more expensive per ton/unit but also appear to be more competitive than British rival exports, suggesting that a relationship exists between export competitiveness and higher value per ton on the one hand and higher unit value on the other. For example, countries such as West Germany, Japan, and France, which manufacture products with higher value per ton/unit, appear to perform better in export trade than the UK with similar products of lower value per ton/unit. In other words, the loss of UK share of world exports has been accompanied by a fall in the value-to-weight ratio and unit value of exports relative to those of competitors. Further, the fact that these measures produced similar results in a number of different product groups — textile and agricultural machinery, fork-lift trucks, and portable power tools — demonstrates convincingly that the value-to-weight ratios (value per ton) and unit value of many of the UK products are much lower than those of its major foreign competitors, even after allowing for shifts in exchange rates and different factor costs. It is therefore, not an unreasonable assumption that value-to-weight/unit is a fair reflection of higher quality and technical sophistication, which leads to the inevitable and disturbing conclusion that many British products are relatively inferior when compared to the rival offerings, with the result that the UK imports 'dear' and exports 'cheap' manufactured products.

An examination of patent data ('a proxy measure for technical change') indicates that Britain maintained a dominant position in patent activities and thus in export trade in the nineteenth and early twentieth centuries, particularly in textiles. (See also the evidence from Albu (1976) on shipbuilding in Chapter 1.) However, further evidence suggests that the UK was unable to capitalize on its growth opportunities due to complacency, lack of foresight, and its inability to transform some of its inventions into innovative realities, and was subsequently overtaken by its major rivals such as Japan, the USA, and West Germany. It is also evident in our analysis that Britain relied heavily on products of low technology content and virtually lost touch with major breakthroughs, design excellence, and technical quality.

A comparison of UK and West German performance in OECD and non-OECD markets reveals that West Germany performed better than the UK in OECD markets on grounds of technical superiority (more technically sophisticated, reliable, and durable, more satisfactory ergonomic and aesthetic features, better performance in operation and design excellence). Although the UK was found to perform better than West Germany in non-OECD markets, a close examination reveals that the UK's higher percentage of exports to non-OECD markets 'represents the exports of relatively unsophisticated products', which appears to signify 'technical backwardness'. Even at that, West Germany appears to be fast penetrating the non-OECD markets with products of higher quality and better value for money, and thus, seems to be rapidly catching up with the UK in these markets.

The analysis of the reasons the UK customers/users prefer foreign to home-made goods points to the inability of many British manufacturers to satisfy the requirements of the domestic users. Indeed, evidence suggests that UK customers are willing to buy from the home source if manufacturers can make products as good as, if not better than, the foreign rival offerings. In other words, domestic users tend to see the home-made products as less appealing in terms of aesthetic and ergonomic features, overall performance in operation, and even price. The customer's viewpoint is important not only because all products must, in the final analysis, 'stand in the dock' for judgement by the customer, but because it appears to validate the use of value per ton, unit value, etc., as useful tools for measuring the extent to which a nation is technically superior in manufactured exports, and thus, internationally more competitive than its rivals. One can reasonably conclude that the reason the UK exports 'cheap' and imports 'dear' is mainly because most of its products are technically inferior when compared to those of the competitors, thereby encouraging the penetration of more foreign products into the UK and reducing the chances of increasing its share of manufactured exports. In effect, this tends to lead to an imbalance of trade, which may adversely affect the British balance of payments (see Table 3.20). In turn, this may

Table 3.20 UK balance of payments current account (£ billion)

	1981	*1982*	*1983*	*1984*	*1985*
Visible balance	3.4	2.3	−0.8	−4.4	−2.1
Non-oil trade balance	0.9	−1.5	−6.8	−9.3	−8.3
Oil trade balance	2.5	3.8	6.0	4.9	6.2
Invisible balance	2.8	1.6	3.9	5.6	5.7
Current account	6.2	3.9	3.1	1.2	3.6

Source: Annual Abstract of Statistics, 1987 edn, HMSO, London.

slow down economic growth, which may result in the loss of revenue, jobs, and lead to a poor quality of life when compared to those of the foreign rivals.

Table 3.20 speaks for itself. In 1981, the UK had a favourable balance (even if marginal when compared to other trade balances) of £0.9 billion in non-oil trade (i.e. manufactured goods), but by 1984, the trade balance was markedly adverse (−£9.3 billion). Of great concern is the accelerating rate of deterioration of such trade imbalance. The effect of this can be clearly seen in the current account balance — from about £6 billion in 1981 to £3.6 billion in 1985. This 'marginal' balance is likely to slow down the growth in the British economy, thus affecting the quality of life when compared with those of the major competing countries, as alluded to earlier.

A close inspection of spending on R & D indicates that the UK does not spend as much as it should to be able to catch up and/or compete effectively with its major foreign rivals. Available evidence suggests that the UK expends less than its foreign counterparts on non-defence-related R & D activities — the main source of technical activities for commercial innovation (innovation involving consumer and industrial products, as distinct from products generated from military R & D applications). Consequently, the UK seems to lag behind its major competitors in technical activities and thus, innovation. A link, therefore, appears to exist between R & D expenditure and technical progress/international competitiveness and/or between R & D expenditure and increases in productivity.

An overview of import penetration into the UK shows an alarming increase of foreign-manufactured imports, suggesting an apparent lack of value for money in home-made products when compared to the rival offerings. The rapidly increasing penetration of foreign-manufactured products into the UK can be attributed to the inability of much of UK industry to meet the requirements of the home users, particularly in those product areas where they ought to be excelling their competitors. As a result of this, firms that cannot compete must seek to merge with and/or acquire other companies, diversify, divest, or close down factories in order to survive; or even liquidate as a last resort! The consequences are clear. Economic growth will slow down, revenue will be lost, and unemployment will increase.

Much of this chapter is based upon earlier studies that may now seem dated, particularly when one considers the apparent turn-around in the UK economy since 1980. At the time of writing (mid-1987) Britain has one of the fastest (if not the fastest)—growth rates in the Western world, well in excess of 3 per cent per annum. And the London Business School is forecasting that it will continue to do so for the next four years. Similarly, British productivity has been growing at between 4 per cent and 4.5 per cent since 1981, the Stock Market is at record levels and many apparently hopeless case industries like British Steel have been invigorated with new life. *But* as Table 3.20 reveals, our dismal balance on visible trade in physical goods is only sustained by windfall profits from sales of a non-replaceable asset — oil (Lord Stockton's 'family silver') and invisible or service industries. The symptoms may have been alleviated but the underlying cancer is still gnawing away at the heart of our economy — our manu-facturing base. If Britain is to be restored to full health then it must address the fundamental causes of its decline — a failure to offer customers, both at home and abroad, true value for money.

Our analysis indicates that value, like beauty, lies in the eye of the beholder and is a complex concept comprising many dimen-sions. To provide value we must understand user needs and then design and develop products that will deliver the desired satis-factions. In the next chapter we are going to address how this might be done.

Chapter four

The characteristics of commercially successful firms

Introduction

Although the preceding chapters have tended to paint a gloomy picture of British manufacturing industry, it is quite clear that the failings identified do not apply to all firms in every industry. On the contrary, there are many companies that compete very effectively in both domestic and foreign markets. If, therefore, one is concerned with seeking to improve the overall performance of our industry, it would seem sensible to start by seeking to determine what are the characteristics of commercially successful firms. If we can define policies, practices, or behaviour that are present in successful firms then we can compare these with less successful firms as a basis for seeking to improve the performance of the latter group.

To this end this chapter is concerned with defining successful firms in terms of their technical progressiveness and innovative ability for, if nothing else, our earlier analysis has clearly shown that value implies more and better ways of satisfying needs for a given price. In turn, this indicates that successful firms are those that take a lead in adding value through continuous improvement and new-product development. Having defined technical progressiveness, we then review previous studies that have aimed at discovering factors that characterize successful innovators and technically progressive firms.

Based on the characteristics identified, we present the findings of our own survey of Queen's Award Winners, who may be regarded as commercially successful firms, and who were asked specifically to report on their own behaviour *vis-à-vis* these characteristics. This analysis is supported by a review of factors associated with lack of success (failure!) on the grounds that there is ample evidence of firms containing the 'success factors' that are performing badly. Thus, many of the firms cited as examplars in Peters and Waterman's (1982) study *In search of excellence* were experiencing difficulties in the marketplace within two or three years of the book's publication. It follows that, in pursuing success, one must also guard against failure and these are not necessarily mirror images of each other.

The technically progressive firm

Britain's declining competitiveness in international markets has been apparent for many decades now and the subject of numerous inquiries into its causes and consequences. Among the most influential studies is that carried out by Charles Carter and Bruce Williams on behalf of the Science and Industry Committee in the mid-1950s. The three reports resulting from their inquiry — *Industry and technical progress, Investment in innovation* and *Science in industry* — are required reading for all serious students of industrial performance and competitive success. Carter and Williams (1957) define a technically progressive firm as 'one which on a necessarily subjective judgement, is keeping within a reasonable distance of the best current practice in the application of science and technology'.

Since this study is based upon the performance of recipients of the Queen's Award to industry for technology aimed at producing goods for sale as well as for export achievement, the criterion for technical progressiveness is a commercial one, and will be defined from the context of the criteria set by the Queen's Award Committee (1985).

A technically progressive firm is, therefore, defined as one that has made a significant advance, leading to increased efficiency in the application of technology to a production or development process or the production for sale of goods that not only incorporate new and advanced technological qualities, but also show evidence of commercial success.

Generally speaking, commercial success occurs when an innovation (new product) or firm is profitable, i.e. when 'an innovation obtains a worthwhile market share and profit'. Based on this seemingly general notion, the term 'technically progressive firm' is further broadened to include the firm that has made significant inroads in international markets (i.e. markets where two or more competitors from different countries compete) through sale and sustained increase in sale of good-quality goods/services (i.e. goods/services acceptable to customers/users in international markets) when compared with those of its competitors. It also includes improved marketing organization and/or new initiatives to cater for export markets; and this, for the purpose of this study, will be our operational definition of the term 'commercially successful firm'.

Factors characterizing successful innovators and technically progressive firms

Judging from the context of our definition of technically progressive firms, all the Queen's Award winners are perceived to be technically progressive and/or commercially successful. However, it is relevant to note right from the outset that the fact that a firm has not won the Queen's Award does not necessarily render it unprogressive or not commercially successful. There are several reasons for this. First, there are other similar national awards, the Design Council Award for technological achievement, for example. Second, there are international awards such as the International Export Association Triple Gold Award for exports. Thus any firm that wins one of the above recognized awards can be said to be technically progressive and/or commercially successful.

Further, there appears to be no evidence to suggest that the winner of the Queen's Award is guaranteed the Design Council award for technology or the International Export Association's award for exports and vice versa. Thus, while some firms may be interested in national awards, others may be interested in international awards.

There has also been a 'flood' of protests from non-winners of the Queen's Award on grounds that 'the criteria on which winners are chosen are vague'. For example, although the Queen's Award Office has the 'discretion to give general guidance to firms and to indicate factors which might have adversely affected an application', sometimes, 'such guidance is often limited in practice — since in many cases, there are no obvious reasons why the firm was not chosen other than the strength of the competition.' This is evident in a letter written by D.J. Sidebottom (Chairman, Glasdon Ltd) to the Secretary of the Queen's Award Office asking for some guide as to how his company fell short of the exalted standard required. The answer basically said: 'Well done but other people did better, try again next year.' Obviously this sort of response could discourage firms (existing and prospective applicants) from applying for the award. Add to all of these, the probability that bias might influence the selection itself and some progressive firms may be dropped during selection.

Carter and Williams' Study

Carter and Williams (1957), pioneers in this field, identify twenty-four characteristics of technically progressive firms:

1. High quality of incoming communications. It is found that technically progressive firms read high-quality 'technical literature' that enables them to keep in touch with current ideas on technological innovation and to come in contact with experts — engineers, scientists, technologists, managers, etc. Conversely, unprogressive firms are found to indulge in reading 'low quality trade journals' or not to read at all. Consequently, they appear to be slow at contacting people with experience or keeping abreast with current ideas in new product development and design.

   However, other evidence confirms that there is a limit in the reading of printed materials (high or poor quality) beyond which the understanding of technological complexities require the physical presence of someone with a technical know-how or expertise to direct the operation, hence technically progressive firms appear to lay more emphasis on technological transfer 'on the hoof' rather than 'on paper', i.e. transfer by individuals moving physically from one firm or department to another rather than through published materials.
2. A deliberate survey of potential ideas.
3. A willingness to share knowledge.
4. A willingness to take new knowledge on licence and enter joint ventures. This association is manifest in the technically progressive firms' ability to adopt and adapt to the fast-changing environment through licence/patent, joint ventures, and suchlike business activities.
5. A readiness to look outside the firm. This suggests that progressive firms are more outward-looking in action than non-progressive firms. It appears to show that progressive firms search for ideas outside their firms; emulate their rivals; draw upon the standards set by rivals; and, in turn, set achievable standards in accordance with the size of the firm, while at the same time comparing these standards with those of competitors.
6. Effective internal communication and co-ordination. This appears to explain that progressive firms are characterized by:

   (a) Effective team work during new-product design and development and beyond.

(b) Clear definition of objectives or formulation of goals to avoid role ambiguity; clear responsibility backed up by authority; and better collaboration between and among all functional areas.

(c) Differences (i.e. usual internal conflicts (intra-firm conflicts)) between functional departments appear not to disrupt the overall corporate objectives or goals, and 'Board decisions (where required) are given without undue delay'.

7. High status of science and technology in the firm. Technically progressive firms tend to give special preference to science and technology because it accords them a 'higher status'. Consequently, more funds are appropriated on research, design, and development.
8. A consciousness of costs and profits in the research-and-development departments (if any).
9. Rapid replacement of machines.
10. A sound policy of recruitment for management. Progressive firms are more management-quality conscious, and as a result of this, seem to recruit better-qualified and better-trained management staff than backward firms.
11. An ability to attract talented people. Progressive firms do not only have the capability to attract talented and able personnel, they also provide adequate incentives (intrinsic and extrinsic) to motivate them.
12. A willingness to arrange for effective training of staff.
13. The use of management techniques. Progressive firms tend to use management techniques, such as method study, work measurement, production planning, and control.
14. An ability to identify the outcome of investment decisions. Progressive firms tend to calculate the probable outcome of investment decisions.
15. High quality in the chief executive(s).
16. Adequate provision for intermediate managers. Progressive firms tend to have enough middle managers to handle day-to-day company problems, giving senior managers more time for planning, policy formulation, and decision-making. Based on similar size and environment, backward firms are observed to have fewer managers, thus, compelling senior managers to spend less time on their main tasks, as referred to previously, in order to attend to minor issues that otherwise could have been handled by middle managers.
17. Good quality in intermediate management.

18. An ability to bring the best out of managers.
19. The use of scientists and technologists on the board of directors. Absent in unprogressive firms, this factor has been found to characterize technically progressive firms.
20. A readiness to look ahead. This implies that progressive firms are forward-looking, long-term planners, and forecasters, while unprogressive firms tend to be myopic — always planning in the short-term and medium-term, particularly the former.
21. A high rate of expansion (the rate of increase of assets) appears to characterize technically progressive firms while serving as a stimulus to would-be progressive firms. Non-progressive firms seem to lack this element.
22. Ingenuity in getting round material and equipment shortages. Progressive firms, unlike backward firms, are quick at overcoming problems concerned with shortages of material and equipment.
23. An effective selling policy. This suggests that progressive firms are good at marketing efforts, while unprogressive firms tend to assume that good products sell themselves. Consequently, unprogressive firms tend to relax their marketing efforts. This has been confirmed by Cooper (1979a).
24. Good technical service to customers. This characteristic implies that technically progressive firms, *vis-à-vis* unprogressive firms, are customer-orientated. They tend to interact with, and get feedback from customers/users. This enables them to continuously redesign or reinnovate existing products or design new ones in order to compete or sustain competitiveness in world trade.

In summary, Carter and Williams' (1957) study suggests that factors characterizing technically progressive firms are multivariate. They range from the ability and willingness to communicate and learn from other firms, the deliberate search for and updating of innovative ideas, a readiness to exchange these ideas and knowledge with other firms, through intra- and inter-firm collaboration, forward-lookingness, market awareness, effective training and the creation of an organizational climate to fit the people, to the recognition of the importance of research, design, and development of new products as factors contributing to success.

Overall, Carter and Williams' work has stood the test of time. It is an influential study upon which many researchers have drawn and since Carter and Williams' study, an army of researchers has relentlessly continued in this field in an attempt to identify more factors that characterize technically progressive firms or successful innovators.

Robertson (1977) summed up some of the important conclusions on innovation studies in his review of 'Case study based research on innovation', which seems to suggest that technically progressive firms or successful innovators possess the following characteristics:

1. Need recognition. About 50 per cent of the cases show that clear identification of market/customer need is a vital ingredient for technical progressiveness.
2. Key individuals. The presence of key individuals (top people or others prepared to commit themselves to the new-product design and development or corporate goal(s) of the firm are commonly noticed in technically progressive firms.
3. High-quality management personnel. Progressive firms are ready to spend, recruit, and train the right managers (top, middle, low) at the right time, in the right place, as well as to create an organizational climate to fit.
4. Intra- and inter-firm co-operation. Forward firms tend to create an atmosphere conducive to co-operation between functional departments. They also collaborate with other firms with which they exchange new design/innovation ideas via inter-firm visits, meetings, conferences, commercial agreement, etc.
5. Ability to adapt to change. Progressive firms are known for their preparedness to adapt to new environmental changes (changes in the market, society, etc.) and seem to be quick at adopting new technologies, new designs, processes, and new techniques.
6. Ability to select viable projects. Progressive firms appear to select the most viable projects to embark upon, fully justified on economic grounds. Such innovators tend to be cost-conscious. They are, therefore, more likely to use such techniques as CAD/CAM, FMS, value engineering, quality circles, and so on, to reduce costs and improve quality.
7. Good communication within and outside the firm. Successful innovators tend to state clear objectives, define responsibilities, and ensure that communication within the organization flows without disruption. In addition, they

tend to create good lines of communication with the outside world so as to keep in touch with current events — new technologies, competitors' activities, changes in tastes/preferences, new ideas, etc., which are important for the design and development of new products.

8. Involvement of customer/user in new-product/process design. Successful innovators tend to involve customers/users right from the first stage of new-product design and development.

These conclusions are particularly important to this study because they appear to be consistent with the findings in later studies.

Project SAPPHO

Project SAPPHO (Scientific Activity Predictor from Patterns with Heuristic Origins) is another influential study in this field. It is a study designed to discover the differences between successful and unsuccessful innovations.

Using a 'paired comparison' technique (where a successful innovation is compared with an unsuccessful one), Rothwell *et al.* (1974) investigated forty-three pairs — twenty-two in chemical processes and twenty-one in scientific instruments. The aggregated results of both industries are shown in Table 4.1 in a ranking order. As can be seen, all the twenty-three factors characterizing progressive firms are highly significant (with a probability of chance occurrence of less than 0.1 per cent for the first fifteen variables and less than 1 per cent for the last eight variables). It is also interesting to note that four variables had equal ranking in the two separate studies (i.e. project SAPPHO, phases 1 and 2), lending credence to the consistency and importance of these factors to the success of firms (see Table 4.1, nos 1–3 and 6). Table 4.1 also indicates that fifteen variables identified in SAPPHO, phase I, as factors characterizing progressive firms were confirmed as highly significant by the results of SAPPHO phase II (see nos 1–15).

However, Rothwell *et al.* (1974) have grouped the combined principal results of both projects (i.e. SAPPHO, phases 1 and 2), under five major underlying dimensions — defining major areas of competence of successful firms, as shown in Table 4.2. Among such areas of fundamental importance to the success of firms are: strength of management and characteristics of managers, understanding user needs, marketing performance, efficiency of development, and communications. Again, Table 4.2 shows that the characteristics of successful firms range from encountering less

Table 4.1 Aggregate results for both (process and scientific) industries

(a) Variables for which the probability of chance occurrence is less than 0.1%

Ranking[a]	*Variable*	*S > F*	*S = F*	*S < F*	*Binomial test*[b]
1(1)	Successful firms understand user needs better	33	10	—	1.2E-10
2(2)	Successful innovations have fewer after-sales problems	1	13	31	7.7E-09
3(3)	Successful firms employ larger sales efforts	22	21	—	2.4E-07
4(5)	Successful innovations have fewer 'bugs' in production	1	17	25	4.0E-07
5(7)	Successful firms have better coupling in specialised areas	23	19	1	1.5E-06
6(6)	Successful firms pay more attention to educating users	23	19	1	1.5E-06
7(4)	Successful innovations need less adaptation by users	0	28	15	3.1E-05
8(12)	Successful innovations need fewer unexpected adjustments in production	2	20	21	3.3E-05
9(15)	In successful firms there is less opposition to the innovation on commercial grounds	3	20	20	2.4E-04
10(21)	Successful firms give more publicity to the innovation	17	24	2	3.6E-04
11(10)	Successful firms benefit from outside technology during production	25	12	6	4.4E-04
12(8)	Successful firms drop products as a result of the innovation	11	32	—	4.9E-04
13(11)	Successful firms seek the innovation more deliberately	18	22	3	7.4E-04
14(19)	Successful firms have better external communications	18	22	3	7.4E-04
15(13)	The executive in charge of the successful innovation has more power	20	19	4	7.7E-04

Table 4.1 continued

(b) Variables for which the probability of chance occurrence is less than 1%

Ranking[a]	*Variable*	$S > F$	$S = F$	$S < F$	*Binomial test*[b]
16	The executive in charge of success has more responsibility	18	20	4	1.3E-03
17	The innovation is made as a result of marketing decisions rather than production decisions	9	34	—	2.0E-03
18	The executive in charge of success has more diverse experience	20	18	5	2.0E-03
19	The executive in charge of success has more enthusiasm	14	27	2	2.1E-03
20	The executive in charge of success has higher status	18	21	4	2.2E-03
21	In successful firms there is less opposition to the innovation on technical grounds				
22	Successful firms have better internal communications	13	28	2	3.7E-03
23	Successful innovations are more radical for world technology	16	23	4	5.9E-03

Source: Rothwell *et al.* (1974) SAPPHO updated — project SAPPHO phase II, *Research Policy* 3, pp. 261–2.

Note: a. The quantity in brackets gives the rank order of the variables as derived from the results of the original 27 pairs (i.e. Project SAPPHO, phase I)

b. For convenience, 10^{-x} is written as E−x.

c. A further 17 variables emerged as significant at the 5% level.

Table 4.2 Principal results of projects SAPPHO and the underlying factors

Factor	*Symptom*
1. Strength of management and characteristics of managers	Successful firms (1) encounter less opposition to the innovation on commercial grounds (2) seek the innovation more deliberately (3) encounter less opposition to the innovation on technical grounds (4) make the decision to innovate more for marketing reasons (rather than production ones) (5) perform more of the R & D in house, sometimes completely (6) have R & D chiefs with greater seniority (7) employ larger teams at the beginning of development (8) overspend less than unsuccessful firms
	The business innovator[1] responsible for successes (9) has more power (10) has more responsibility (11) has had more diverse experience (including overseas experience) (12) has more enthusiasm for the innovation (13) has a higher status than his counterpart in the unsuccessful firm
	Successful firms (14) have someone who plays the role of 'product champion'[2] whereas failures do not
2. Understanding user needs	Successful innovations (15) need less adaptation by users (16) need fewer modifications resulting from user experience after commercial sales than unsuccessful innovations
	Successful firms (17) understand user requirements better (18) see user problems earlier than unsuccessful firms

Table 4.2 continued

Factor	*Symptom*
3. Marketing performance	Successful firms (19) employ greater sales efforts (20) devote more effort to educating users (21) give more publicity to the innovation than unsuccessful firms
4. Efficiency of development	Successful innovations (22) have fewer after-sales problems (23) have fewer technical 'bugs' in production (24) have fewer unexpected adjustments in production (25) are less modified during development than unsuccessful innovations Successful firms (26) rate prospects of technical success lower at the outset (27) make more use of development engineers in planning for production than unsuccessful firms
5. Communications	Successful firms (28) have better coupling with the external scientific and technical community in the specialised areas concerned (29) benefit from dependence on outside technology during production (30) have better external communications (31) have better internal communications than unsuccessful firms

Source: SAPPHO updated — phase II (1974), *Research Policy*, vol. 3, pp. 265–6, 291

Notes: 1. Business innovator: 'The individual who was actually responsible within the management structure for the overall progress of this project.' He/ she could be the technical director, research director, sales director, technical innovator, chief engineer or the chief executive.

2. Product champion: 'Any individual who made a decisive contribution to the innovation by actively and enthusiastically promoting its progress through critical stages.'

a. Technical innovator: The 'inventor' or individual who made the major contribution on the technical side to the development and/or design of the innovation. He would normally, but not necessarily, be a member of the innovating organisation. He would sometime, but not always, be the 'inventor' of the new product or process.

b. Chief executive: The individual who is formally the head of the executive structure of the innovating organisation, usually but not necessarily with the job title of 'managing director'.

(Notes a and b as per Rothwell *et al.*'s definitions in SAPPHO updated (1974) *Research Policy*, vol. 3, p. 291.)

opposition to innovation on both commercial and technical grounds, understanding user requirements better, through greater sales efforts, making more use of development engineers in planning for production, to having better internal and external communications.

Nevertheless, after thorough analysis of the data collected from both studies, using different statistical techniques (factor, univariate and component analyses), project SAPPHO came up with the conclusion that there are five major areas of difference between success and failure. These main areas of difference, aggregated for both industries, are briefly summarized as follows:

1. Successful innovators were seen to have a much better understanding of user needs.
2. Successful innovators pay more attention to marketing and publicity.
3. Successful innovators perform their development work more efficiently than failures but not necessarily more quickly.
4. Successful innovators make more use of outside technology and scientific advice, not necessarily in general but in the specific area concerned.
5. The responsible individuals (technical innovator, business innovator, chief executive and product champion) in the successful attempts are usually more senior and have greater authority than their counterparts who fail. See 1., 2., a, and b in notes for Table 4.2.

One can draw the conclusion, based on this review, that progressive firms are, among other things:

(a) customer/market-orientated
(b) design/technology-driven
(c) forward-looking, and
(d) have experienced leaders.

Project NewProd

Project NewProd, undertaken by Cooper (1979a) is an investigation into new-project outcomes. It represents one of the pioneer works in this field. The study appears to be similar to project SAPPHO in the sense the 'paired' comparison of innovations were also made — 'one a commercial success and the other a failure'. There are, however, some areas of difference in terms of sample size and statistical methods used as well as questions as to how far one can generalize from a study based solely on Canadian experience.

As the basis of his study Cooper reviewed previous work in the subject and as can be seen, Table 4.3 shows twenty-one variables discriminating between success and failure, grouped under five main underlying factors. Of particular interest are five factors that occurred more frequently than others (i.e. variables that repeatedly occurred in not less than five of the seven separate studies), suggestive of the significance of these characteristics to successful innovation. They are:

(1) Extensive customer–producer interfacing;
(2) Market pull (idea derived from market);
(3) Better internal and external communication;
(4) Fewer after-sales problems;
(5) Better planning and systematic approach to innovation.

NewProd results

Cooper (1979a) identified six groups of variables from his 'conceptual descriptive model' and formulated eight hypotheses designed to show the 'expected' effect each of the six variables would have on new products as shown in Table 4.4. These six underlying dimensions contained seventy-seven variables carefully designed to test the eight propositions in Table 4.4. Using factor analysis, Cooper was able to collapse the seventy-seven variables into a more useful and manageable eighteen factors that were further reduced to eleven factors that were found to discriminate between 'new-product successes and failures'. They encompass:

(1) Introducing a unique but superior product;
(2) Having market knowledge and marketing proficiency;
(3) Having technical and production synergy and proficiency;
(4) Avoiding dynamic markets with many new-product introductions;
(5) Being in a large, high-need, growth market;
(6) Avoiding introducing a high-priced product with no economic advantage;
(7) Having a good 'product/company fit' with respect to managerial and marketing resources;
(8) Avoiding a competitive market with satisfied customers;
(9) Avoiding products 'new to the firm';
(10) Having a strong marketing communications and launch effort;

Table 4.3: Recent findings from success versus failure investigations

Variables discriminating between success and failure	*Research study*[a]	*Type of variable*
Understanding user needs	1, 2, 3, 6	
Extensive customer–producer interfacing	1, 2, 3, 4, 6	New product process
Efficient performance of development process (e.g. few 'bugs', fewer modifications, better planning, etc.)	1, 2, 3, 6	
Sales-forecasting carried out	1, 2, 6	
Strong selling and marketing effort	1, 2, 3, 6	Launching effort and product offering
Product and/or price advantages	4, 7	
Strong promotional, user education effort	1	
Fewer after-sales problems	1, 2, 3, 4, 6	
No initial marketing difficulties	4	
Market pull (idea derived from market)	1, 2, 3, 4, 5, 6	Nature of venture
Technology push (for major successes)	5, 6, 7	
Good fit with company potentials, resources	3	
Close to current market (familiarity with market)	1, 3, 4	
Better internal and external communication	1, 2, 3, 4, 5, 6	
Better co-ordination of R & D, marketing, production	1, 2, 3, 6	Organisational
Product champion or top management support	1, 2, 4, 6	
Better planning and systematic approach to innovation	1, 3, 4, 5, 6	Descriptors
Industry maturity (affects nature of venture)	4, 6	
Business cycle (influenced efforts)	5, 6	External industry, environment
Government role (induced effort)	4, 5	
Government support insignificant (except for major innovations)	1, 4, 6	

Source: Cooper (1979b) 'The dimensions of industrial new product success and failure', in *Journal of Marketing*, vol. 43, summer, p. 95. (Part-report of Project NewProd., 1979).

Notes: a. Studies refer to:
1. Rothwell, 1972, Rothwell *et al.*, 1974
2. Rothwell, 1974 (Hungarian SAPPHO)
3. Kulvik, 1977
4. Utterback *et al.*, 1976
5. Gerstenfeld, 1976
6. Rothwell, 1976
7. Davidson, 1976

Table 4.4 Research propositions

New-product success is expected to be positively related to:

1. Products which are superior, have a differential or economic advantage, or are unique relative to competing products; 2. Products where the other elements of the commercial entity — selling, distribution, production, etc. — are proficient;	Commercial entity
3. Projects where considerable technical and market knowledge is acquired;	Information acquired
4. Projects where the technical, marketing, and evaluative (process) activities are proficiently undertaken:	Proficiency of process activities
5. Products entering mass, large, growing, dynamic and uncompetitive markets, with a high but unsatisfied need for such products;	Nature of marketplace
6. Projects where a high degree of resource comparability exists between the needs of the project and the resource base of the firm; 7. Familiar projects to the firm (do not involve new technologies, new markets, etc.);	Resource base of firm
8. Market-derived projects (product ideas come from the marketplace).	Nature of the project

Source: Cooper (1979) 'The dimensions of new product success and failure', in *Journal of Marketing*, vol. 43, summer, p. 96. (Article based on result of Project NewProd, 1979.)

Note: In addition, other characteristics of the project, for example, whether a custom product or not; whether a true innovation or a 'me too' effort, etc. were expected to impact on product outcomes, but in a moderating way.

(11) Having a market-derived idea with considerable investment involved.

The overall review of Cooper's (1979) work suggests that successful innovators:

1. Emphasize superior overall performance. For a product to be overall superior, its design must be right. By this is meant that a superior product should incorporate all or a combination of design attributes — reliability, durability, performance in operation, flexibility, ease of use, safety in use, ease of maintenance, etc., all of which make one product superior to, or better than the other. The design should also encapsulate some elements of unique

differentiation to provide choice for consumers.

2. Have a thorough knowledge of both market and customer. This implies that progressive firms use market research to identify who buys their products, why and where they buy the products, and when. Again, with the feedback from market research, progressive firms will be able to modify their products via redesign or designing new ones to satisfy various customer needs and wants.
3. Develop an effective new-product-launch strategy, implying that progressive firms develop strong marketing efforts — prompt delivery, efficient after-sales service, effective advertising and promotion, effective distribution, and personal selling. As discussed in Chapter 2, all these are design-related factors in the sense that a customer does not only buy the physical product, he also buys the accompanying services (prompt delivery and efficient after-sales service). Besides, if a customer is not made aware of the existence of a product, he may not enquire, let alone buy, so the product should be advertised and promoted. Even where a product is advertised, but is not made readily available in retail outlets, customers may not find it to buy, so it should also be effectively distributed — hence it is argued that delivery, after-sales service, advertising and promotion, etc., are important non-price factors.

Since project NewProd, Cooper has furthered his studies in this area. In his study of 'the strategy-performance link in product innovation' Cooper (1984a) found that the strategies leading to high performance were characterized by the following dimensions:

1. An aggressive technological orientation: strong R & D orientation and proactive in acquiring new technologies.
2. A venturesome, offensive programme that was viewed as a leading edge of corporate strategy.
3. A market orientated programme, featuring a strong effort to identify customer needs, and a proactive search effort for new-product ideas.
4. The development of products with a differential advantage, which met customer needs better than competitive products, and had a marked impact on customers.
5. The use of sophisticated technologies, but with a high degree of synergy with the firm's resource base.
6. A relatively diverse new-product programme: products,

technologies and end uses not necessarily closely related to each other.

He concluded: 'What we witness is not a single strategy, but a packet of strategies that differentiated these high performers from the rest of the firms. A marriage of technological prowess, a strong marketing orientation, the search for a differentiated advantage, and a willingness to accept risk appears to be the key to a high performance program.'

Other related studies

Other studies portraying the characteristics of technically progressive firms and successful innovators abound. In his review of the results of some important innovation studies carried out in Hungary, Belgium, Holland, the US and the UK, Rothwell (1977) highlighted twelve highly significant success factors, as shown in Table 4.5.

In addition to the major variables, shown in Table 4.5, other types of variables such as resource availability, improvements, small-step innovations, government support, and successful implementation of technical change were also found to contribute to successful innovation and technical progress. In summary, the review suggests that progressive firms:

1. enter into collaborative arrangements with customers/users;
2. co-ordinate and co-operate between and among departments;
3. communicate effectively both internally and externally;
4. are in search of technical excellence;
5. have clear objectives;
6. are customer/market-driven; and
7. have qualified and experienced staff.

Overall, this review is particularly important because of its consistency with the findings of previous studies — Carter and Williams (1957), Myers and Marquis (1969), Langrish *et al.* (1972), and Projects SAPPHO (1971, 1974), as discussed earlier. In turn, these findings underpin Ughanwa's (1986) study of Queen's Award winners that we now report.

In the review of literature, a number of factors that determine the competitiveness of firms in international markets were identified. The firms in our survey were therefore asked to indicate the

Table 4.5 Characteristics of successful innovators and technically progressive firms

Variables characterizing success	*Research study*[a]	*Type of variable*
Good internal and external communication important to success.	1, 2, 3, 4, 5	Good communication and effective collaboration
Successful innovators collaborated with external agencies/customers from early stage in the innovation process.	1, 3, 5	
Made good contacts with scientific and technological organisations.	2	
Harmonious co-operation between all functional areas (research, development, production, financial and marketing departments).	1, 2, 3	Innovation as a corporate-wide task
Successful firms performed their development work more efficiently than failures (eliminated technical bugs before commercial sales).	1, 2	Effective development work
Most innovations were formally budgeted for at the start (budgets for market and development available).	1, 2, 3, 4, 5	Planning and management techniques
Made use of systematic sales forecasting to add innovation to the firm's product lines.	1, 2	
Employ careful production costing and control procedures, as well as engineers with considerable experience in development and production.	1, 2	
Have explicitly formulated policy towards innovation (planning was more highly structured and sophisticated in successful cases).	1, 4, 5	
Majority of the chief executives in successful firms possess technical qualifications — many of them graduates.	1, 4	Quality of management, personnel policy, and management
Open-minded progressive managers play an important role in success.	1, 2	
Successful innovators have ample scope for training.	4	
The presence of an effective middle management structure is associated with success.	1, 2	
Successful firms perceived purposeful innovation as taking positive action to promoting key personnel's creativity.	4	

Table 4.5 continued

Variable characterizing success	*Research study*[a]	*Type of variable*
Successful innovators are customer-orientated (market/user need recognised before technical solution)	1, 2, 3, 4, 5	Marketing and user needs
Successful firms provide efficient and reliable after-sales maintenance services as well as user education.	1, 2, 5	After-sales service and user education
If in marketing the innovation, the enterprise was able to penetrate new fields, the chances of success improved considerably.	2	
Success is promoted by enthusiastic top management	1, 2, 5	Key individuals
No lack of new materials, components, parts, and labour force hindered the launching of mass production in success.	1	The impact of resource availability
Successful innovators are associated with formal R & D department (average monthly expenditure on successful projects was, on average, twice that on unsuccessful projects).	1, 2	
Successful radical innovators, on average, have three times the employment of incremental innovators.	1	
Progress occurs through a combination of both radical and incremental innovation.	1	Role of improvement/small step innovations in technical advances.
Government support fostered success.	1	Government support
Innovation perceived as of great importance in successful firms.	1, 2, 4	Contribution of successful implementation of technical change

Source: Rothwell (1977), 'The characteristics of successful innovators and technically progressive firms', *R & D Management*, vol. 7, no. 3, pp. 192–9 (adapted version)

Note: a. Studies refer to:

1. Textile Machinery Study: investigated the factors surrounding the generation of twenty radical and fifteen incremental innovations (all commercial successes) in the textile machinery industry. Looked also at the factors associated with eighteen failures (ten incremental and eight radical). The project included the detailed study of some twenty enterprises and the sample was international.
2. Hungarian Study: an adaptation of Project SAPPHO pair-comparison technique to twelve success/failure pairs in the Hungarian electronics industry.
3. Belgian Study: studies innovation strategy and product policy in twelve Belgian enterprises over a ten- to fifteen-year period. Success criterion commercial (profit margin better than 7%).
4. Dutch Study: studied the factors affecting the innovation potential of forty-five Dutch companies in the metal-working sector between 1966 and 1971. The success criterion

$$\frac{\text{1971 turnover of innovations marketed since 1966} \times 100}{\text{total turnover 1971.}}$$

5. MIT Study: investigated the factors affecting success and failure in innovation in five industries (automobiles, industrial chemicals, computers, consumer electronics, and textiles) and five countries (France, F.R. Germany, the Netherlands, Japan, the UK) — total sample consisted of 164 innovations.

extent to which these factors have contributed to their competitiveness in international markets. The aim was to identify the most important factors that characterized the commercial success of the Queen's award winners in manufactured exports, with a view to generalizing such findings. Our findings were that commercially successful firms:

1. *Have clear objectives.* The Queen's Award winners set clear and achievable objectives. They know where to go and how to get there. Of the eighty-seven firms who responded, 87 per cent (75 out of 87) said it was an important element in determining their competitiveness, in marked contrast, only 2 per cent rated it as not important. This is an important finding because a company without clear objectives could be likened to a ship without a pilot. Thus, a clear definition of objectives tends to make for a better understanding of the design/innovation concept that, in turn, enhances the management of design and thereby promotes international competitiveness. There is, therefore, the need for innovators to set clear and achievable objectives in their organizations to achieve good results.
2. *Attach great importance to the effective management of design and innovation.* Of the sample, 97 per cent rated this factor (87 out of 90); 89 per cent perceived it as important as against 2 per cent who did not believe it was important. As revealed in the literature and as will be seen later in Chapter 7, the management of design and innovation is of paramount importance in the success of new products. The effective management of design encourages, among other things, better definition of design, continuous design evaluation and review, co-operation and co-ordination of efforts between and among functional areas, creation of design/cost awareness throughout the organization, and so forth. This implies that firms should pay attention to the effective management of design and innovation to enhance commercially successful products.
3. *Encourage closer interaction between R & D, production, and marketing.* Of the sample, 96 per cent (87 out of 90) rated this factor, 85 per cent believed in its importance in determining their competitiveness in international markets, while 3 per cent were opposed to this view. A commercially successful product can hardly be achieved without closer interaction between R & D, production, and

marketing. This is because the development of a new product is basically dependent on market information in relation to customer needs/wants, tastes/preferences. The marketing department usually provides this information. Similarly, before developing a new product, the production department must be consulted by R & D to ensure that a suitable process is available for its manufacture. The role of the marketing department comes into play again soon after the product rolls off the production line. The cycle starts all over again when the marketing department sends feedback to R & D. This means that firms should attempt to encourage a closer co-operation between R & D, production, and marketing to promote commercially successful products.

4. *Make extensive use of ideas derived from the market.* Of the eighty-seven firms who responded, 79 per cent (69 out of 87) rated it as important, 6 per cent did not support this view.
5. *Make extensive use of ideas derived from technology.* Of the sample, 99 per cent (89 out of 90) rated this factor, of which 76 per cent said it was an important determinant, compared to 5 per cent who perceived it as not important.

   In a dynamic environment where tastes/preferences change fast, firms require to keep in constant touch with the market in order to find out what the needs and wants of customers/users are, with a view to improving upon the existing products or designing new ones to satisfy those needs. Often the market needs (need-pull) triggers search for new technologies (technology-push), aimed at meeting such needs, and vice versa. Thus, ideas derived from both the market and technology are often complementary. In other words, while searching for market needs, firms should strive to find suitable technologies with which to exploit them.
6. *Indicate readiness to carry out in-depth research before entering new markets.* Of the eighty-eight firms who rated this factor, 76 per cent (67 out of 88) believed it was an important determinant, 10 per cent believed the contrary. It was evident in the literature that one of the reasons for the UK's lack of competitiveness was lack of in-depth study before entering new markets. This is particularly important because a 'shallow' study of any market tends to present sketchy information about that market. In turn, the firm may find it difficult to define

specific needs to satisfy. Consequently, the company may not be able to compete on an equal footing with the rival who has carried out in-depth research of the market. The high response given to this factor also seems to reflect upon its importance, suggesting that innovators/firms should carry out in-depth research before entering new markets.

7. *Indicate readiness to adopt modern techniques.* Of the sample, 94 per cent (85 out of 90) rated this factor, 65 per cent supported the view that it was an important determinant of competitiveness in international markets, 12 per cent opposed it. The fact that 78 per cent (70 out of 90) of the sample introduced significant changes to their manufacturing processes during the period 1981–5 strongly suggests that significant gains can be made from such changes. The reasons given by respondents for introducing changes such as increase in profits, increase in productivity, reduction in costs, increase in added value, etc., bear testimony to this. This implies that firms should adopt modern techniques to enhance the effectiveness of manufactured products.
8. *Indicate readiness to look far into the future.* Of the respondents, 66 per cent (57 out of 90) said it was important. The Queen's Award winners were prepared to make long-term plans, compared to 9 per cent who did not see it as an important factor. In this highly competitive environment, long-term planning seems to be an important strategy for sustaining and maintaining competitiveness. It may be a recipe for disaster, for instance, to concentrate on short-term plans nowadays, in light of the leap-frogging nature of design and innovation, suggesting that firms should look ahead at all times and embrace changes as they emerge, or they may be left steps behind their major competitors.
9. *Show willingness to enter into collaborative arrangements.* Of the eighty-eight firms who responded, thirty-seven (42 per cent) found this factor to be an important determinant of their competitiveness in international markets, as against 34 per cent (30 out of 88) who felt it was not important. Perhaps the lower rating accorded to this factor may be borne out by the fact that firms tend to drag their feet when it comes to compromising with fellow competitors on projects, for fear that rivals might capitalize on their weak points. This is not disputed. However, collaborative arrangement does not strictly imply collaborating with

competitors, firms can also collaborate with users. Even if it means collaborating with rivals only, there still seems to be more gains than losses to be made from such an arrangement, otherwise such collaborative arrangements between Austin Rover and Honda, Bedford and Pontiac, and Vauxhall and Isuzu, may not have been contemplated in the first place. Furthermore, a collaborative arrangement tends to speed up the cross-fertilization of innovative ideas between and among firms. It also enhances the design and development of commercially successful products, particularly when users are party to the arrangement. It seems clear from the above discussions that collaborative arrangement is an important competitive tool to which innovators' attention is drawn.

10. *Seek support (financial, political, etc.) from government.* The average rating of 'government support' was lowest of the responding firms. Only 24 per cent of the respondents rated this factor as important: in marked contrast, about twice this figure (46 per cent) said it was not important. This may be explained by the findings in Chapter 3 that the UK government spends a larger proportion of its GNP on defence *vis-à-vis* civilian industry, and that the government appears to be less keen on giving financial support to firms, particularly medium and large firms. For example, the loss of the Bosphorus Bridge (Turkey) contract to a Japanese firm by Cleveland Consortium (UK) resulted from the government's unwillingness to help financially. In marked contrast, the Japanese winner of the contract was financially supported by the Japanese government.
11. *Pay more attention to design and R & D.* In 77 per cent (70 out of 91) of the cases, the design function was located at the top-management level, whilst in 23 per cent it was located at the middle management.

    Similarly, 72 per cent (18 out of 25) of the sample classified the R & D function under top management, compared to 28 per cent who located it at the middle-management level. In both cases, no firm located design or R & D in the lower management cadre, suggesting the importance attached to design and R & D as major contributors to success in manufactured products.
12. *Pay more attention to choice of materials.* Of all the design/manufacture interface considerations (range of colour, existing manufacturing processes, suitability of existing manufacturing processes, and reduction in rather

than addition to the number of components per set), 'choice of materials' was perceived by the Queen's Award winners as the most important factor when taking decisions on the design and manufacture of new products.

13. *Pay more attention to technical qualities of a product.* Of all the factors that influenced the sales of the Queen's Award winners' products against those of their competitors in international markets, 'performance in operation' received the highest rating with 93 per cent saying it was influential compared to a meagre 5 per cent who did not support this view. 'Reliability' was second with 86 per cent rating it as influential as opposed to 3 per cent who said it was not.

    Attention to technical qualities also manifests itself in the management of design where the technical director was assigned the responsibility for managing design in 45 per cent of the cases — more than other directors. Similarly, the technical director was found to have the highest proportion representation of the R & D function at the board level in sixty-four firms investigated.
14. *Believe in effective leadership.* The Queen's Award winners believe that effective leadership is the key to commercial success. Consequently, 'was directed by an individual with outstanding authority and power (top person)' was rated as the most important factor responsible for the successful design and development of the award-winning products. Of the firms, eighty-six out of eighty-nine considered this factor, of which 74 per cent rated it as important compared to only 4 per cent who said it was not.

    Similarly, 'top management support' received the highest rating as the most important determinant of competitiveness in international markets. Of the firms, eighty-eight considered this factor and eighty (91 per cent) said it was important as against 4 per cent who believed the contrary.
15. Of the Queen's Award winners, 78 per cent (70 out of 90) made *significant changes to their manufacturing processes* between 1981 and 1985. Among the major changes are: automation (flexible manufacturing system, materials requirement planning, automatic testing equipment, computer numerically controlled machine, printed circuit board) (55 per cent), CAD/CAM (28 per cent), modification of existing processes (22 per cent), robotics (14 per cent), and replacement of existing processes (12 per cent). This result largely portrays the rapid growth of

the use of computers in commercially successful firms. It also strongly suggests that the use of computers to aid the effectiveness of design and manufacture of new products has come to stay and firms who ignore it do so at their peril.

16. *Incremental innovation strategy.* The majority of the Queen's Award winners pursued incremental innovation as a powerful design strategy for achieving success in manufactured products. 'Major innovation (innovation leading to major changes in or major additions to an existing product)' and 'improvements (innovation leading to minor changes in or minor additions to the existing product)' were each considered by 78 per cent (70 out of 90) of the sample. However, whilst 67 per cent of the respondents ranked 'major innovation' as most important, 46 per cent said 'improvements' was the most significant design strategy. Only thirty-six firms considered 'radical breakthrough (innovation leading to an entirely new technology)' of which 19 per cent ranked it as most important.

    It is important to note that incremental innovation and radical breakthrough strategies are not mutually exclusive. Some series of incremental innovations (i.e. major innovations and improvements) may not be feasible without the initial major breakthrough. However, while efforts are concentrated on incremental innovation in the short and medium terms, in the light of the advantages they have (e.g. it takes shorter time to achieve, costs less, etc.), attempts at radical breakthroughs as a long-term strategy should not be relaxed.

17. *Pursue proactor (leader) strategy.* The majority of the Queen's Award winners were proactive, rather than reactive, in their approach to innovation. Of the sixty-three firms that pursued the strategy of 'introducing new products into specific market segments', (proactor (leader) strategy), 70 per cent ranked it as the most important innovation strategy, compared to 26 per cent who ranked 'introduce products in response to competitive pressure' as most important or 17 per cent for the strategy of 'protecting existing products via process innovation (e.g. cutting maufacturing costs)', both of which are classified as reactor (follower) strategies. By introducing their new products into specific market segments, the Award winners are able to gain deeper knowledge into the particular needs

of the specific segment of the market. Also, through this innovation strategy, they are able not only to satisfy these needs profitably, they satisfy some of the criteria for winning the Queen's Award such as:

(a) A substantial and sustained increase in total exports over a period of three years.
(b) A substantial and sustained increase in the percentage of total export sales to total business over a period of three years.
(c) A percentage of exports to total business that is considerably higher than the average for the applicant's sector of industry.

Nevertheless in today's competitive world, there is the need to balance leaders and followers' strategies to maintain and sustain competitiveness in international markets.

18. *Have a broad conception of design.* The award-winning firms tend to have a broad conception of design. That is, they do not consider design as an embodiment of a single factor, rather they perceive design as an amalgam of factors comprising design, design-related, and design-influenced elements that are market-led. Of the four major factors selected by respondents as best encapsulating the concept of design (see Chapter 6, Table 6.1A) 'making products that make profit' was ranked as most important by 51 per cent of the sample (sixty-eight firms), followed by 'making products that sell' (35 per cent). 'Superior overall performance' and 'making products that function technically', were ranked third and fourth by 36 per cent and 24 per cent of the responding firms, respectively. As can be seen, there seems to be no outright majority for any one factor. Besides, the responses overlap in places. This means that one factor was ranked by more than one firm, which lends credence to the thesis that the concept of design encapsulates a combination of factors.

On the other hand, the link between the design concept and market orientation is borne out by the fact that the four major factors selected by respondents as best encapsulating the concept of design mainly depend upon the customers/users' evaluation of value/quality and so underline the fact that successful firms *must* be marketing-orientated.

Further, to make profit, a firm must sell its products; and to sell, a product must have desirable qualities. The remainder of our findings will demonstrate how the respondent firms have achieved these qualities in their manufactured products to win the Queen's Award. In other words, the findings will present the characteristics of the Queen's Award winners that enabled them to achieve success in international markets.

19. *Encourage continuous involvement of customers/users in new product development.* The award winning firms perceived the continuous involvement of customers/users during the process of design and development as a powerful tool for achieving and sustaining commercially successful products. Of the sample, 96 per cent involved customers in the development of their new products compared to 4 per cent who did not. In this way they are able to interact closely with customers/users, discover and understand their needs, and satisfy them profitably. In fact, 'close interaction with customers/users' was found to be the most significant source of new ideas used by the Queen's Award winners for the design and redesign of new and improved products. However, there is evidence in the findings to suggest that 'close interaction with customers/users' is by no means the only important source of new ideas for the design and development of new products. The commercial success of new products is a function of the right mix of new ideas.
20. *Employ more engineering designers than industrial designers in-house.* Of a total of 19,904 designers employed in-house by seventy-eight Queen's Award winners, 97.7 per cent were engineering designers compared to a meagre 2.3 per cent who were industrial designers. In another investigation, 50 per cent of the sample satisfied their industrial needs through engineers rather than through industrial designers themselves. The remainder of the needs were satisfied by the use of external consultants and industrial designers.
21. *Strive to build products with higher technological contents than rival product offerings.* Of the sample, 96 per cent (85 our of 89) considered this factor, of which 63 per cent stated that it contributed significantly to the successful design and development of the award-winning products, as against 11 per cent who considered it to be not significant.
22. *Make continuous product reviews during and after product design and development in the light of changes in the environment.* The Queen's Award winners tend to review their products from time to time in line with changes in the

market environment. Of the respondents, 67 per cent (55 out of 83) believed this accounted for the successful design and development of the award-winning products; 12 per cent (ten firms) believed to the contrary.

23. *Make more use of new and improved manufacturing techniques.* The winners of the Queen's Award tend to introduce new techniques to their manufacturing processes (e.g. CAD/CAM, FMS, MRP, etc.) and seek to improve upon the existing ones to increase efficiency and productivity. Of the sample, 94 per cent (84 out of 89) reacted to this factor of which 60 per cent said it contributed significantly to the successful design and development of the award-winning product. In contrast, only 12 per cent said it was not important.
24. *Encourage designers to see new products through to commercialization.* Over half of the sample (57 per cent) encouraged their designers to see products they have designed through to commercialization, compared to 14 per cent who did not. By doing so, designers are able to get feedback directly from customers/users as to the performance of the products. The products are then improved based on the feedback, or new ones designed, to provide better value for money.
25. *Ensure that products are designed and developed by a team of qualified engineering and industrial designers.* The Queen's Award winners combined the skills of engineers and industrial designers in the design and development of their new products, and 61 per cent said that this accounted for the successful design and development of the award-winning products; 16 per cent were opposed to this view.
26. *Have the reputation for the use of design, design-related, and design-influenced factors to achieve international competitiveness.* The winners of the Queen's Award attached great importance to:

(a) Design factors (technical)	Performance in operation Reliability Technical sophistication Durability
(b) Design-related factors (marketing)	Efficient delivery Quality of after-sales service Advertising and promotion Sale price

(c) Design-influenced factors (ergonomics)	Ease of use Safe to use Easy to maintain Parts availability Flexibility and adaptability in use Operator comfort
(d) Design-influenced factors (aesthetics)	Attractive appearance/shape Style/fashion

They tend to make the best fit between and among these aspects of design dimension to enhance their competitiveness in manufactured products/exports.

In the subsequent section, the major factors underlying design success in international markets will be outlined.

Major factors underlying design success in international markets

In all, thirty-four factors underlying design success in international markets were identified and labelled (see Table 4.6). Factor analysis (SPSSX Routine — varimax rotation) was performed to reduce the thirty-four variables to their underlying factors (all eleven factors generated from a five-point interval scale), as presented in Table 4.7. All eleven factors had eigenvalues in excess of 1.0; one (factor 1) had an eigenvalue greater than 8.0, two (factors 2 and 3) had eigenvalues greater than 3.0 and 2.0 respectively, whilst the eigenvalues of the remainder (factors 4–11) were greater than 1.0.

The eleven principal components explained 75.5 per cent of the variance in the original thirty-four variables after rotation, and thus seem to underpin the major factors characterizing design success in international markets. This is attested to in the variable loadings that were found to be high, mostly over 0.60.

We shall now discuss the eleven dimensions (factors) in detail, to demonstrate how they influence the design success in international markets.

1. *Managerial and technical synergy and proficiency.* This dimension describes the managerial and technical orientation that underpins the design success in international markets. The managerial functions include:

Table 4.6 Factors underlying design success in international markets

Factors determining the competitiveness of firms in manufactured exports

1. Top management support.
2. Had clear objectives.
3. Effective management of design and innovation.
4. Closer interaction between R & D, production and marketing.
5. Made extensive use of ideas derived from the market.
6. Made extensive use of ideas derived from technology.
7. Readiness to carry out in-depth research before entering new markets.
8. Readiness to adopt modern techniques (e.g. CAD, CAM, automation, etc.).
9. Readiness to look far into the future.
10. Willingness to enter into collaborative arrangements.
11. Government support.

Factors determining the successful design and development of the award-winning products

12. Was directed by an individual with outstanding authority and power (top person).
13. Interacted closely with customers/users in design and development stages.
14. Had higher technological content than rival product offerings.
15. Made continuous product reviews during and after product design and development in the light of changes in the environment.
16. Made more use of new and improved manufacturing techniques.
17. Designer(s) saw product through to commercialization.
18. Product designed and developed by a team of qualified engineering and industrial designers.

Factors influencing the sale of award-winners' products against their rival offerings in international markets

19. Performance in operation.
20. Reliability.
21. Sale price.
22. Efficient delivery.
23. Quality of after-sales service.
24. Technical sophistication.
25. Durability.
26. Easy to use.
27. Safe to use.
28. Easy to maintain.
29. Parts availability.
30. Attractive appearance/shape.
31. Flexibility and adaptability in use.
32. Advertising and promotion.
33. Operator comfort.
34. Style/fashion.

Table 4.7 Major factors underlying design success in international markets

Factor name (% variance explained)	*Variables loading on factor*	*Type of variable*	*Variable loadings*
1. Managerial and technical synergy and proficiency	Effective management of design and innovation	Management	.790
	Readiness to look far into the future	Management	.764
	Top management support	Management	.753
(24.4%)	Had clear objectives	Management	.745
	Had higher technological content than rival product offerings	Commercial entity	.716
	Closer co-operation between R&D, production, and marketing	Project	.688
	Readiness to adopt modern techniques (e.g. CAD, CAM, automation, etc.	Project	.604
	Easy to use	Ergonomic	.441
	Performance in operation	Technical	.436
	Made extensive use of ideas derived from technology	Technology push	.390
	Made continuous product reviews during and after product design and development in the light of changes in the environment	Project	.372
2. Aesthetic/ergonomic, and communication synergy	Easy to use	Ergonomic	.395
	Style/fashion	Aesthetic	.823
	Attractive appearance/shape	Aesthetic	.771
(10.1%)	Advertising and promotion	Marketing	.721
	Operator comfort	Ergonomic	.703
	Flexibility and adaptability in use	Ergonomic	.426
	Durability	Technical	.363
3. Performance in use	Had clear objectives	Management	.313
	Easy to use	Ergonomic	.366
(7.6%)	Easy to maintain	Ergonomic	.805
	Parts availability	Ergonomic	.735
	Flexibility and adaptability in use	Ergonomic	.480
	Quality of after-sales service	Marketing	.378
	Technical sophistication	Technical	.426

Table 4.7 continued

4. Marketing knowledge and proficiency (5.5%)	Made extensive use of ideas derived from the market	Need-pull	.849
	Interacted closely with customers/users in design and development stages	Project	.807
	Readiness to carry out in-depth research before entering new markets	Marketing	.589
	Designer(s) saw the product through to commercialisation	Commercial entity	.309
5. New product development synergy and proficiency	Readiness to adopt modern techniques (e.g. CAD/CAM, robotization, etc.)	Project	.459
	Readiness to carry out in-depth research before entering new markets	Marketing	.396
(5.4%)	Made more use of new and improved manufacturing techniques	Project	.790
	Efficient delivery	Marketing	.688
	Quality of after-sales service	Marketing	.622
	Made extensive use of ideas derived from technology	Technology push	.386
	Designer(s) saw the product through to commercialization	Commercial entity	.314
6. Product uniqueness/superiority	Performance in operation	Technical	.405
	Quality of after-sales service	Marketing	.303
(4.9%)	Safe to use	Ergonomic	.737
	Durability	Technical	.613
	Made extensive use of ideas derived from technology	Technology push	.567
	Reliability	Technical	.566
	Technical sophistication	Technical	.495
7. Commercial production synergy	Product designed and developed by a team of qualified engineering and industrial designers	Project	.808
(4.5%)	Made continuous product reviews during and after product design and development in the light of changes in the environment	Project	.624
	Designer(s) saw the product through to commercialization	Commercial entity	.565

8. Design/manufacture, management synergy and proficiency (3.6%)	Top management support	Management	.330
	Closer co-operation between R&D, production and marketing	Project	.327
	Easy to use	Ergonomic	.329
	Technical sophistication	Technical	.315
	Designer(s) saw the product through to commercialization	Commercial entity	.368
	The design and development of the award-winning product was directed by an individual with outstanding authority and power (top person)	Project	.794
9. Collaboration and research magnitude (3.5%)	Readiness to carry out in-depth research before entering new markets	Marketing	.373
	Willingness to enter into collaborative arrangements	Management	.835
10. Government effort (3.1%)	Government support	Government	.900
11. Flexibility/adaptability advantage (3.0%)	Flexibility and adaptability in use	Ergonomic	.420
	Sale price	Commercial entity	.846

management of design/innovation, looking far into the future and making relevant long-term plans, giving support (financial, expertise, moral, etc.) to economically viable projects, setting clear and achievable objectives, reviewing and evaluating products in line with customer needs/wants and changes in the environment as well as introducing modern techniques to enhance the effectiveness of new products.

Technically, products are expected to perform well in operation and be easy to use. Often, this involves the application of new technology to improve the quality of new products, and in this context, to enhance the functional performance and ease of use of new products, both of which are unique selling points (USP).

2. *Aesthetic/ergonomic, and communication synergy.* Factor 2 draws attention to the aesthetic, ergonomic, and communication advantages that account for the design success in international (export) markets. The aesthetic features include: style/fashion and attractive appearance/shape, whilst 'easy to use', 'operator comfort', 'flexibility and adaptability in use', etc., exemplify ergonomic features. Aesthetic and ergonomic features can substantially influence the sale of one product over another. As discussed elsewhere, bulldozers, tractors, etc., are now produced in glittering colours, even though the manufacturers know, as we do, that these implements will soon soil in use and may never be cleaned. However, to increase awareness and sales, new products should be advertised and promoted.
3. *Performance in use.* Factor 3 describes the performance in use dimension. Customers/users expect products to be easy to use and maintain. In some products, flexibility and adaptability in use (e.g. mains and battery radio) and technical sophistication (e.g. remote-controlled TV) constitute important buying influences. A farmer may prefer to purchase a technically sophisticated agricultural machine that can combine two or more functions (e.g. planting and harvesting crops) rather than one with only one function.

   However, not all customers or users are skilled in or have time for carrying out repairs or routine maintenance of products in use. Therefore, there is the need to provide adequate after-sales service, particularly for sophisticated products. Thus, firms that have the capability of incorporating these various customer requirements in their new-product designs are likely to be more competitive than those who have not.

4. *Marketing knowledge and proficiency.* The fourth factor, which explains 5.5 per cent of the variance points to a marketing orientation as an important dimension underlying design success in world markets. It stresses that a firm with adequate knowledge of the market — a factor that may be brought about by carrying out a deep study of the market — often makes extensive use of ideas derived from the market when designing and developing new products. Close interaction with customers/users is also a particular characteristic in marketing knowledge orientation. It enables firms to ensure that the new ideas derived from the market are built into new products in accordance with customer/user specifications.

   Further, marketing knowledge is enhanced through the firm's encouragement of its designer(s) to see new-product design through to commercialization. The rationale is to get feedback on product performance upon which the existing products may be improved/modified or new ones designed.

   In sum, a firm with adequate knowledge of the market tends to identify and understand clearly the specific needs of the market and is thus, better able to satisfy them (i.e. to provide better value for money).
5. *New-product development synergy and proficiency.* Factor 5 describes the company and its proficiency in new-product development. The readiness of firms to find out exactly what the markets needs and wants are, before designing and developing new products; adopt modern techniques (e.g. CAD/CAM); and/or make more use of new and improved manufacturing techniques complemented by efficient delivery and effective after-sales service, all combine to create commercially successful manufactured products.
6. *Product uniqueness/superiority.* A product that performs well in operation, is durable, reliable, safe to use, technically sophisticated, and supported by an adequate after-sales facility is truly unique or superior; and is likely to be more competitive than its rival offerings.
7. *Commercial production synergy.* Design success in international markets is more likely to be achieved when a product is designed and developed by a team of relevant professionals (e.g. engineering designers and industrial

designers), rather than by a single individual. It is an old adage that two heads are better than one. In other words, a team of qualified people working on a design project is more likely to produce a synergistic (2+2=5) effect, than that executed by one person.

None the less, to maximize this effect, the product must be continuously reviewed and evaluated in line with the market requirements to ensure that the project incorporates the prevailing needs of customers/users. Thus the successful commercialization of new products appears to depend on the extent to which adequate market information/feedback is obtained to enable designers to create products that people would want to buy.

8. *Design/manufacture management synergy and proficiency.* The design and manufacture of commercially successful products is enhanced when it is supported and directed from the top by someone with enthusiasm, outstanding authority and power. As alluded to earlier, the management tends to encourage designers of products to interact with users aimed at soliciting on-the-spot comments about the overall performance of products while in use. The user records of the good and bad aspects of the product performance are noted and subsequently used to improve the existing products or make new ones to satisfy customers/users in a better way.
9. *Collaboration and research magnitude.* Factor 9 explains 3.5 per cent of the variance, and addresses the collaborative and research orientation. Design success in international markets is stimulated and triggered by the extent to which firms are willing to collaborate with 'outsiders' (users, universities/colleges, competitors, etc.) in design projects. Such collaborative arrangements often inject new innovative ideas into design projects that, in turn, help to create acceptable and competitive products. Examples include collaborative design project arrangements between Austin Rover (UK) and Honda (Japan); Bedford (UK) and Pontiac (USA); Vauxhall (UK) and Suzuki (Japan).

Nevertheless, the success of collaborative design projects appears to depend on the extent to which the market needs have been clearly identified and understood. In other words, the success or failure of new products largely depends on the extent to which firms carry out in-depth research of markets to find out what the specific user needs/wants and tastes/preferences are, before designing and developing new products.

10. *Government effort.* The role of government in the marketplace can be of considerable importance to the success of firms in manufactured products. Many firms, particularly small firms, cannot compete in some product areas — 'high-tech' sector, for example — either because they have no technical expertise or have expertise but not sufficient funds to support design projects or undertake research aimed at discovering new uses, modifying existing products, or designing new ones. In both respects, the government can be of immense help by providing support (financial and/or expertise) that will enable these companies to develop their potentials and expand. The UK government, for example, operates a 'funded consultancy scheme' whereby firms are provided with a range of consultancy services on design projects, free of charge. Under the Funded Consultancy Scheme — administered on behalf of the Department of Trade and Industry by The Design Council — the Government helps pay for the consultant design work needed to develop a new product or improve an existing one. The scheme has a dual purpose. First, is to help British companies, in the short term, to get well-designed products on the market quickly. Second, is to persuade companies, in the longer term, through their own practical experience, that investment in design pays off — in fuller order books and healthier profits. By 1984 'more than 1,700 firms have already taken advantage of the scheme, which is available for most kinds of product-related design in most kinds of company'.
11. *Flexibility/adaptability advantage.* A product that is flexible and adaptable in use is more likely to generate sales than one without these qualities. In sum, firms who are able to build flexibility and adaptability into their new products are likely to be more competitive in international markets than those who do not possess this capability.

That these descriptors emerge in a study of new-product design/innovation is not new. Indeed, virtually all have been referred to in one form or another in previous works — Myers and Marquis (1969), Rothwell *et al.* (1974); Rothwell, (1976a,b, 1983,) Cooper (1975), Kulvick (1977), and Walsh and Roy (1983), to name a few. The dimensions identified help to clarify and simplify the set of variables that underline design success in international markets. Instead of working with thirty-four or more interrelated design characteristics, the problem has been reduced to a more manageable proportion — eleven independent dimensions that characterize design success in international (export) markets.

It will be observed from the 'variables loading on factor' (see Table 4.7) that a number of factors characterizing design success in international markets interrelate and overlap strongly, suggesting that these factors should not be treated in isolation. Rather, attempts should be made to make the best fit between and among them, in the light of customer/user specifications/requirements, as well as changes in the environment.

These findings, therefore, have one major underlying implication for innovators, managers, and marketers alike. *They clearly signal the dangers of concentrating on one factor as the sole influencer of design success in international markets. This study indicates that the success of new-product designs in international markets is dependent on the matching of the appropriate or relevant interrelated factors.*

Efforts have, so far, concentrated on identifying factors that characterize success in international markets. However, if we know only factors that characterize a firm's success, we may not be able to guard against factors that characterize failure. In the following section, we shall briefly review the literature on barriers to innovation and then report our research findings in this area.

Innovation/design barriers

Rothwell (1977b) furthered his study on successful innovations by investigating factors contributing to failure* and delay to innovation or new-product design covering four major studies — Project SAPPHO, Hungarian SAPPHO, Queen's Award, and textile-machinery. Grouping the failure and barrier factors into four major components — marketing/market-related factors, technical factors, management factors, and mixed/other factors, Rothwell (1977b) identified four main dimensions associated with delay in innovation and the consequent failure, as follows:

1. Poor communication (internal and external);

* An unsuccessful innovation is one which, in the eyes of the customer/user, fails to satisfy the technical and/or economic requirements/specifications, and/or make any worthwhile profit as expected by the firm, and consequently has left the market prematurely, i.e. has been withdrawn from the market after launch or before the end of its estimated life cycle.

2. Poor quality of management and bad management practice;
3. Paucity of marketing effort and failure in interaction with potential customers;
4. Poor development work.

These factors are particularly important because their opposite numbers, that is, good communication (internal and external), good-quality management and good management practice; good marketing effort and frequent interaction with customers (existing and potential), and efficient development work have been shown to characterize successful innovators and technically progressive firms (see Table 4.5).

The conclusion to be drawn from this is that factors associated with success and failure in innovation seem to be amenable to manipulation by progressive firms through effective management of design/innovation and other related factors. This implies that 'while chance and uncertainty can upset even the best-laid schemes, responsibility for the success or failure of designs/innovations rests firmly in the hands of the innovating firms' own management'.

Survey results

It was felt that the investigation of barriers to innovation in the award-winning firms would help to throw more light on or explain the UK's laggardness in manufactured exports when compared with its major foreign competitors, and thus help in drawing reasoned conclusions on how to maximize success while minimizing failure. Respondents were asked to indicate factors they considered to be major barriers or delay factors to innovation (or new-product design) projects in their firms. The answers to this question are presented in Table 4.8, from which the following observations emerge:

1. Market uncertainty was the most frequently mentioned factor with more than half (52 per cent) of the sample considering it as a major barrier to innovation. This is an important finding and it strongly confirms an earlier work by Zeid (1981) who found that 75 per cent of the responding firms put the major reason behind the steady decline in the UK's export competitiveness to 'fluctuation in exchange rates'. The remedy is, clearly, to improve one's marketing intelligence and research activities.
2. 'Shortage of resources' (38 per cent) and the 'lack of financial return on R & D in today's competitive world'

Table 4.8 Barriers (delay factors) to innovation

	No. of responses	%[a]
No market or need	18	21[d]
Lack of financial returns on R & D in today's competitive world	31	37
Some other technology not sufficiently developed	16	19
Resistance to new ideas (i.e. over-attachment to old ideas)	20	24
Resistance to change from trade unions	3	4
Lack of in-depth research	16	19
Lack of technical know-how	11	13
Lack of qualified manpower/expertise	27	32
Shortage of resources	32	38
Market uncertainty	44	52
Lack of creativity/inventiveness	10	12
Others[c]		
Total	228[b]	271

Note: a. Based on 85 firms who answered this question.
b. A number of firms considered more than one factor as major barriers to innovation.
c. Others are:
(1) Lack of backing from senior management.
(2) Rigid control structure.
(3) Individual confidence — predisposition to risk.
(4) Management business strategy not well developed.
(5) Lack of investment in marketing.
(6) Shortage of graduates in production.
(7) Need ideas that lead to practical course of action.
(8) Lack of funds.
d. To be read: 21% of the resondents stated that 'no market or need' was the major barrier to innovation in their firms.

(37 per cent) were the second and third most frequently mentioned barriers/delay factors to innovation, respectively. The difficulty here is to break the circle as the lack of financial returns on R & D is likely to lead to the shortage of a company's financial resource which, in turn, may lead to the lack of investment in R & D. Once again the diligent application of Baker's (1975) composite model of buying behaviour, as described earlier, will help to solve or minimize this problem. The model simply requires the innovator (firm) to weigh up the pros and cons of courses of action. In this case, to do the cost/benefit analysis of investing in R & D in the light of competitive activities in

that product area on the one hand, and the exploration of an alternative investment in the light of the available company financial resources on the other. Armed with this formula, the innovator will be able to come up with an overall judgement as to which project to choose in the light of how it measures against those of the competitors. In this way (provided the necessary data are available) the innovator will be able to take a correct decision from the outset whether or not to start, suspend, or abort the project, rather than invest a huge sum of money on a project only to blame 'poor financial returns on R & D' for a failure that probably could have been avoided or minimized by applying this particular model.

3. Although only 24 per cent of the respondents perceived 'resistance to new ideas (i.e. over-attachment to old ideas)' as a major barrier to innovation, this appears to be a worrying figure in the light of strands of literature stressing the parochial mindedness or conservative attitude of British management towards innovation. It is feared that with the leap-frogging nature of design in this highly competitive environment, the 'resistance to new ideas/over-attachment to old ideas' on the part of British management, may detract from the competitiveness of British-made products in international markets. However, given the respondents are unlikely to hold these attitudes perhaps their own perception is unsurprising.
4. Also mentioned as major barriers/delay factors to innovation are: lack of qualified manpower (32 per cent), lack of technical know-how (13 per cent), and lack of creativity/inventiveness (12 per cent). This appears to negate our contention that Great Britain does not lag behind its major competitors in these areas but is probably more a reflection of the sample.
5. Only 4 per cent (three out of eighty-five) of the respondents considered resistance to change from trade unions as a major barrier to innovation. Again this may reflect the experience of firms that have overcome these problems (which is why they've succeeded) as well as the marked change in the industrial-relations climate during the Thatcher years.
6. A number of respondents considered other factors as barriers to innovation. These factors are stated under 'others' (see Table 4.8). As can be seen, many of these factors strongly support the main findings. For instance,

> 'rigid control structure' and 'lack of backing from senior management' are associated with 'resistance to new ideas/over-attachment to old ideas'; 'predisposition to take risks' correlates with 'market uncertanty'; 'shortage of graduates in production' supports the 'lack of qualified manpower'; whilst 'need ideas that lead to practical course of action', and 'lack of funds' corroborate 'no market or need', and 'shortage of resources', respectively.

Conclusion

The overall review of factors characterizing commercially progressive firms appears to suggest that there is no Royal road to, or 'cookbook recipe' for success. To assert that there is one, may be grossly misleading in the light of the dynamic nature of the environment in which design/innovation takes place, as well as the 'competitive nature of innovation and the uncertainty inherent in the process'.

There seems also to be no clear-cut or distinct line of demarcation between success and failure. This is because, as Baker (1975) strongly argues, 'it is difficult to define a common denominator with which we may judge success and failure'. That is to say, that today's failure could be tomorrow's success. This lends credence to our earlier argument that some of the factors associated with success and failure are amenable to manipulation.

Although Project SAPPHO's (1971) assertion, that 'success is not always self-evident', is not disputed, there always appears to be that glimmer of hope for success in every firm and at every introduction of a new product/process, as no firm goes to the marketplace with the intention to fail. What happens after the introduction is largely a function of many exogenous factors — market uncertainty, competition, government policy and legislation, luck, and, in fact, the effect of general economic and technological environment.

This can be illustrated with Project SAPPHO's (1971) analogy with a football game:

> Both in football and in competitive industrial innovation (new product design) there will always be winners or losers. This is the nature of the game. Knowledge of those factors which are conducive to success may lead some managers and players to succeed more often, if they are able to apply this knowledge in practice. But their opponents will also learn, so that circumstances are constantly changing and the end-result may

> simply be an improvement in the standards of management all round. The managers of football teams mostly know what their teams ought to do in order to win, but the factors which they are striving to control are not easy to manipulate and they certainly cannot guarantee success in any particular game. What can be much more positively asserted is that a team which has not learnt to adapt its tactics and level of fitness to contemporary standards will find itself at the bottom of the League, (or 'knocked out' if it is that type of competition). To learn the 'rules' of innovation management may sometimes lead to well-earned success, depending upon the relative efforts of competitors and an element of chance. Not to know the 'rules' or how to apply them often means disaster.

Although the great majority of our sample considered market uncertainty as a major barrier to innovation, it is not in itself sufficient a factor to explain the UK's lack of competitiveness in manufactured exports. This is based on the proposition that market uncertainty is an uncontrollable (external) factor that universally affects every competitor in the marketplace.

Market uncertainty can only be minimized, it cannot be stopped. One way of minimizing market uncertainty is suggested here. First, the innovator (firm) must be aware of the presence of market uncertainty in our environment. Second, the innovator must have confidence in him/herself and must be predisposed to take risk. It is often said that 'the greatest risk is not taking risk at all'. Both self-confidence and risk-taking may be strengthened and sustained by setting clear and achievable objectives. Last but not least, the innovator must take precautions in the light of changes in the environment and must be ready to move with time. Once again, the application of Baker's model will help to facilitate this move. That is to say, it will enable innovators to 'model' their own thinking towards the appropriate precautions (or courses of action) necessary to guard against or minimize the market uncertainty.

However the model is a greatly simplified representation of an exceptionally complex reality and the real challenge is to operationalize it. In the chapters that follow, we attempt to provide some pointers and inputs that will facilitate this. In Chapter 5, we examine ways in which the firm can tailor its output to match more closely the needs of its intended customers by adding value through effective management of technology and marketing. Our analysis will show that design and design management have an important role to play and these will be analysed in greater detail in Chapters 6 and 7.

Chapter five

Sharpening the competitive edge

Introduction

Previous chapters have chronicled the continuing decline of British manufactures in world markets. While the factors underlying loss of market share are many and multifarious, fundamentally they all come down to the same thing — a customer's perception that a competitive product offers better value for money. In turn it has been established that 'perceived value' is quite different from 'lowest price'. Value represents an overall judgement of the total bundle of benefits in terms of performance factors, service, and risk that the user will receive in exchange for a given amount of money.

In the case of Britain it would seem that both complacency and myopia are responsible for our declining economic fortunes. Complacency because we were the first major economy to industrialize and so, by definition, were more technologically advanced than our competitors. Given this advantage, together with the huge captive markets of the Empire — a source of both raw materials and enormous potenial demand — it is unsurprising that by the middle of the nineteenth century the UK accounted for almost half the world's trade. In that no country could ever account for more than 50 per cent it was inevitable that we should lose share as other countries industrialized and began to trade in manufactures. On the other hand, given the rapid increase in world trade as a whole, this loss of share was compensated for by ever increasing sales volume and herein lie the seeds of our decline. Given that quantity is emphasized to the neglect of quality, many British firms have taken sales and volume as their watchwords rather than quality and value added.

It is a well-known economic fact that with very few exceptions the quantity demanded will increase as the price is reduced. It is also the case that price elasticity of demand tends to vary between potential users so that most industry demand schedules would appear like that in Figure 5.1.

What Figure 5.1 tells us is that those with the strongest demand will pay very high prices to obtain a supply, that as price falls demand will expand rapidly to a point where demand is saturated and further price reductions will yield very few new orders. Given that Britain is poorly endowed with natural resources, it makes sense, as we have argued earlier, to add the maximum value to those materials we have to import by harnessing the inventive genius of our scientists and technologists to the skills of a highly trained workforce. This implies operating at the top end of the demand curve, not slugging it out in the volume market where users' expectations — and disposable income — are much lower.

Just as Henry Ford dominated the US car market in 1920, so the UK dominated many world markets for basic manufactures. Both became complacent because of their success. Their myopia prevented them from recalling the basis of their original success. They had captured demand and diverted consumers' disposable income from the purchase of other goods by offering something perceived as *better*, i.e. giving more satisfaction per unit of cost than other alternative goods. But, having introduced consumers to cars or whatever, you have expanded their expectations and hori-

Figure 5.1 Typical demand schedule

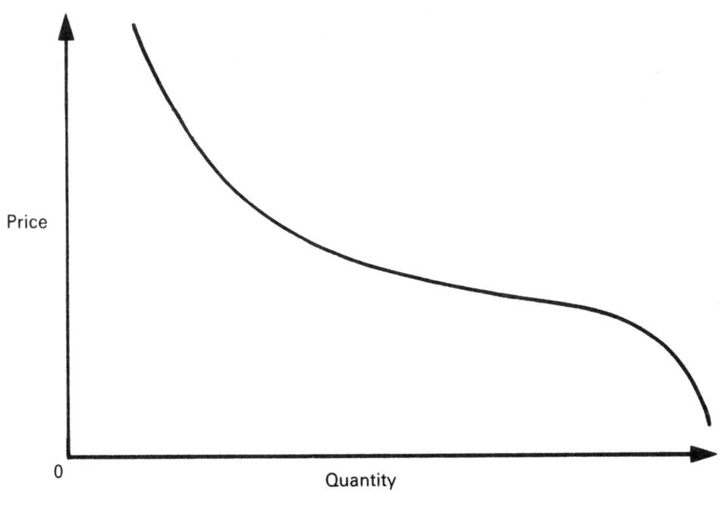

zons. Either you cater for these through a process of continuous improvement in quality and value for money, or someone else will do it and attract 'your' customers away just as GM did to Ford in 1923/4 when it adopted the concept of a product range with different product configurations designed to match the heterogeneous demand of consumers.

What is needed then — confirmed over and over again in the preceding chapter, which looked at the characteristics of successful firms — is a marketing orientation through which one establishes the nature of demand(s) precisely and then sets about designing and developing products to match these demands at prices consumers are willing to pay.

Design and marketing resemble each other in that both comprise a philosophy and a function. As philosophies, the pursuit of good design and a marketing orientation are states of mind that can be held and shared by all members of an organization. In this sense the mission of all organizations could be expressed as achieving their specific objectives *by satisfying customers' needs profitably through the supply of products and services designed to match their precise needs and wants.* In doing this the marketing and design *functions* will have a particular role to play alongside R & D, production, personnel, finance, etc., but the combined efforts of the functions will be optimized only if they subscribe to and seek to implement the common mission.

Given that the primary focus of this book is design rather than any of the other functions, this will be the primary focus of the next two chapters. Before seeking to look at design in the particular (Chapter 6), it will be helpful to put it into context and to this end this chapter looks first at some of the technological and management factors that influence design. Next we examine the contribution of marketing to new-product design and development before reporting upon the attitudes and practices of our survey companies towards these factors.

Factors influencing design

The specific concern of this section is to examine some of the factors that influence design, discussed under three major subheadings, namely, technological, management, and marketing factors.

Technological factors

There are many technological factors that tend to influence design, a few of which will be discussed here. Among them are: innovation, value analysis, aesthetics, and ergonomics.

Innovation

> We live in a business world that increasingly worships a great tribal god innovation, lyrically hailing it not just as desired, but as a necessary condition of a company's survival and growth.
>
> (Theodore Levitt, 1960)

Innovation has received extensive analysis in literature and has been defined in a variety of ways including the following:

Schumpeter (1939): 'The process by which invention is first transformed into a new commercial product or process.'

MacLaurin (1949): 'When an invention is introduced commercially as a new or improved product or process it becomes an innovation.'

Carter and Williams (1957): 'Invention we define as the creation and development of a new idea, whereas innovation is the act of bringing it into practical use.'

Levitt, (1965): 'Generally speaking, innovation may be viewed from at least two vantage points:

(1) Newness in the sense that something has never been done before, and

(2) Newness in that it has not been done before by the industry or by the company now doing it.'

de Farrenti (1966): 'The function of transferring the progress made in science into new technical advances.'

Schmookler (1966) Schon (1967), Ray (1972), and Lorenz (1979), all define innovation as a process whereby invention is commercially exploited.

Langrish (1968): 'Innovation is used by different writers in three different senses:

(a) A new 'thing' which has been used.

(b) A process separate from the process of invention (innovation in this sense is the next step after invention). It is usually associated with an 'entrepreneur' who invests money in an invention.

(c) A process involving research aimed at a discovery. The making of a discovery or invention and the use of the discovery or invention. In this sense innovation includes invention and is not separate from it.'

Robertson (1970): 'The word "innovation" is understood to

mean the process of taking an apparently saleable notion (an invention) and developing it to the point where it achieves a profitable measure of market penetration.'

Tinnesand (1973): studied 108 innovations and came up with six different definitions: '(1) new idea (2) introduction of a new idea (3) invention (4) introduction of an invention (5) an idea different from existing forms and (6) introduction of an idea disrupting prevailing behaviour.'

Holt (1977): 'Innovation is a process which covers the use of knowledge or relevant information for the creation and introduction of something that is new and useful.'

Pilditch (1978): 'The whole process by which new ideas or processes are translated into the economy.'

Saren (1980): 'The process by which knowledge is first transformed into a new commercial product or process.'

Rothwell and Robertson (1973) 'Technological innovation is the technical, industrial and commercial steps which lead to the marketing of new manufactured products, and to the commercial use of new technical processes and equipment.'

Orr and Van den Hoven (1981) 'Innovation is by definition the commercial exploitation of a new idea from whatever source.'

OECD (1982): 'Innovation may be defined as the successful launching of a new product on the market or the development of a new manufacturing process in the firm.'

Twiss (1980): 'The process by which an invention or idea is translated into the economy' (an adaptation of the US Department of Commerce definition (1967)).

Rothwell *et al.* (1983): 'The commercial exploitation of basic idea or concept, i.e. the exploitation of specific product characteristic at a particular price.'

Johne (1984): 'The process of commercially exploiting the invention.'

Witcher (1985): 'By innovation I mean the commercial introduction of a new process, product or service to the society.'

Drucker (1985): 'Innovation can be defined the way J.B. Say defined entrepreneurship, as changing the yield of resources. Or, as a modern economist would tend to do, it can be defined in demand terms rather than in supply terms, that is, as changing the value and satisfaction obtained from resources by the customer.'

Combining all of these definitions, innovation is broadly defined here as the commercial introduction of anything new to the company, customer/user, and the society, be it product, process, knowledge (e.g. published knowledge — articles, reports, etc.), or service (for example, a change in distribution practice or advertising may result in a new market).

However, despite this common thread, it is clear that innovation can take many forms and that firms have a wide range of policies towards it. Some basis for classifying innovation will therefore be useful to ensure that one is comparing like with like when analysing corporate innovative behaviour.

Robertson (1970) classifies innovation (from the point of view of users) into three: (a) continuous innovation, (b) dynamically continuous innovation, and (c) discontinuous innovation.
Langrish *et al.* (1972) group innovation into four dimensions:

(1) Radical breakthrough innovation: innovation leads to a new technology, that is, a new textbook with a new title is required. An example of this is open-end spinning.
(2) Major innovation: the innovation makes several chapters of the standard book out of date or requires the addition of a new chapter or chapters. Examples of this are shuttleless weaving and electronic patterning in knitting.
(3) Incremental innovation: the innovation requires alteration to a chapter or additions of a few paragraphs to the book.
(4) Improvements: the innovation makes zero or very slight difference to the standard book.

Orr and Van den Hoven (1981) classify innovation into three dimensions:

(1) It is the synthesis of a particular consumer need.
(2) It is the technical means of satisfying that need.
(3) The synthesis must take place at a time when the market need is clearly expressing itself.

Drucker (1985) puts forward seven sources of innovative opportunity — the first four are within the industrial sector, whilst the remainder are outside it:

(1) The unexpected — the unexpected success, the unexpected failure, the unexpected outside event.
(2) The congruity between reality as it actually is and reality as it is assumed to be or as it 'ought' to be.
(3) Innovation based on process need.
(4) Changes in industry structure or market structure that catch everyone unawares.

(5) Demographics (population changes).
(6) Changes in perception, mood and meaning.
(7) New knowledge, both scientific and non-scientific.

In his study of the textile machinery industry Rothwell (1976a) classified innovations in terms of the type of change involved (Table 5.1) and the intentions of the firms adopting the innovations (Table 5.2).

Table 5.1 Types of change

	Classification of innovation	
	Classes 1 and 2	*Class 3*
Change based on new technological development	9[a]	3[b]
Change based on new scientific development	4	—
Non-mechanical technological change	2	—
Complex mechanical change based on established technology	—	2
Simple mechanical change	—	5
	15	10

Source: Rothwell (1976b) 'Innovation in the UK textile industry: the results of a postal questionnaire survey', *R & D Management*, vol. 6, no. 3, p. 132.

Note: a. Five of these innovations took the form of complex mechanical change.
b. One of these innovations took the form of complex mechanical change.

Table 5.2 Classification of innovation: intention of firm

Classification of innovation	*Primary intention*	*Secondary intention*
1 & 2	1 To provide a completely new application for a machine. 2 To produce a completely new machine 3 To increase the productivity of an existing machine. 4 To increase machine productivity	1 To improve the quality of user's product. 2 To increase the machine's efficiency. (reduce down-time) 3 To reduce labour requirement.
3		1 Reduction in labour requirements. 2 Reduction in down-time. 3 To improve user environments (i.e. the reduction of noise and/or fly)

Source: Adapted from Rothwell's (1976b) 'Analysis of the classification of innovation in the UK textile machinery industry', *R. & D. Management*, vol. 6, no. 3, p. 132.

Taking into account the points of view of both manufacturer and user, Utterback (1971) classified innovation into two — radical and incremental innovation.

Rothwell (1976b) appears to agree with Utterback's grouping of innovation, but seems to prefer to call radical innovation 'large step revolutionary technical change' and incremental innovation, 'small step evolutionary technical change'. However, in another article, Rothwell (1977) describes radical innovation as 'major innovation' and incremental innovation as 'minor innovation'.

Nevertheless, Johne (1984) notes that 'the classification of innovation as radical or incremental depends entirely on the background of those involved in applying an invention for either product or process innovation purposes.' He cites, as an example, the invention of a microprocessor unit that provides producers of electronic products with an opportunity for incremental innovation of the product type, whilst the similar invention offers manufacturers of less-sophisticated electrical products the opportunity for radical product innovation because such sophisticated electronic components are very new to them. Again, this demonstrates how design (and in this case the design of a microprocessor) can be influenced by incremental or radical innovation.

Bright (1969) goes as far as classifying radical innovation into four: scientific (i.e. search for knowledge); engineering (i.e. reduction to practice), entrepreneurial (i.e. introduction to society), and managerial (i.e. optimization of usage). However, for our purposes it will suffice if innovation is classified under two broad dimensions — 'Breakthrough innovation' (i.e. radical, major, large-step or discontinuous innovation), and 'improvements' (i.e. incremental, small-step, or minor innovation).

Irrespective of whether it is a breakthrough innovation or an improvement on an existing product, all innovation involves design changes. By the same token one could argue that all design changes result in innovation and this interdependency of the two concepts is particularly apparent in the case of product improvement. In turn, many product improvements are initiated by dissatisfaction with some aspect of the performance of an existing product or an underlying belief that as Curtis (1971) has observed, 'The inquiring mind is never satisfied with things as they are. It is always seeking ways to make things better and do things better. It assumes that everything and anything can be improved.'

It is this attitude of mind that has led to the development of such formal techniques as will be discussed in the subsequent sections.

Value analysis/engineering

The term value analysis was coined by an American — Lawrence D. Miles and was first introduced in the US in 1947, and later in the UK in the late 1950s and early 1960s. It is a technique aimed at maximizing the value in a product at a minimum cost and has since been widely accepted as a useful tool for cost reduction in the design and redesign of new products as well as in the purchasing of raw materials/components. Value engineering is another term often used for the same purpose of minimizing costs in order to maximize the added value in a product/process. Thus, both terms are essentially the same and will be used interchangeably in this book, although a number of authors have attempted to draw a distinction between the two terms.

For our purposes, value analysis or value engineering is defined as a technique designed to minimize costs in existing and/or new products/processes in order to maximize their value content to the benefit of both customer and manufacturer. Value, as implied here, is the embodiment of all the benefits that a customer stands to gain from or lose in a product (i.e. those qualities upon which the desirability, utility, and, in fact, the entire worth of a product depends). Examples include: attractive appearance, reliability, durability, maintainability, ease of use, safety in use, and so on and so forth. As can be seen, all of these value factors are characterized by design from which it follows that value analysis/value engineering influences design. This can be demonstrated.

The achievement of maximum value consistent with price for the customer/user on the one hand, and minimum cost, (probably resulting in economies of scale, high profit, etc.) for the manufacturer on the other, requires that a number of vital questions be asked by the designer/value engineer when considering changes in the existing design or selection of materials/components for the design of new products or improving existing ones. This will enable the right product to be designed with the right materials/components for the right people (consumers). Leech (1972) puts forward thirteen questions:

1. What is it for? (i.e. Will this product fit the purpose for which it is built? What use or value is this material/component in the design of this product that will benefit both the customer and manufacturer?).
2. Do we need it?
3. Is someone buying the same function for less money?
4. Is any part of it redundant?

5. Can it be made of cheaper materials?
6. Can it be made by a cheaper process?
7. Can standard parts be used?
8. Can we use cheaper or better tools?
9. Can we buy the same materials cheaper?
10. Can we save some paper work?
11. Can we widen tolerances?
12. Can we buy in better quantities?
13. Can inspection be simplified?

Similarly, Baily (1978) considers ten questions:

1. Can we cut it (e.g. part) out altogether?
2. Can we simplify it; does it need all its features?
3. Is there anything better for the purpose?
4. Can we replace it or parts of it, with a standard part?
5. Is it made on proper tooling considering the quality required?
6. Would some cheaper materials do as well, or a dearer material be better?
7. Would some cheaper production process give a satisfactory part?
8. Do material, labour, overheads, and profit add up to its price?
9. Can we find a cheaper supplier who will be satisfactory?
10. Is anyone buying it for less?

He illustrates question 6 with Figure 5.2, which shows a cost-saving of 50 per cent by using glassfibre in place of steel material for a fan shroud. It also indicates a saving in weight of 75 per cent consequent upon the use of glassfibre instead of steel. It is pertinent to point out that changing material from steel to glass-fibre is likely to affect the design of the fan shroud in terms of appearance, shape, style, etc. Again, this strengthens the argument that value analysis/value engineering influences design.

However, the questions asked during value-analysis study may vary from firm to firm, probably depending on the focus of investigation. For instance, emphasis on cost reduction could attract different questions from stress on quality improvement. The designer of a car engine, for example, may have the option of improving the engine combustion to increase power or opt for lower fuel consumption and thus concentrate on weight reduction. Nevertheless, the ultimate objective in each case is usually to maximize value in the product while minimizing cost.

Figure 5.2 Value analysis: alternative materials for a fan shroud

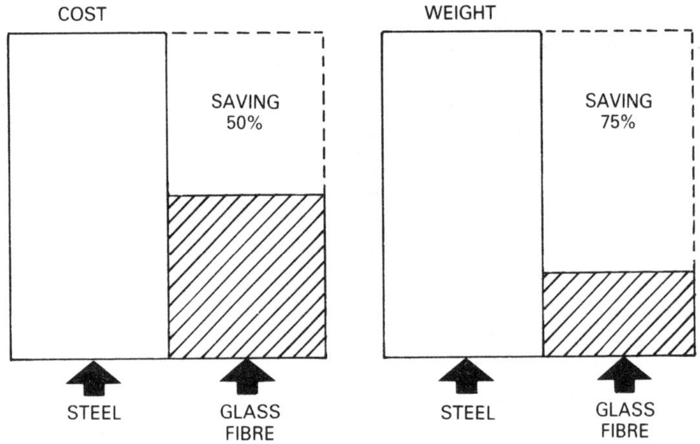

Source: Baily (1978) *Purchasing and Supply Management*, 4th edn, Chapman and Hall, London, p. 240

It is said that many people not directly involved in value-engineering (VE) work are of the view that the application of value-engineering techniques appears to 'cheapen' a product and lowers its performance capabilities. To confirm or infirm this hypothesis, the American Ordinance Association (AOA) (1964) carried out a survey of representative randomly selected items that had been value-engineered. The objective of the AOA study was to verify the benefits resulting from the value-engineering effort. Listed in Table 5.3 are benefits evaluated in the study and their definitions. Almost all the benefits are design attributes. It was shown that the advantages gained from design changes far outweigh the disadvantages during the VE effort, once again demonstrative of the influence of value engineering upon design.

Further, the AOA (1964) has also demonstrated the influence of value-engineering on the design of selected products. For example, before the VE effort, the 'old design' of a 2.75 rocket motor tube cost $5.08. However, after the design changes (i.e. the elimination of gold plating) during the VE exercise, the cost of the 'new design' was reduced to $3.73 with an annual net saving of $760,522. In addition, a saving of $30,000 in development costs was made on the production of 585,572 units.

Another example shows the results of the elimination of the gold plating of an intervalometer during a value engineering study.

Table 5.3 The benefits of value engineering and their definitions

Benefit	*Definition*
1. Reliability	Ability to meet performance requirements for a determined number of times.
2. Maintainability	Relative ease of repair or replacement.
3. Producibility	Relative ease of repetitive manufacture.
4. Human factors	Acceptability of change related to necessary education or dexterity.
5. Parts availability	Relative ease in obtaining or manufacturing simplified or standard parts.
6. Production lead time	Elimination, standardization or simplification of operations or materials.
7. Quality	Characteristics of parts to meet everything specified consistently
8. Weight	Lighter in weight.
9. Logistics	Quantity and complexity of parts needed for field support of end items.
10. Performance	Ability of the change to carry out the intended function at time of initial test or qualification.
11. Packaging	Relative ease of protecting parts until ready for use.

Source: The US Department of Defense (1969) Principles and applications of value engineering, Washington DC, pp. 16–17.

First, the unit cost was reduced from about $761 to $392. Second, 60 per cent weight and 49 per cent volume reductions were made. Third, 137 parts were eliminated and an annual savings in cost of about $454,000 made. Once again the influence of VE upon design in terms of weight reduction, parts elimination, etc. is demonstrated. Similarly, the fuel tank of a semi-trailer was reduced into half, from four compartments to only two, without change in capacity. However, during the VE study, the original specifications were changed, 'incorporating new welding procedures, new materials and new suspension system'. These new specifications meant changes in the original design hence four compartments were reduced to two, suggestive of the influence of VE on design.

In one final example twenty parts required for the design of a Minigun barrel clamp were reduced to only three after it (the minigun barrel clamp) had been value engineered. In addition, the following benefits accrued: improved accuracy (i.e. reliability); longer life (i.e. durability); and easier assembly (i.e. maintainability/ease of use) all of which are largely characterized by design, hence it is reasoned that value engineering influences the design of new products.

The need to reduce fuel consumption in cars and in fact, other fuel-powered machines appears to have helped to speed up the application of value engineering. Griffiths (1985) reports that British Leyland and Alcan of Canada have applied for patents in aluminium automobile bodies. Both companies have jointly developed a technology that allows the volume production of cars with aluminium frames, the components of which are primarily adhesive-bound rather than welded. Further, the car will have a plastic body panel that is designed to be glued to the aluminium frame. According to Griffiths, the use of aluminium and plastics in place of steel for building future generation cars will have the following advantages:

(1) The feasibility of a 50 per cent weight-saving compared with a conventional spot-welded steel structure has already been demonstrated.
(2) High corrosion resistance.
(3) Durability.
(4) Good crash resistance.
(5) Eliminates the need for paint primers, i.e. paint being needed only for cosmetic purposes, not to perform a protective function.
(6) Eliminates joint sealers (i.e. seam sealing to prevent leaks will no longer be required).
(7) Eliminates anti-corrosion treatments.
(8) Reduces capital investment in spot-welding facilities.
(9) Because the vehicle is not prone to rust, it is said to have an 'expectable enhanced resale value'.
(10) Although a cost premium of about £100 has been estimated compared to the cost of a conventional steel car, it has been equally estimated that this cost is likely to be recovered in less than 20,000 miles through improved fuel consumption brought about by the 'lighter weight' of the car.

It is the contention of the Design Council (1984a) that 'The British motor industry could build an all-plastic car now, but at the moment it would be expensive to change production methods and the strength of steel still makes it preferable.' This is where the value-engineering study appears to be crucial. Since a steel car is heavier and consumes more fuel, and an all-plastic car consumes less fuel but may be too light and less strong, could it not be possible to use part of each material (i.e. steel and plastic) to achieve a balanced strength while reducing weight to optimize fuel consumption? The answers seem to be positive. For example, the Design Council (1984a) reported that a technical solution has been found whereby 'layers of plastics and steel can be used together to give strength and lightness of weight'. According to the Council the 'lightness, aerodynamic styling and efficient engines are all going to cut down the amount of fuel used'. The Council points out that 'new cars bought in the UK now use 15% less petrol on average than they would have five years ago'. It goes on to say: 'This trend is likely to continue, and the possible introduction of "lean-burn" engines which use a less rich mixture of petrol and air, would boost the economy drive.'

All of the aforementioned are examples of value-engineering activities that should be taken into consideration when designing new products or improving existing ones, such as cars, engines, etc. This strengthens the contention that value engineering influences design and that value analysis/value engineering influences the design of new products. This implies that in designing new products, thought should be given to the principal benefits of value engineering (i.e. maximization of value and minimization of cost of new and existing products). To achieve this (i.e. to increase the value for money and reduce the cost of production per unit), means that changes in design (such as changes in specification, style, shape/appearance, safety, etc.) should be made, and by so doing, value engineering is said to influence design.

Relating this to the focus of this chapter, namely how to improve the competitive edge, it follows that an efficient application of value-engineering/value-analysis techniques will lead to both *cost* and *value* advantages that will enable firms to either reduce price or charge a premium. With these two benefits (i.e. charging lower price due to lower production costs and higher price due to the higher value content of the product), a firm will not only be competitive, it can also fight off competition, particularly in highly competitive and saturated markets such as electronics. It follows that value-engineering studies should be undertaken on a regular basis throughout the product life cycle to ensure a satisfactory result (i.e. achievement of maximum value at a minimum cost). In turn, this means that new-product designs should be continuously reviewed to ensure that they are updated in line with benefits sought by customers/users. Put another way, products should be continuously reviewed and updated so that customers/users get the maximum value for money (superior quality, reliability, safety in use, etc.), whilst manufacturers provide such value at a minimum cost (i.e. at a profit). This seems to agree with the US Department of Defense, which argues that 'there is some point in the product life cycle after which the cost to do VE and implement VE changes may exceed the potential savings'. Consequently, it argues that 'VE should be applied at all stages in the product life cycle before this point of diminishing returns is reached'.

It is believed that value-analysis/value-engineering techniques when utilized throughout the product life cycle can ensure that maximum value is achieved in both technological and aesthetic appeals of a product. Value analysis and value engineering are, therefore, not limited to technical product specifications. Often aesthetic product benefits can accrue from such applications. The following section examines the aesthetic components of a product, and the contribution of aesthetics to market-led designs.

Aesthetics

The Concise Oxford English Dictionary defines aesthetics as 'the theory or philosophy of the perception of the beautiful' (1985). The beauty of an object (product) *per se* tends to be perceived in different forms. For example, some customers may be attracted to a product because of its appearance, style, shape; others, by colour, texture, etc. Consequently, different authors have used such terms as appearance, style, colour, etc. to represent aesthetics. Thus, whichever terminologies are used — style, appearance, and so on, they all express virtually the same thing — the visual beauty of a product.

Cain (1969) defines appearance as 'the outward expression of the quality of the product and is the first point of contact with the user'. This will be used as the operational definition of aesthetics here. The outward expression encapsulates shape, style, proportion, colour, etc. As Cain strongly argues, 'No product can be entirely dissociated with the quality of appearance at some stage of its life, and it is therefore essential to determine the correct nature and degree of priority of appearance aspects of any product as a step in the design process.' He argues that appearance (aesthetic feature) 'has a more profound effect on the user than is generally realised by engineers'.

Since appearance influences customers/users who, in turn, influence the design of new products (e.g. via specifications or changes in preferences), and since the quality of appearance cannot be dissociated with any product, as alluded to earlier, then it is reasonable to argue that appearance (aesthetics) influences design. In the words of Murdoch and Flurscheim (1983) 'In the past, form or shape (or appearance) evolved principally through straightforward functional considerations, this is no longer the case.'

Styling

The object of the styling, as said earlier, is 'to develop products that satisfy the aesthetic requirements of customers/users as well as possible within the technical and economic constraints'. In other words, it is a useful tool for differentiating a company's products in order to satisfy the varying tastes or preferences of consumers. This can take the form of colour, shape, pattern, balance, proportion, and so on. To meet these differing tastes and preferences of customers/users implies that changes may be made, in terms of colour, shape, pattern, proportion, etc., while designing new products or improving the existing ones, hence it is hypothesized that styling (aesthetics) influences design.

In most consumer goods, such as motor cars, the pop-up toaster, television, radio, clothing, etc. styling is an evolutionary process that tends to change design probably to accommodate developments in technology, engineering, production techniques, social influences, etc. This may help to explain the reason Cain (1969) described style as 'an elusive quality, yet is based on hard reasoning. It is the result of doing the job the right way, yet with impeccable taste. If the same principles of design are consistently applied through a range of products a common style is created.'

Murdoch and Flurscheim (1983) cite two instances of the influence of style on new-product designs. One example illustrates the style of the 1920s, which is now being reintroduced in the modern motor-cycle industry. They point out that 'tilting the engine of a motor cycle forward to the extent that it can be employed as the front member of the frame was used in the 1920s and has now been developed by Honda in an arrangement that gives an impression of forward movement and speed using a purely functional shape'. The second example is the cars of the 1950s, most of which were characterized by 'the tail fins springing out of the rear wings (reminiscent of the then American cars (Cadillac, Chevrolet, etc.)) suggestive of aircraft performance'. This style appeared to be short-lived. The oil crisis of the 1970s meant that more attention was given to fuel economy. This appears to have virtually changed the design of new cars with emphasis being laid on what Karen (1984) describes as 'aerodynamic styling', that is, 'the paring down of a vehicle's design into a simple uncluttered shape, via, for example, the use of lightweight materials, such as plastics'. The overall objective is to reduce the weight of motor vehicles and lower fuel consumption. However the Mini Metro was nearly a non-starter and had to be redesigned about eight times as discussed in Chapter 1, because the engineers prioritized functional performance to the neglect of styling — the mono-sided structure as originally conceptualized by the BL engineers meant that the total body side of the Metro car, including door aperture, was made as a one-piece pressing, leaving no room for styling.

In the words of Bache (1980), a specialist in aesthetics who was employed by British Leyland in 1975 to upgrade the quality of the Metro's styling: 'I started to grumble about the lack of form of the thing. It was essentially sane, but a little too functional... The original packaging brief had laid down both interior and exterior widths to the extent that there was virtually no room for me to style the body.' The result was that Bache and his team reworked the mono-sides, changing the front wings, side doors, rear quarter sections, and rear hatch door. The refinements to the rear door extended the back end and tail downward, adding another 65 mm onto the car, thus making room for both exterior and interior styling. In this way, styling (aesthetics) is said to influence the design of new products.

Colour

Colour is another aspect of aesthetics that tends to influence the design of new products. Although the term colour has received adequate treatment in engineering literature, its application to products appears to have received little attention. However, the detailed discussion of colour and its technicalities is outside the scope of this study.

Perhaps, the limited attention paid to colour may be explained by the fact that selection of colour itself seems to depend mainly on intuition, emotion, experience, and taste. According to Whitfield and Wiltshire (1983) 'When it comes to selecting new colours (by designers) for a product that people will like, it is largely a question of guesswork'; the reason being that, 'there are no known 'absolute' laws of taste ...' For example, even some general connotations often associated with colours such as orange/ yellow for warmth/cheerfulness; pink for softness/femininity; green for growth/freshness; white for purity; red for aggression/ masculinity, etc. may not be strictly relied upon for two reasons. First, different individuals tend to perceive colours differently, that is, according to preferences or tastes. Second, the perception of colour and meaning attached tend to vary from culture to culture or from one ethnic group to another. Thus in designing new products, these factors tend to play significant roles when choosing colour.

Selecting colour also goes hand in hand with selecting the right materials to match. In turn, the type of material chosen tends to determine the type of design technique to use. For instance, the choice of a plastic material may require a different design approach from that of wood, steel, or a combination.

Gloag and Keyte (1957) refer to colour as 'a capacity to express form or breakdown of surface'. Hardy (1966) describes colour as having the 'ability to accentuate a satisfactory three-dimensional structure or to correct unsatisfactory proportions by altering their apparent form and creating centres of attention'. Finally, Whitfield and Wiltshire (1983) define colour as 'a means of influencing the perceived structure of an object in a controlled and predictable manner'. They point out however, that 'colour has also been investigated as a means of minimizing strong brightness contrasts between sources of illumination and their surroundings, and as an aid to controlling the level and distribution of illumination in a room'.

Combining all of these definitions, colour is defined as a means of identifying a product through perception.

The *structural* function of colour (i.e. the extent to which colour can influence the way in which a product or item is perceived by individuals) can influence design in three main ways:

(1) Centre of attention. Products are usually coloured to suit their surroundings. However, where the objective is to draw special attention, say, of customers, to the product, then a centre of attraction may be created by the use of colour that strongly contrasts with the surrounding background.
(2) Separation/association. Products are usually composed of many components. Depending on what image the product is supposed to portray to the consuming public, colour can be manipulated (particularly at the boundaries between areas of colours) to have an effect of 'making the product appear visually more unified or fragmented, more simple or complex'. In this way, colour tends to influence the design of new products.
(3) Proportion and orientation. It is possible to change or increase the visual proportions of a product by breaking the surface into areas of colour. Examples include chessboard, ludo, and dartboards. The same thing can be said of the

product orientation, that is, the extent to which colour can be manipulated to make a product appear, say, vertical, horizontal, or triangular in shape in the eyes of the observer.

Further, the influence of colour on military hardwares is well known. For example, colour can be used to camouflage virtually all military hardware (armoured vehicles, aircraft, machine guns, etc.) and even human beings (soldiers) to be compatible with the fighting surroundings, particularly in a battle.

The *ergonomic* influence of colour on new-product designs is associated, among other things, with safety, visibility, and durability of appearance of a product. Safety is one of the most important attributes of new-product designs. A product that is not safe to use is not worth buying at all. Thus, to ensure safety in potentially dangerous products, such as electrical equipment, colour is often used to draw the attention of the the user to such dangerous part(s).

Colour does not only draw the attention of the user to a potentially dangerous product or parts of a product, it also makes such parts more *visible* to users and non-user alike. In addition, the colouring of control mechanisms in, say, manually operated machines, enables the operator to see clearly what and how to manipulate, thus improving productivity. It also enhances the identification of various parts or functions of a hardware. For example, the control panel of an agricultural machine may have red, green, etc. buttons to indicate various functions. Similarly, coloured parts/components also help to quicken the dismantling and assembling of a product, saving time, energy, and costs. In this instance, it can be argued that colour aids maintenance, which, in turn, strongly influences the design of new products, such as motor cars.

The appearance of some products tends to deteriorate as they are exposed to dirt, oil, grease, corrosive materials, etc. Consequently, the designer may have the choice to:

(1) Select the matching colour of the anticipated contaminant, such as grease, 'so that the product's general appearance does not actually alter much in use'.
(2) Select a grey or dull green colour, typical of such machinery, such as tractors, fork-lift trucks so that when discoloured by the contaminant, the appearance will still be fair (i.e. will not be totally unpleasant to the operator to live with).
(3) Select a colour that will stress the effect of the contamination *per se*, aimed at 'encouraging users/operators to keep it clean'.

On the other hand, the selection of colour tends to depend on the type of material, which, in turn, appears to influence the design of new products.

The aesthetic influence of colour

In spite of the fact that confusion still appears to exist in literature as to the relevance of aesthetic features in new-product designs, particularly engineering products, the recognition of the aesthetic function of colour appears to be gathering momentum (Whitfield and Wiltshire 1983). Typical examples include agricultural machines, tractors, bulldozers, etc., 'which are being produced in glittering colours, to appeal aesthetically to customers/users, even though manufacturers know that such machines will soon soil in use and may never be cleaned'. Further, colour may be used to 'maximize what may in reality be minor differences in product range', such as minor alterations in a product manufacture, and may be used to disguise some outward defects, depending of course, on the type of product.

Perhaps, the use of colour to depict aesthetics is even more marked in the design of fashion clothing, carpets, wallpapers, etc. For example, the blend of colours a garment requires to fit appears to determine how the garment is to be patterned or designed. In designing new garments, therefore, designers often look for materials that will marry well with the chosen colour, so that the benefit of the full brightness of colour can be gained. In other words, the use of the right colour can be exploited to influence the overall perception of a new-product design.

According to Whitfield and Wiltshire (1983) 'a product can be made to appear warmer, lighter in weight and even smaller in size by the appropriate use of colour. It can also be made to appear expensive, subtle, and sophisticated ...', once again demonstrative of the influence of colour (aesthetics) over design.

One can conclude, based on the previous review, that aesthetic features — style, colour, and the like, influence the design of new products. As Whitfield and Wiltshire put it, 'a good-looking machine will have a sale advantage over a not-so-good-looking machine ... an attractive appearance should help offset minor practical disadvantages'. They conclude: 'In effect there would seem to be no advantage in neglecting the aesthetic appearance of a product, only disadvantage.' If this carries conviction, then considerable attention should be given to aesthetic features when designing new products, hence it is argued that aesthetics influences design and improves competitive performance.

Ergonomics

'Throughout centuries, machines had of course been designed to be used by men and women, but their manual operational and control requirements were elementary, and were met by *ad hoc* arrangements and by skilled craftmanship rather than by ergonomic analysis of need' (Flursheim, 1983). However, the introduction of high technology into our society has placed both men and women in a position where mistakes in the operation of some machines and turnkey installations (such as nuclear reactors), spacecraft, electrical appliances, etc., may have far-reaching consequences. Thus, it becomes necessary to provide adequate control and safety mechanisms to avoid accidents.

Similarly, with regard to mechanically driven products such as, tractors, motor vehicles, agricultural machines, and the like, there appears to be some concern amongst users, for the apparent lack of ease of maintenance, ease of use, comfort, as well as the rising costs of repairs and operation.

As a result of such problems, designers often seek, *inter alia*, to design products that are safe to use, easy to use, easy to control and maintain and comfortable, in order to satisfy customer needs (i.e. to increase the value for money in new-product designs). All of these are ergonomic factors that tend to interface between man/woman and the machine during operation, suggesting the importance of ergonomics in the design of new products.

A review of definitions of ergonomics appears to suggest an agreement, in terms of concern for human factors. These include, among other things, ways and means of ensuring that a product, say, a combine harvester, is safe to use, easy to use, and easy to control and maintain, and that the operator feels comfortable as well as protected from such environmental factors as noise, vibration, smell, etc., while in operation. (The operator is the man or woman in charge of a manual- or mechanical-control system and to some extent responsible for the monitoring, maintenance, and repair of a wholly-automated system). This is also true of protecting non-users, say, local people, from the effect of ergonomic factors such as pollution, noise, vibration, etc., created by machines such as cars, power tools, and the like. Thus, ergonomics is defined here as the study of the man/machine interface aimed at minimizing human problems (be it at workplace, home, etc.), such as safety, control, operator comfort, maintenance; psychological and physiological problems such as fear, claustrophobia, fatigue, perception; and environmental problems such as noise, pollution, smell, vibration, and so on.

Minimizing human problems

Ergonomically, human problems can be minimized in a number of ways. The design of the Massey-Ferguson's tractor (2000 series) is one case in point. According to Gilligan and Myerson (1983), ergonomically, the tractor is safe and easy to maintain and refuel. It has several purpose-designed features that enhance alertness and increase productivity. For example, whereas its slim corner pillars and tinted-glass panels ensure all-round visibility, the soft, adjustable seats with foldable armrests provide comfort. This is particularly important because farmers tend to spend most of their time in the cab twisting round to look out of the back to monitor the progress of the implement in the soil, and this can seriously injure their backs. As for the control system, the tractor's 'well-positioned, easy-to-identify controls is just as good as an average family car'. They conclude: 'All in all, it is a far cry from the days when farmers straddled the tractor on a tin seat covered with straw matting and finished the day's ploughing standing up, sick from fumes, and shaking with vibration.'

According to Watkin (orthopaedist and designer) (1984), most British wheelchair manufacturers appear not to be ergonomically driven. He argues that most UK producers are 'very backward' at wheelchair design, describing British-made wheelchairs as 'an anatomical insult', adding, 'we need a marriage of the talents of people like Flight Equipment and Engineering of Chesham who are doing revolutionary things in the aircraft industry, with someone who understands the human body'. He declares: 'Be sure to include actual wheelchair users and you will have the kind of think tank that will come up with the answers we need.' Two deductions can be made. First, it appears to support the argument that designers should give thought to body dimensions (standing, sitting, bending, reaching positions, etc.) while designing new products. Second, it lends weight to the hypothesis that ergonomics influences design. For example, by taking specifications from wheelchair users, designers will be able to make changes to the existing wheelchairs or design new ones.

Reaching a high shelf or top drawer of a filing cabinet from a wheelchair is an almost impossible task for the handicapped, particularly those confined to a wheelchair for life. However, with ergonomic consideration in mind the Trunkers of West Germany appear to have solved this problem when they designed a wheelchair called a 'Butler Chair', with an in-built, hydraulic-lift device that enables the handicapped to place items on or take items from any level of shelf or cabinet in the kitchen or office. As Mara

(1984) describes it: 'The chair is comfortable, orthopaedically designed, and self-propelled, with a smart executive look.'

In the case of body dimensions, designers should ask such questions as: 'What is the nature of the operator's work? Does it involve standing, sitting, bending, walking about, or a combination of these? How frequently is the operator exposed to noise?' The answers to these questions will enable designers to ensure that the control mechanisms, visual display of information on the keyboard or elsewhere in the workplace, the process of maintenance and repair, etc., are simple enough to allow the operator to carry out his/her functions with minimum discomfort or inconvenience, in situations most likely to be encountered under normal operating circumstances. By designing and redesigning new and existing products, respectively, to suit distinct user purpose, ergonomics is said to influence design.

The next section will examine the management factors that influence the design of new products.

Management factors

A number of management factors tend to influence the design of new products. Among them are: quality circles, timing, and leadership style.

Quality circles

> Quality of design, production and marketing wins markets. Only satisfied customers will repeat orders and make British goods and services their first choice.
>
> Margaret Thatcher, 1984

The concept of 'Quality circles' (QC) was first developed in Japan in 1946 by NEC, 'then a telephone company rebuilding from the ruins of World War II'. According to Barratt (1984c), 'to rationalize command communications the allied forces needed good telephones, but the Japanese telecom network had been almost entirely destroyed and what was left of the producing industry was turning out goods of very uncertain worth. So experts from the QC section of GHQ taught NEC about statistical quality control (SQC).' Two deductions can be made. First, QC can be said to be born out of a military necessity. Second, QC appears to be applied initially in a narrow sense to solving the production quality and reliability problems.

However, through the influence of an American, Dr Deming,

NEC was able to introduce QC in a broader sense to encapsulate management components. Consequently, in 1965, NEC established a zero defect programme to enable employees throughout the company to participate in quality management.

Thus, the concept of QC in a wider perspective was introduced in Japan in the 1960s with the overall objective of improving product-quality standards via the improvement of leadership abilities of supervisors, encouraging self-development, and increasing workers' morale as well as creating an environment in which every worker is conscious of quality and the need for its continuous improvement. As Barratt (1984c) puts it, 'to supply a satisfactory product to the user the entire company must work together'. In fact, he quotes the Japanese plants he visited as stressing that 'quality does not just concern production of products: it involves commitment at all levels of the company to making the right decisions and implementing them properly . . .'

The application of QC in Japanese industries was so successful and the benefits so substantial that it prompted interest in the West, particularly in the USA and the UK. In 1974, Lockheed Missiles and Space Company of California pioneered the use of QC in the USA. Similarly, in 1978, Rolls Royce introduced QC into the UK. Later in the UK, Wedgwood (the ceramics manufacturer) and Jaguar joined the circle 'race' (Cox, 1984).

Although we prefer to use the term 'quality circles', it is relevant to note that different organizations have used different names to denote quality circles. Among them are: 'qualitivity teams; participation teams; performance circles; employee involvement program; productivity teams; excel teams; and employee participation groups' (O'Donnell and O'Donnell, 1984). Others include: quality control circles (QCC); total quality control (TQC); (Barratt, 1984c); and last but not least, statistical process control, Williams (1984) and Creighton (1985).

According to O'Donnell, the rationale for using a name other than quality circles seemed to be attributed to two principal factors: a desire to dissociate the name from Japanese origin, or a desire to broaden the emphasis from only quality considerations to include many other program objectives.

However, whatever the terminology that may be used, the fact remains that the ultimate objectives appear to be one and the same thing, that is, to improve the production quality and reliability of an organization's product, process, or service through the application of good management principles, including the use of statistical techniques by production engineers to evaluate quality problems and trace their sources. The overall purpose is to satisfy customer needs and wants and quality circles are defined here as a management/production technique designed to improve the production quality and reliability of a company's product, process, or service with the overall objective of achieving a zero-defect result (i.e. zero returns) in mind.

Further, since the major objective of quality circles is to improve the quality of a product, and since quality cannot be inspected into a product, then conventional wisdom suggests that a good-quality product can only be achieved through design or redesign of new and existing products, respectively.

Quality circles in practice

The Jaguar car

> Only three years ago, Jaguar's once glorious image as the champion of British engineering and design lay in shreds. The car's defects and failings were an owner's sick joke. In the US, easily the Jaguar's biggest single market, it became known as the 'Friday night/Monday morning' car; . . . Never let your wife out in a Jaguar after dark they said, because the car might strand her, you never know where.
>
> (Isaac, 1984)

This was the state of Jaguar before John Egan took over as chairman, when it was split off as a separate company from British Leyland in April 1980. His main objective was to improve the quality, particularly the reliability, of the Jaguar car. This was tackled on three fronts.

First, he organized a survey of owners of Jaguar and rival (BMW and Mercedes) cars to find out how Jaguar compared with them in terms of quality and warranty statistics, as well as to establish the number of faults. A total of 150 faults were found, 60 per cent of which originated from suppliers of its components. Next, Egan sent his senior managers to visit Mercedes and BMW plants in West Germany as well as some Japanese car manufacturers. The aim was to learn about their arch-rivals' quality control and management techniques. The chosen technique was the 'most innovative and the much publicised Japanese-style Quality Circle' (Isaac, 1984).

Finally, the chairman set up a communication system to tell everyone in the company what the problem was and how it could be tackled. The company's slogan was 'In pursuit of perfection'. Representatives of management, supervision, workforce, quality control, purchasing, marketing, etc., were brought together in task forces to identify and solve quality problems. The result was significant.

According to Isaac (1984), in 1980 the number of quality inspectors dropped from 677 to 360. Its market warranty statistics improved by 40 per cent. Its quality index also improved considerably. For example, in 1980, the average score for Jaguar paint finish was 'a miserable 30', but by 1984, the score had risen to 70. Furthermore, the body of the current saloon, which used to be made up of over 560 pieces, now contains only 330.

Today, Jaguar has about sixty circles, 'each with around a dozen members (i.e. about 10% of the workforce) operating in areas ranging from cylinder-block and cylinder-head production to engine assembly and reconditioning' (Isaac, 1984). The overall result is that Jaguar has become more reliable, selling fast in overseas markets, particularly in the US. This results from the application of QC which, in turn, has brought about changes in design, hence it is argued that QC influences design.

The Coventry Climax. The Coventry Climax (producer of Climax fork-lift trucks) was once very competitive in terms of quality and after-sales service, and its engineering widely recognized as first class. However, in 1982, it became clear that this company had serious quality and reliability problems. When the problems were diagnosed, it was found that Climax fork-lift trucks produced irritating faults that annoyed its customers during the first three months or so after purchase. Among such faults were leaks in hydraulic hoses and minor electrical faults that apparently took too much of the company's time in an attempt to repair or put them right — faults that ought not to have occurred in the first place!

To combat these problems, the company introduced the quality circle. A weekend seminar was organized. Participants were made to understand that 'quality management is not just a matter of exhortation, rigorous inspection, and setting-up a quality department; it involves a fundamental appraisal of every job, starting at the top'. In particular, engineering, design, and marketing were made to understand that quality failures could no longer be shrugged off as someone else's responsibility — quality involves everyone, not just the quality department.

Brian Sharpe, the Managing Director, met with some strong opposition at the shop floor following the introduction of the quality circle on the grounds of personal accountability that usually characterizes QC. Afraid of unfair blame and possible victimization, the shop floor rejected the suggestion that they should accept personal responsibility for the quality of their work — their reason being that a lot of faults were not created by them but were rather forced upon them.

However, the problem appeared to be short-lived. The foremen accepted the responsibility for the quality of their work. The first impact was that everyone became quality-conscious throughout the organization; then came a spate of complaints from the shop floor about new-product designs that were not made with ease of manufacture in mind — a fact that appeared to confirm their fears. Insistence on loading these faulty designs on to the production line meant that the number of scraps/rejects and rate of rework would increase. Consequently, designers were compelled to redesign the Climax fork-lift trucks — designs that no one had ever questioned before. It is on this premiss that it is once again argued that quality circles influence design.

Overall, the introduction of QC in this company appears to have produced impressive results. For example, in 1982, forty-two faults were detected, but by 1984, this figure had been reduced to seven. It was also found that rework dropped by 50 per cent every six months since the QC was introduced. Moreover, 'in 1983, Coventry Climax became the first and only manufacturer of fork-lift trucks to be registered by the British Standard Institution (BSI), and the only British capital goods manufacturer' (Dale and Mortiboys, 1984). Above all, Coventry Climax now offers warranty about six times as long as its competitors — three years instead of the customary six months — an evidence of reliability. All of these significant results can be traced to improvements made on Climax fork-lift trucks via redesign of existing trucks or designing new ones that were prompted by the application of quality circles.

Other examples illustrating the influence of quality circles over design abound. A survey conducted by Binder Hamlyn Fry and Co. (1981) in Japan's iron-and-steel industry showed that the five subjects that dominated discussions during quality-circle sessions were: 'Cost (19.3%), safety (19.3%), equipment (16.6%), quality (16.1%) and efficiency (11.7%).' Similarly, Dale and Ball (1983) studied eighty-six manufacturing companies in the UK and found that the five major subject areas for quality-circle projects were: 'quality (17.7%), cost reduction (15.4%), production process (13.8%), productivity improvements (11.8%) and wastage (10%).

Timing

Timing refers to the design, development, and launch of the right product at the right time in the right market. A survey by the National Industrial Conference Board in the USA in 1964 revealed that 'poor timing' was the most frequently cited cause of new-product failure — a finding confirmed by Hopkins (1981). If the timing is wrong, that is, if a product is developed and introduced too early (i.e. earlier than scheduled) or too late (i.e. later than planned), it may result in lack of market acceptance for a number of reasons. First, the product may be a victim of new technology, as a result of late introduction. Second, it may have been overtaken by the first-to-market rival who has established a much stronger foothold in the market with a better product, presumably again as a result of late launch. Third, an attempt by a supplier to pre-empt the action of a competitor by introducing a new product too soon may result in the product being defective.

In each of these cases, in order to remain in the market (i.e. to ensure market acceptance) or to compete effectively, the product may have to be recalled for modification (redesign) or for the design of a new one. Even in the absence of redesign/new design, any attempt to use design-related factors such as delivery, after-sales service, advertising, etc., may still require changes/improvements in the existing method, such as increasing efficiency in delivery, after-sales service, or advertising in order to increase the added value or customer satisfaction.

This argument corroborates Alder's (1966) finding, which states that 'marketing executives are aware of the advantages that accrue to the new product that hits the market first. They are also aware of being too late — hitting the downside of a product cycle; becoming the victim of changing consumer habits, tastes and styles; or of having an idea fail because it is ahead of its time.' However, he reasons that 'there is one temporal aspect that is often underrated. This is the length of time it takes to develop new products. To put it another way, one of the prime causes of product failure lies in underestimating the time ...' He asserts: 'The fear of wrong marketing timing leads to faulty development timing. That is, anxiety about timing in the market makes time an enemy in the laboratory.' This is particularly important because often management tend to introduce new products in a hurry, that is, earlier than scheduled.

Perhaps, a few practical examples will help to further demonstrate how timing influences the design of new products. At the end of 1978, the Imperial Group (manufacturer of Lambert & Butler king-size cigarettes) decided that the time was right to go after a share of the 'high quality market' (Murray, 1980). Consequently, the Lambert & Butler (L&B) pack was redesigned, and in 1979, L&B was launched 'in a three-row, silver and black pack bearing the now familiar L&B logo'. According to Mike Haine (the marketing operations manager), the importance of the pack design was stressed 'since in this (i.e. cigarette) market it (i.e. packaging design) constitutes most of the added value perceived by the customer' (Haine, 1980). As a result of this, the company based its unique selling proposition on 'strong visual appeal' — an aesthetic factor that largely influences design as discussed earlier. In fact, this enabled the company to charge, at that time, a premium of seven pence on a pack of L&B king-size cigarettes.

In launching its new product (Bird's 'Whisk and Serve'), General Foods 'cannily waited to see the effect of Unilever's pioneering of instant custard powder before coming in with the added value of its own famous Bird's brand'.

In the mid-1950s, the Unilever Research and Development team simulated a variety of meats by using first groundnut (peanut) protein and later soya bean protein. The research resulted in the production of simulated ham, luncheon meat, and chicken breast that were proclaimed to be 'nutritious and very realistic'. However, they appeared to get the timing of the design, development, and introduction of these new products wrong. This is attested to by the joint statement made by Sir David Orr and Van den Hoven (Chairmen of Unilever) which reads: 'they were far too early and even now the public is not quite ready to accept substitutes at worthwhile prices' (Orr and Van den Hoven, 1981). A decision was taken and the product withdrawn, the reason being that the R & D team got the timing wrong, and had to either rethink for a possible redesign to meet the consumer tastes and preferences or probably wait until the time is right for reintroduction.

After what Unilever described as 'some very good research study', its design-and-development team developed a formulation that enabled detergent powders to bleach clothes at medium temperatures instead of at the boil. There was clear evidence of energy saving. However, like the simulated ham example, the introduction of the detergent powder was too early. In other words, the launch appeared to be wrongly timed, for it was not until the oil crisis of the 1970s that the energy-consuming public became aware of the need to conserve energy, and the time was right for the introduction of the detergent powder.

Renault Truck Industries of France held back the introduction of its medium trucks by four years (Gooding, 1985). The medium-truck version was originally planned to be introduced between 1988 and 1989 but has now been rescheduled to be launched between 1991 and 1992. The reason for the delay was said to be that three vehicle projects (two for heavy trucks and one for a medium truck) for the UK plant were running concurrently and that 'it would be impossible to launch three product ranges at the time'.

Gooding quotes the company as saying, 'it was simply a question of project priorities, and a decision has been taken to allow the heavy truck projects to go first'. Among the reasons underlying this decision to follow the 'strategy of the fast second (watchful-waiting strategy)' was to enable it to gather more market information about the type of medium vehicles its major competitors are likely to introduce in the 1990s, rather than introduce its version at the wrong time only to regret later (i.e. to recall it for redesign). The decision also reflects moves made by its major competitors.

The failure of Sir Clive Sinclair to introduce his Z88 lap-top portable computer in May 1987 as scheduled was the fifth in the series of late introduction of his new products. According to Miles (1987) 'Sinclair has failed to meet delivery dates at least four times before. In 1967, it was his first pocket-sized TV. In 1980, customers waited months for his Z80 personal computer. Again, in 1982, his ZX Spectrum computer was delayed before being supplied to mail order buyers, followed by more delays with the QL computer, launched in January 1984. Now the Z88, which in February 1987 was announced at the retail price of £199.95 has been held up. When it does appear it will probably retail at £250, a significant price hike.' Two observations can be made:

First, it is possible that when the Z88 model does finally appear on the market at the predicted retail price of £250, and its rival computer of the same quality sells at a lower price, rational buyers will go for the lower priced computer. Second, the Z88 computer may be overtaken by new technology due to its late launch and so become obsolescent. In either case, Sinclair will have to face two hard choices:

(1) To withdraw the Z88 computer for redesign/modification, aimed at matching the rival brands. Hence we maintain that timing influences the design/redesign of new/existing products.
(2) To match the competitive price, which may result in selling below the production cost. Nevertheless, as categorically stated in Chapter 2, 'if two products of similar price and different quality are available, all rational consumers will prefer the better quality product'. Thus, if Sinclair's Z88 model is of lower quality than those of his competitors, it may not sell which, in turn, may warrant its withdrawal from the market, for modification or redesign. Alternatively, the

> Z88 project may be abandoned and a new one embarked upon in a costly venture.

Kraushar (1969) notes that 'timing is a particular problem'. He contends that 'because a product fails at a particular point in time it does not follow that it cannot succeed perhaps five years later or indeed that it could not have succeeded five years earlier'. For example, we are much readier today to accept convenience food products than we were about thirty years ago. Similarly, Alder (1966) argues that 'new product development is crucial to the success of many companies ... and likewise the time dimension is crucial to successful new product development'. Table 5.4 shows how timing influenced the design of seven products. It appears to indicate that almost all the seven products were modified (redesigned) at one time or another (either because they were introduced too early or too late), to be successful and/or retain market share.

Of particular interest is the Red Kettle product (dry soup mixes). According to Alder, the innovators withdrew the product because they 'were dissatisfied with the processes and packages available'. This is not disputed. Nevertheless, it is reasonable to point out that if both the product and timing were right (i.e. if the right product was introduced at the right time in the right market), the question of innovators/suppliers being dissatisfied say, with the packaging should not have arisen – the reason being that it is customers/users not suppliers who judge or have the final say about the product in the marketplace. If the product satisfies the needs and wants of consumers, and is introduced at the time they want it, then the product will be accepted and vice versa. It is, therefore, reasoned that Red Kettle was recalled not so much because of the innovators being dissatisfied with the packaging but because the timing of the launch was presumably wrong. Consequently, Red Kettle lacked market acceptance.

One can conclude, based on the aforementioned review, that timing tends to influence the design of new products or redesign of existing products. This often occurs when an innovator, in their anxiety to be first to market, to pre-empt the action of its competitors, introduces a product too early. Alternatively they may introduce a product too late, fall victim to new technology, probably as a result of late introduction, and so on. This conclusion seems to be consistent with the earlier findings. According to Scanlon (1980) 'Many revolutionary products suffer from being ahead of their time, as British aviation can testify with the Comet, TSR-2 and Concorde.'

Table 5.4 New-product development time

Product	*Company*	*Date development started or idea born*	*Test of initial markets*	*Large-scale or national launch*	*Elapsed time*	*Remarks*
(1) Ban (roll-on deodorant)	Bristol-Myers	About 1948	1954	March 1955	6 years	Roll-on idea came from an outside inventor, hence presumably predates 1948. Bristol-Myers developed product that failed in test markets in 1951. Company researchers worked on plastics, finally assigned outside company job of making plastic marbles. Consumer-panel studies favourable in fall, 1953. Final test markets, summer, 1954. National advertising, March 1955
(2) Calm powder deodorant in aerosol can	Alberto-Culver	1959	N/A	February 1964	5 years	Non-spray powder deodorants were tried about 1948-50, did not 'get off the ground' then
(3) Marlboro (filter Philip Morris cigarettes)	Philip Morris	May 1953	March 1955	N/A	2 years	Malboro had previously existed as a premium non-filter cigarette; development of filter, hard package, flip-top were new; red ('beauty') tip derived from earlier non-filter product
(4) Red kettle (dry-soup mixes)	Campbell Soup	Before 1943	N/A	August 1962	19 years plus	Campbell first tested dry noodle soup in 1943-4, withdrew product because 'we were dissatisfied with the processes and packages available'. Resumed testing in 1959

Table 5.4 continued

(5) GE (electric tooth brush)	General Electric	1958-9	October 1961	April 1962	3 to 4 years	Electric tooth brushes available for 30 years; wall-socket recharging, new
(6) Talon (zippers)	Corporate predecessor of Talon, Inc.	1883	1913	1918	30 years	Zippers first thought of as shoe-fastening device in 1883. In 1894, product first developed for use in shoes. First modern zipper concept emerged in 1908. First successful mass production 1913; and first applications to clothing in 1913-18. In 1918, an ex-GI suggested use of zippers for money belts, and this was really the first commercial success
(7) Flav-R-Straws	Frontier Foods Corporation and others	1953	April 1956	Early 1957	3 years	Inventor sold idea to Frontier in 1955. By January 1956, Frontier was in trouble, and product taken over by others

Source: Alder, (1966) 'Time lag in new product development', in *Journal of Marketing*, vol. 30, January, pp. 18-21 (adapted version).

Finally, Peter Drucker (1985) concludes that 'those entrepreneurs who start out with the idea that they'll make it big — and in a hurry — can be guaranteed failure. They are almost bound to do the wrong things.'

Leadership style

> There is no progress without leadership, and that has to come from the very top, from the board of management itself.
>
> John Egan (Chairman, Jaguar plc (1985))

Leadership style has received extensive treatment elsewhere and it is not intended to discuss it in detail here. Consequently, it will be examined briefly to buttress the argument that the style of leadership in an organization influences the design of new products. The term leadership, for our purpose, includes all managers (top, middle, and low) who are charged with the responsibility of the management of design projects in an organization.

'The attainment of cost-effective quality is not just the result of operating a quality department, defining policy, using statistical quality-control techniques, collecting quality costs, etc., ... the overall important factor is the drive by top management to permeate quality through every aspect of company life in a bid to achieve zero defects; quality is what top management want it to be ... The chief executive in any company must take the lead' (Dale and Mortiboys, 1984).

This drive to 'permeate through evey aspect of company life in a bid to achieve zero defects' appears to depend largely on the way in which effort is piloted at the top. In other words, it seems to depend on the style of leadership in the organization. A few questions come to mind. What style of leadership influences design? Is it democratic/consultative, autocratic or participative leadership, or a combination of these?

A democratic/consultative leader is more likely to arrive at decisions by consensus, even if occasionally punctuated with some elements of bureaucracy or 'red-tapism' in implementation — design projects, for example. This may cause delay in the design of a new product that, in turn, may result in haste or a late launch. Products introduced in haste or late may fail to achieve market acceptance for two reasons. First, a hasty launch may lack finish. Second, late introduction may suffer from technological obsolescence. Consequently, the product may have to be redesigned. Since the manager (leader) takes the decision to introduce a product earlier or later than planned, one can argue that any product failure resulting from such a decision may lead to the modification (redesign) of the product or the designing of a new one.

On the other hand, an autocratic leader is more likely to impose his/her will on others or pass down instructions to subordinates for execution (top-down approach) probably without consultation, without listening to or considering ideas contributed by others, particularly designers. Such leaders may be described as parochial-minded in the sense that they tend to exhibit unwillingness to delegate power and/or authority; and seem to be much more likely to cling to old design concepts only to realize at a late hour that they are wrong and their subordinates (designers) are right (Ames (1970); Manasian (1980)). This may cause the existing product to be redesigned or abandoned, or a new design project embarked upon — another outcome of leadership style.

In his own style, a participative leader tends to show the willingness to allow his workers to participate in decisions affecting their individual projects and take responsibility — reminiscent of quality-circle projects (bottom-up approach). This type of leadership is more likely to listen to others, accept and encourage new ideas and may be much more willing to implement design projects such as modifying (redesigning) the existing product or designing new ones in order to satisfy the customer requirements. It follows that leadership style influences design.

This strongly supports Sir John Harvey-Jones's (1985) argument that a good manager (leader) is one who has 'the ability to listen, sensitive to the views of others and ... willing to embrace change and welcome the new rather than the more usual attempt to resist movements and cling to the past'.

Similarly, Horrocks (1985) (Group Chief Executive, British Leyland) contends that a good manager is one who has 'the courage to see decisions through if he believes them to be correct — and perhaps even greater courage to accept he might occasionally be wrong and to change direction accordingly, setting aside personal embarrassment and concentrating on what is good for the company'. Typical examples include John Egan (Chairman, Jaguar Ltd) and Brian Sharpe (Chairman, Coventry Climax Ltd), both of whose leadership styles have led to various improvements (redesign) on Jaguar cars and Climax trucks which, in turn, have resulted in significant increases in quality. They seem to listen to others, particularly customers, accept and encourage new ideas from subordinates, all of which probably may have prompted the introduction of quality circles. In turn, this has enabled them to detect faults and eliminate defects, as alluded to earlier.

However, in some cases, a leader could combine any of the above leadership styles such as combining autocracy with participative or democratic style, depending, of course, on the situation (stick-and-carrot approach). Nevertheless, whichever style of leadership is adopted, the success or failure of many new-product designs seem to depend mainly on the manner in which a decision is taken at the top to start, continue, suspend, or abort a design project. By so doing, leadership style is said to influence design.

Drawing inference from Egan's dictum that 'there is no progress without leadership', one can conclude that there may be no successful product (particularly design-driven products) in this competitive world without good design. In turn, there may be no good design without good leadership, hence it is argued that leadership style influences design, a conclusion supported by numerous studies.

The US National Academy of Sciences (1966) for example, refers to 'the technical entrepreneur (i.e. innovation champion) as a missionary — the man who carried the torch for a new idea — is often the catalyst of technical progress ... his courage and tenacity are frequently a vital element of successful innovations'. Similarly, Miles (1967) concludes that 'most innovations appear to be stimulated, triggered, shepherded, and nurtured by some active person or groups' (i.e. leader(s)/champion(s)). Finally, Olins (1985) declares: 'Design won't get anywhere without understanding and commitment at the top ...'

Marketing factors

As Rothwell *et al.* (1983) have pointed out 'The first and basic stage either in the improvement of an existing product or in the creation of a new one is to discover what the customer wants or will want by the time sales will begin. Only then can a specification be drawn up and the product described in terms understandable by a designer.'

This section contains two main parts. The first part reviews the 'concept of marketing', which is the essence of the attention to user needs. The second part examines the attention to user needs based on Rothwell's 'need dimensions' model, which is explored under four headings:

(1) Need elements: focuses on economic and technical needs of customers/users such as sale price, delivery dates, reliability, etc.
(2) Need intensity (felt need): attempts to balance the built-in

benefits with the price the customer is prepared to pay.

(3) Need stability: specifically concerns itself with the degree of individual need and the willingness of a customer to pay a premium to acquire a particular value designed into a product.

(4) Need diffusion: explores how design enhances the acceptability of new products in the marketplace.

The marketing concept

User needs are the essence of the marketing concept. This means that a review of the marketing concept is essential to provide a spring board for a better understanding of discussions on 'attention to user needs'. The marketing concept *per se* has received extensive analysis in literature and only a selection of better-known works in this area will be examined to illustrate our case. Baker's (1979a) assertion that 'virtually every text on marketing starts with a different definition' of the marketing concept, confirms that the marketing concept has received extensive treatment in the literature.

The application or practice of the marketing concept itself can be traced back to the eighteenth century when Adam Smith (1776) wrote:

> Consumption is the sole end and purpose of all production; and the interest of the producer ought to be attended to, only so far as it may be necessary for promoting that of the consumer.

In the words of Wilmshurst (1978), 'this maxim is so perfectly self-evident that it would be absurd to attempt to prove it'.

Stanton (1971) reasons that 'the marketing concept is based on two fundamental beliefs. First, all company planning, policies, and operations should be orientated toward the customer; second, profitable sales volume should be the goal of a firm. In its fullest sense, the marketing concept is a philosophy of business which states that the customer's want satisfaction is the economic and social justification of a company's existence.' 'Consequently', he continues, 'all company activities in production, engineering, and finance, as well as in marketing, must be devoted first to determining what the customer's wants are and then to satisfying those wants while still making a reasonable profit'.

Drucker (1978) argues that 'it is the customer who determines what a business is. It is the customer alone whose willingness to pay for a good or service converts economic resources into wealth, things into goods.' He declares: 'What the business thinks it produces is not of first importance — especially not to the future of the business and to its success. What the customer thinks he is buying, what he considers value, is decisive — it determines what the business is, what it produces and whether it will prosper.' It is his contention that 'what the customer buys and considers value is never a product. It is always utility, that is, what a product or service does for him ...' What 'a product or service does for a customer/user largely depends on the extent to which the innovator or designer pays attention to user needs or requirements/ specifications.

In his study, Levitt (1960) maintains that 'Every major industry was once a growth industry. But some that are not riding a wave of growth enthusiasm are very much in the shadow of decline. Others which are thought of as seasoned growth industries have actually stopped growing. In every case the reason growth is threatened, slowed, or stopped is not because the market is saturated. It is because there has been a failure of management. The failure is at the top. The executives responsible for it, in the last analysis, are those who deal with broad aims and policies.'

Citing a typical example of the lack of attention to user needs he asserts:

> Thus, the railroads did not stop growing because the need for passenger and freight transportation declined. That grew. The railroads are in trouble today not because the need was filled by others (cars, trucks, airplanes, even telephones), but because it was not filled by the railroads themselves. They let others take customers away from them because they assumed themselves to be in the railroad business rather than in the transportation business. The reason they defined their industry wrong was because they were railroad-oriented instead of transportation-oriented; they were product-oriented instead of customer-oriented.
>
> (Levitt, 1960)

Taking a wider view, Kotler (1972) argues that the core concept of marketing 'is that of market transactions'. He contends that 'marketing is specifically concerned with how transactions are created, stimulated, facilitated and valued'. He describes this as 'the generic concept of marketing'. Three crucial questions emerge:

(1) How can a product be created?
(2) How can a product be valued?
(3) How can a product be stimulated and facilitated?

To create a product (i.e. to make a product) the innovator has to carry out market research to find out who the customers/users are, and what they want or need (i.e. what their requirements or specifications are). That done, the designer then designs to specification, that is, designs for marketability or commercialization, a product that can be valued by the customer/user.

To recap our earlier definition of value: 'Value is an embodiment of all the benefits that a customer/user stands to gain or lose in a product.' Among them are reliability, durability, maintainability, safety in use, etc. To put value to a product, the customer's view must be sought to find out exactly what he/she wants. As argued earlier (Chapter 1), no one product is likely to encapsulate all the values. Besides, what is valuable to A may not be valuable to B. In other words, A's preference may be B's distaste. Thus, it is essential to seek the user's opinion, hence attention to user needs is all important in the design of new products.

Kotler (1972) puts forward four ways in which value could be created in a product: 'he (i.e. the innovator) can try to design the social object (i.e. product) more attractively (configuration); he can put an attractive term on the social object (valuation); he can add symbolic significance in the social object (symbolization); and he can make it easier for the market to obtain the social object (facilitation)'. Again, none of these dimensions can be executed successfully without consultation with the customer/user. For example, to design aesthetic appeals into a product, the customers' preference must be sought in terms of style, colour, shape, size, texture, etc.

'Stimulation' as applied here denotes advertising and promotion whilst 'facilitation' connotes effective distribution. To advertise, promote, and distribute a product effectively, some essential questions must be asked:

Who buys the product?
Why does he buy?
How does he buy?
When does he buy?
Where does he buy?

The answers to these questions will enable the innovator to:

(1) know the target/market audience;
(2) formulate an advertising theme/message;
(3) target the advertising message to the right audience;
(4) identify customers' buying behaviour;
(5) set up an effective distribution system.

Once again, this will require interaction with the customer/user in order to carry out the above tasks successfully.

Finally, as basic needs become satisfied, people turn to more and more complex products and services. This implies that a more sophisticated approach may have to be adopted by innovators/firms, including deeper probes into customer needs and wants; and greater interaction and collaboration between customers/users and suppliers. This means that as more and more choice becomes available, as the steep increase in demand tails off or perhaps turns into a decline, then manufacturers need to consider more and more what kind of product their customers truly need. This is particularly important because products developed jointly with customers/users are more likely to succeed commercially.

All of these foregoing arguments strongly support the thesis that the marketing concept is the essence of the attention to user needs. Applying the marketing concept to design, it is suggested that innovators should design what customers/users would want to buy, not try to sell what they like designing.

The dimensions of user needs and their influence on design

Rothwell *et al.* (1983) contend that 'the obvious objective of any firm is to make things (products) or to provide services that will sell profitably in their intended market, which means that it is vital for the needs of potential customers to be understood. Once these needs are understood, they must be reflected in the design of the product.' To understand customer/user needs properly, the different dimensions of user needs must also be understood in order to design products to suit them. This may involve changes in the existing design or developing a new design. Rothwell and his colleagues (1983) identify four need dimensions: need elements, need intensity, need stability, and need diffusion.

Need elements

Customers often seek to buy products that combine economic factors (sale price, quality of after-sales service, etc.) with technical factors (reliability, aesthetic appeal, performance in operation, safety in use, and so on). By the same token, designers often take this combination into consideration by attempting to match technical and economic needs or requirements of users during new-product designs. For example, although many prospective car owners might like to purchase reliable and aesthetically appealing cars, they might also like to combine reliability and aesthetics with affordable price. Thus, when a designer incorporates both technical and economic specifications in a product according to the customer requirements, attention is said to be paid to user needs. In this way, attention to user needs tends to influence the design of that product.

This argument strongly supports Rothwell *et al.* (1983) who contend that 'the customer bases his purchasing decision on a trade-off between price and non-price factors in making his decision'; they continue, 'these factors (i.e. technical and economic) are balanced according to the specific requirements needed and preferred and according to the money available' — a proposition central to Baker's composite model of buyer behaviour (see p. 67).

It is their contention, as pointed out in Chapter 2, that although the basic function of the manufacturer is 'to identify optimum characteristics that best match the needs of a particular group of users, it is the function of the designer working for that manufacturer to translate this set of user specifications into a working product or process that best satisfies these needs'. This implies that to produce a workable and/or saleable product, the needs of the targeted group of customers/users must be designed into the product. If so, then attention to user needs influences design.

Need intensity

Building benefits into a product is one thing, ensuring that customers are prepared to pay for the value of such benefits is another. Thus, it becomes necessary to segment markets so that each target market can be served according to the level of need, which should be balanced with the price customers/users can afford. In other words, the price a customer/user is prepared to pay for a product may largely depend on the intensity of need or level of need felt (felt need) by him/her. This can be illustrated thus.

A farmer who has only 100 acres of land that takes seven days to plough with an agricultural machine may not be willing to pay more for a modified version of the existing machine that ploughs the acres in six days (a saving of only one day, which may not be worth the additional cost). On the other hand, a farmer managing 1,000 acres of land may be much more willing to pay more for a modified version of the same machine at the same rate of cultivation on grounds of economies of scale. For instance, whereas the owner of 100 acres saves one day, as referred to earlier, his counterpart saves ten days. This confirms Rothwell *et al.*'s (1983) contention that 'when a user is willing to pay a considerably higher price for some new specification that is added to an existing product or is embodied in a new product, the need intensity is high. On the other hand, where customers are only prepared to pay a relatively small price for the new modification, the need intensity is low.'

A good car is expected to be driveable, reliable, durable, attractive, comfortable, economic to run, to be easy to maintain, to be safe in use, and so on. In most cases, no one car contains all of these benefits. This means that prospective buyers should have to prioritize their needs. That is, the customer may go for reliability, durability, fuel economy, speed, safety, or attractiveness, or a combination but not all the values at one and the same time, the reason being that it may not be possible to satisfactorily encapsulate all the benefits in one car. Thus, depending on the need intensity (felt need), the grand-prix driver may opt for a car with the highest speed. On the other hand, a man requiring a high-status symbol may prefer an aesthetically appealing car to speed, regardless of price. This means that as customer/user priorities change, so do designs (through modifications of existing products or design of new ones). Again, this sustains the argument that attention to user needs influences the design of new products.

Need stability

The preceding section discussed the degree or level of individual needs and the willingness of a customer to pay a premium to acquire a particular value designed into a product. This section specifically concerns itself with how these needs change over time and their resultant influence on design.

Because of the unpredictable nature of buyer behaviour, it is difficult to say or pinpoint what the future customer needs will be or even how long customers will hold onto the products that satisfy their current needs. This may be ascribed to the fact that customers live in a dynamic environment. Consequently, their needs and wants, or tastes and preferences change as the environment changes. In other words, their needs are seemingly unstable. This appears to concur with Abraham Maslow's 'hierarchy of needs' model, which states that as one level of need is satisfied, another higher level of need emerges. Two questions come to mind. How can the designer know when the first level of need is satisfied? How can the designer predict when the next level of need will emerge?

The answer to the former may be difficult to provide because even the level of preferences for the so-called satisfied needs often vary between customers/users, the reason being that the distribution of felt need (need intensity) also varies between the different dimensions of demand.

The answer to the latter question may be even more difficult. For example, the demand for energy-saving products, such as fuel-economy cars could not be predicted or foreseen until the oil crisis of the 1970s during which the OPEC prices rose to a staggering level. This led to many changes in the design of new motor vehicles. Attention that was previously paid to aesthetic shapes and size of cars turned to engine efficiency, smallness, lightness of weight, and aerodynamic styling, the overall objective being to reduce the weight of the car and cut fuel consumption. According to the Design Council (1984a) 'the lower the drag coefficient and the smaller the frontal area of the car, the less fuel the car consumes.' This means that attractive shapes (such as the fin-like shape of the old Chevrolets) that used to appeal to customers is seemingly becoming less popular in the present-day automobile designs. This manifests itself in many cars that now have Citroen-like, frontal, narrow-wedge shapes as opposed to wide fin-like shapes of the 'old' cars. As Tom Karen (1984) (Managing Director, Ogle Design) predicts, the shapes of future cars will be 'much more aircraft-like than a normal car'.

The previous argument strongly supports Jordan's (1984) finding, which reveals that 'until the oil crisis of the mid 1970s, priority in vehicle styling studies was given to aesthetic shapes that would appeal to the buyer but the new demand for fuel efficiency changed the philosophy surrounding styling techniques, and "aerodynamic" considerations moved to the top of the list'.

Thus, it seems clear from these examples that the prediction as to when the next level of customer need will emerge is a difficult task. However, one way of tackling such a problem has been suggested by Rothwell *et al.* (1983). It is their contention that 'to overcome this problem (i.e. the problem of accurately predicting the level of customer needs), the designer must continually update the specification to meet any changes in the market demand'. To update the specification implies that the designer must attempt to incorporate the varying needs/wants of customers/users in new-product designs. Further, Rothwell *et al.* argue that 'the process of design must be treated as a dynamic and continuous process ... and that redesign and reinnovation are essential to keep up with changing market preferences'. They conclude that designers should aim at a moving target and that firms must remain 'plugged in' to their customers/users, continuously monitoring and even anticipating changing user requirements, once again suggestive of the influence of attention to user needs over design.

Need diffusion

The essence of attention to user needs is to ensure that new-product designs satisfy the varying needs of customers. That is, to ensure that newly introduced products will be acceptable to both potential and existing customers/users in the market place. In the words of Rothwell *et al.* (1983) 'Products are made for users and unless these products are wanted, the firm will fail.' For a product to be wanted and/or needed, adequate attention must be paid to the varying customer requirements. This involves 'need diffusion' (increasing or widening of the customer/user preferences and tastes in new products). To achieve this, the designer should continuously assess and monitor current changes in market demands, that is, the designer should pay attention to changes in tastes and preferences of customers, particularly during what may be called 'lead time' (the period between the design of a new product and its introduction).

This corroborates Rothwell *et al.*'s (1983) finding, which shows that 'the continuous nature of the design process is that market preferences may change in the period between original design specification and launch. So while the original design specification may have represented the best fit between the user need profile and the preliminary design studies, it may no longer be appropriate by the time it is ready for production.'

All in all, it seems evident from the foregoing review that attention to user needs does influence the way in which new products are designed. Further, surveys previous to this study strongly suggests that customers/users have largely contributed to the successful design of new products, von Hippel (1979), Rothwell *et al.* (1983). For example, Rothwell *et al.* found that of thirty-three innovations in the medical equipment industry in the UK, twenty-five (75.8 per cent) (out of which twenty-two (88 per cent) were commercially successful), were initiated by the users. Similarly, the findings of Rothwell *et al.* showed that of twenty-five commercially successful innovations in the British textile-machinery industry, 84 per cent of firms collaborated with outsiders, out of which 66 per cent were prospective customers.

In fact, one of von Hippel's findings revealed that in the 'standard and non-standard products' sector, the customer was the sole inventor — 'Customers recognised need, defined functional requirements and specific goods and services needed before contacting potential suppliers' (von Hippel, 1979).

Nevertheless, the more important point is that products developed jointly with customers, as alluded to earlier, are more likely to succeed commercially than products developed solely by the customer or supplier. Close collaboration and interaction between the two parties are most crucial in the process of new-product design and development.

Summary

We set out in this chapter to explore the use of design as an effective weapon for fighting competition. Consequently, we examined factors that influence design *per se* (i.e. factors that bring about changes in or modification of existing products and/or design of new ones). These factors are broadly grouped into three, as follows:

(1) Technological factors (innovation, value analysis/value engineering, aesthetics, and ergonomics);
(2) Management factors (quality circles, timing, and leadership style);
(3) Marketing factors (attention to user needs).

All in all, the dimensions of technological, management, and marketing factors investigated are found to influence the design of new products or the modification of the existing ones. The outcome of the modification of the existing products or design of new ones in accordance with customer/user requirements is usually significant. Sales increase, due to improved value. Lower price consistent with quality is achieved due to economies of scale gained from lower cost of production. In this way, firms are better able to compete, maintain, and sustain competitiveness (particularly in volatile markets).

In sum, a relationship appears to exist between factors (technological, management, and marketing) that influence design and international competitiveness.

We have so far concentrated on the review of the various factors that influence the design of new products and their significant effects on competition. In the subsequent section, we shall report the results of our survey, which seeks to establish the degree of importance placed upon some of the technological, management, and marketing factors by commercially successful firms (Queen's Award winners) aimed at improving the effectiveness of their new-product designs, with the overall view to fighting competition and/or achieving competitiveness.

An empirical perspective

Table 5.5 presents the answers to the question that sought to establish who was responsible for managing design in commercially

Table 5.5: Who is responsible for managing design in your firm?

	No. of responses	%[a]
Managing director	29	32
Marketing director	21	23
Production director	14	15
Industrial design director	2	2
Purchasing director	1	1
Financial director	1	1
R & D director	24	26
Technical director	41	45
Chief engineer	23	25
Others[c]		
	156[b]	170

Notes: a. Based on 92 firms who responded to this question.
b. In a number of firms, design was managed by more than one person.
c. Others are:
(1) Engineering director
(2) Research director
(3) Engineering quality vice president
(4) Design and development manager
(5) R & D manager
(6) Development manager
(7) No-one.

successful firms. This table clearly indicates the large majority (85 per cent) of those who managed design were directors, strongly suggesting the great influence that top management (company leaders) have on design. Since these directors are decision-makers, it follows that they have the power to decide whether or not to start, continue, suspend, or abort a design project. They can also direct that an existing product be redesigned/modified or that a new product be introduced earlier or later than the planned date. These decisions/directions may vary according to the leadership style of each director, hence our proposition that leadership style influences the design of new products is supported.

Further, respondents were asked to rate the degree of importance attached to factors that contribute to the design and development of the award-winning product(s). Again, 'top person' came top of the league table of the influencing factors. Of the respondents, 74 per cent claimed that the successful design and development of the product(s) that received the Queen's Award 'was directed by an individual with outstanding authority and power, "top person"', only 4 per cent of the respondents were opposed to this view. 'Interacted closely with customers/users in design and development stages' was rated second with 76 per cent of the respondents considering it as important, compared to 8 per cent who felt it did not contribute to the successful design and development of the award-winning products. The third factor accounting for the successful design of the award-winning products was 'had higher technological content than rival product offerings' with 63 per cent of the respondents perceiving it as important, whilst 11 per cent said it was not.

In an extensive review of the literature, Britain was found to be inventive but not innovative enough when compared to its major foreign counterparts. As a result of this, the Queen's Award winners were asked to rank the following factors (see Table 5.6) with which British inventions may be translated into commercially successful innovations. As can be seen, 'improvements to own company technology' was ranked by 61 per cent of the respondents as the most important exploitable factor, followed by 'creative innovation' (32 per cent), with 'collaborative arrangements' (25 per cent) in third place. The fact that commercially successful firms ranked 'improvements' and 'breakthroughs' (creative innovation) (both of which are classes of innovation) as the first and second most important exploitable factors for translating inventions into commercially successful products, lends credence to our argument that innovation influences the design of new products.

Similarly, an investigation into the design strategies followed by the Queen's Award winners (see Table 5.7) shows that 78 per cent of the sample adopted 'major innovation' strategy of which 67 per cent ranked it as the most important. 'Improvements' was second in order of importance, with 46 per cent of the respondents ranking it as the most important strategy, followed by 'radical breakthrough' (19 per cent) in third place. As already stated, major innovation, improvements, and radical breakthrough are all classes of innovation that influence the design of new products — once again reinforcing the argument that innovation influences design.

Further, it was hypothesized that 'in commercially successful firms, the introduction of new manufacturing techniques is perceived as a competitive tool for improving the effectiveness of new product designs'. Consequently, firms were asked to state the manufacturing techniques currently used by them. Table 5.8 indicates that CAD (computer-aided design) was the technique currently used by the majority (56 per cent) of the sample, followed by value engineering (38 per cent). Quality circles (25 per cent) was the third most frequently used manufacturing technique. While this indicates that CAD, value engineering, and quality circles influence the design of new products, it also suggests

Table 5.6 Exploitable factors for translating inventions into commercially successful innovations

	Rank in importance									
	1 %	*2* %	*3* %	*4* %	*5* %	*6* %	*Base* %	*n*	*N*	$\bar{x}$
Collaborative arrangements (e.g. with major foreign competitors)	25[a]	35	20	15	5	—	100	20	88	2.40
Improvements to own company technology	61	18	16	4	2	—	100	57	88	1.66
Use of best production methods (e.g. CAD/CAM, automation, robotization, etc.)	23	37	20	14	6	—	100	35	88	2.42
Creative imitation (improving existing products)	14	38	21	17	10	—	100	29	88	2.72
Creative innovation (providing new scientific breakthroughs)	32	35	21	6	3	3	100	34	88	2.20
Use of reliable (high quality) materials/ components	6	27	52	9	3	3	100	33	88	2.85
Others[b]										

Notes: n = the number of respondents who ranked each factor.
N = the number of firms in the sample who responded to this question.
$\bar{x}$ = the mean value attached to each factor.

a. To be read: 25% of the 20 respondents ranked 'collaborative arrangements' as the most important factor for translating inventions into commercially successful innovations.

b. Others are: (1) Improved marketing (ranked first by eight firms, second by one firm, and fourth by one firm ($\bar{x}$ = 1.40).
(2) Proper management of innovation (ranked first by seven firms, second by two firms and third by one firm ($\bar{x}$ = 1.40))
(3) Collaboration with major foreign users (ranked fourth by one firm ($\bar{x}$ = 4.00)

Table 5.7 Design strategies followed by firm

	Rank in importance						
	1 %	*2* %	*3* %	*Base* %	*n*	*N*	$\bar{x}$
Major innovation (innovation leading to major changes in or major addition to an existing product)	67[a]	31	1	100	70	90	1.34
Improvements (innovation leading to minor changes in the existing product)	46	33	21	100	70	90	1.75
Radical breakthrough (innovation leading to an entirely new technology)	19	39	42	100	36	90	2.22
Others[b]							

Notes: n = the number of firms who responded to each design strategy.
N = the number of firms in the sample who followed design strategies.
$\bar{x}$ = the mean value attached to each design strategy.
a. To be read: 67% of the 70 respondents ranked 'major innovation' as the most important design strategy followed by their firms.
b. Others are: via service experience/skill (ranked first by one firm, and third by two firms).

considerable scope for improvement through the use of proven techniques.

Already, it has been established (see Chapter 2, Table 2.1, p. 84) that aesthetic and ergonomic features such as attractive appearance/shape, easy to use, safe to use, etc., influenced the sale of award-winners' products against those of their competitors in international markets — predictive of the influence of aesthetics and ergonomics on new-product designs.

Finally, the essence of the application of technological, management, and marketing techniques, as discussed earlier, is to improve the quality of new products, aimed at satisfying a variety of customer needs and wants. It is also maintained that products developed jointly with customers and users are more likely to succeed commercially than products developed solely by manufacturers — suggestive of the significant role played by customers/users in the design and development of new products. Therefore, it is important to know the extent to which firms involve customers in new product development. Table 5.9 presents the answers to the question that sought to determine the degree of customer/user involvement in the design of new products. As this table clearly indicates, as many as 96 per cent of the sample said they involved

Table 5.8 Manufacturing techniques currently used by firm

Manufacturing techniques	*No. of responses*	%[a]
Computer-aided manufacture (CAM)	2	13[b]
Robotics	—	—
Automation (e.g. flexible manufacturing system (FMS))	1	6
Computer-aided design (CAD)	9	56
Value engineering	6	38
Quality circles	4	25
Parts per million (PPM)	—	—
Others[d]		
Total responses	22	138[c]

Notes: a. Based on 16 firms who responded to this question.
b. To be read: 13% of the responding firms are currently using CAM.
c. A number of firms used more than one technique.
d. Others are:
(1) MRP (material requirement planning)
(2) Subcontracting
(3) Laser
(4) 'Just in Time' component inventory reduction (JIT)
(5) Group technology
(6) Benchmarking against competition.

Table 5.9 Extent of customer/user involvement in new-product design

	No. of firms	%
Extremely involved	19	20
Very involved	23	25
Involved	34	36
Slightly involved	14	15
Not involved	4	4
	94	100

customers/users in the design of new products. In 45 per cent of cases, customers were either 'extremely involved' or 'very involved', whilst 51 per cent of the respondent firms 'involved' or 'slightly involved' customers/users in the design and development of their products. Only 4 per cent of the sample did not involve customers at all.

Table 5.10 Stages of customer/user involvement in new-product development

Stages of customer/user involvement	*No. of responses*	%[a]
First stage (idea generation and drawing)	41	45[c]
Middle stage (development of prototype)	47	52
Last stage (production)	17	19
Commercialization stage (after product has been launched)	33	36
Throughout all stages	18	20
No stage at all	3	3
Total	159[b]	175

Notes: a. Based on 91 firms who responded to this question.
b. A number of firms involved customers/users at more than one stage.
c. To be read: 45% of firms responding to this question said that they involved customers/users in the first stage of new-product development.

Stages of customer/user involvement in new-product development

Involving customers/users in new-product development is one thing, the stages of involvement is another. It was therefore decided to find out the stages at which firms involve customers/users in new-product development. The findings (Table 5.10) show that 45 per cent of the responding firms involved customers/users during the first stage (idea generation, and drawing); more than half of the sample (52 per cent) involved customers at the second stage (development of prototype), whilst 19 per cent of the sample involved customers/users during the last (production) stage. On the other hand, about one-third (36 per cent) of the sample involved customers at the commercialization stage (i.e. after the product has been launched) while 20 per cent (18 out of 91) of the respondents involved customers/users at all stages. Only 3 per cent of the sample did not involve customers/users at any stage at all. Based on the early normative models of new-product development, Booz, Allen and Hamilton, Inc., (1965), A.C. Nielsen Company Limited, and Hamilton *et al.* (1970) to name a few, and strongly inspired by the results of this study, a practical model of customer involvement and its impact on successful manufactured exports has been developed (Figure 5.3).

Figure 5.3 suggests that to produce a competitive product,

Figure 5.3 A practical model of customer involvement in new-product development and its impact on successful manufactured exports. *Source:* Ughanwa (1986)

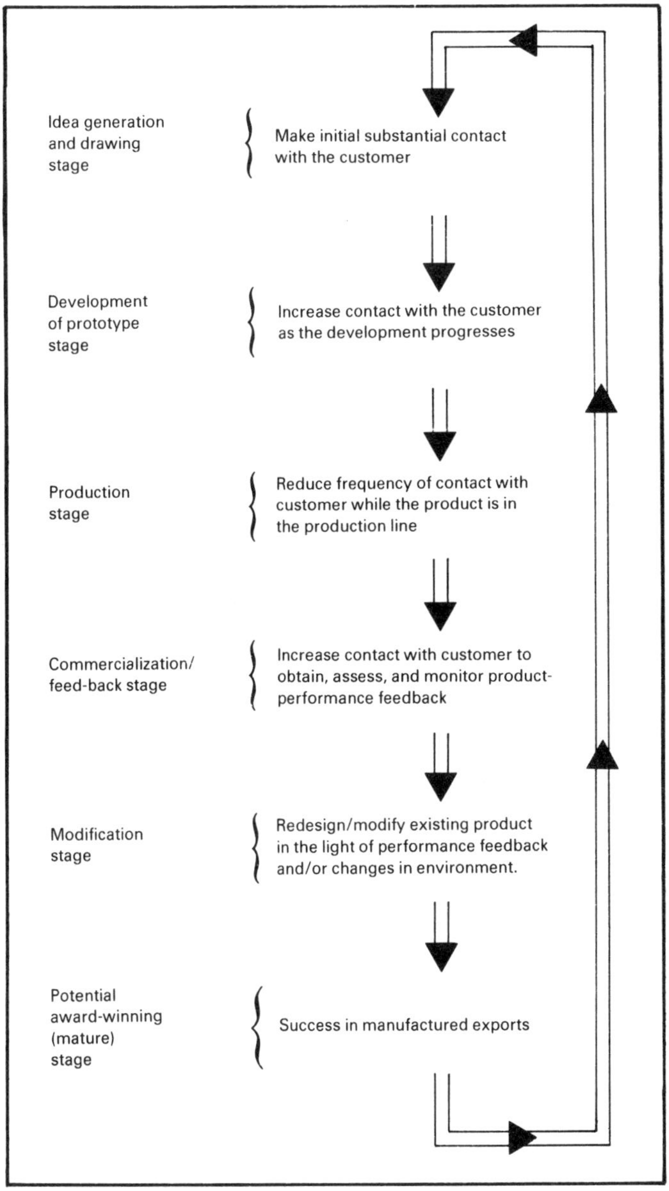

customers/users must be substantially involved at the initial stage of new-product development (i.e. the idea generation and drawing stage). This will enable the innovator (firm) to incorporate the needs and wants of the customer/user.

However, because we live in a dynamic environment in which tastes and preferences change fast, it is possible that the customer might change his/her preference by way of adding to or subtracting from the existing specification before the product is ready for sale. There is, therefore, the need to increase contact with the customer/user at the development of prototype stage to ensure that no details are missed out (i.e. to ensure that the prototype is still in accordance with the customer 'wishes').

Nevertheless, because the customer has little contribution to make while the prototype is in the production line (production stage), little contact with the customer is required at this stage. Rather, what is essential at this stage is the involvement of or close liaison with the designers/developers of the product by the production department to ensure that the product rolls off the production line in accordance with customer/user specifications.

However, there is the need to increase contact with the customer/user as soon as the product is introduced onto the market (commercialization/feedback stage). This is important because the firm (innovator) requires a feedback from customers/users to enable him/her to assess and monitor the product performance. With the performance feedback, the firm can then redesign or modify the existing product or design an entirely new product in the light of performance feedback and/or changes in the environment (modification stage), which in turn, leads to success in manufactured exports (potential award-winning stage). Nevertheless, firms should not stop their innovative activities and rest on their laurels after winning the award. Rather the process should be continuous to achieve more awards. Therefore, this model helps to explain why some firms have received the Queen's Award for export and technology more than once. Short Brothers Ltd (Northern Ireland) for example, has won the award ten times between 1967 and 1981, seven of them (1967–73) consecutively.

Figure 5.4 is the graphical representation of this practical model of customer involvement in new-product development based on the results of our study, and is self-explanatory. The major implication of this model for management, is that firms (be it winners or non-winners of awards) should attempt at all times to enhance their competitiveness in manufactured exports by ensuring that customers/users take an active part in deciding the nature of the product they would like to buy (market orientation), not, as some

Figure 5.4 Graphic demonstration of the impact of customer involvement in new-product development on successful manufactured exports. *Source:* Ughanwa, 1986

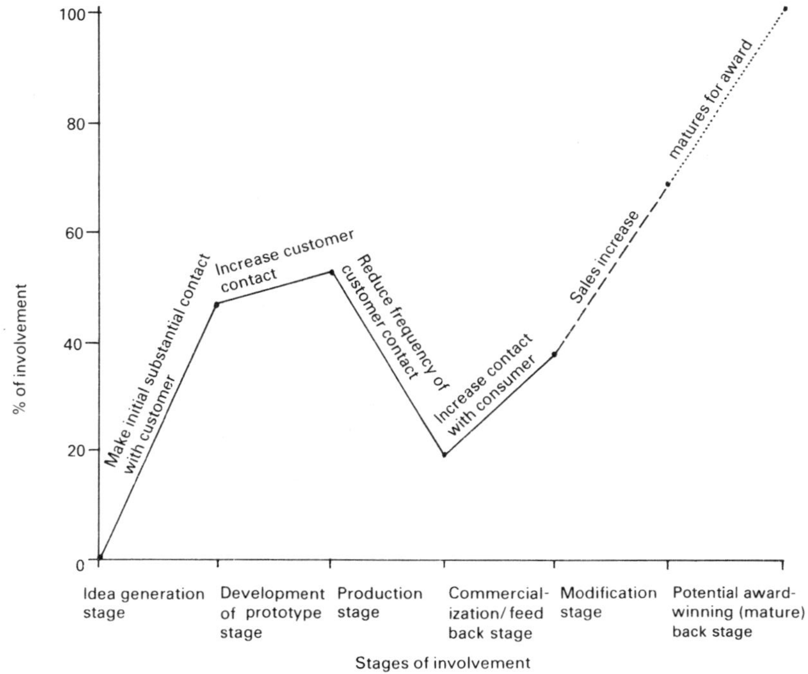

firms usually do, manufacture first, and then look for customers to buy (product orientation). In this way, products can be uniquely differentiated in accordance with the customer/user purpose; in this way firms are better able, not only to fight competition, but also to maintain and sustain competitiveness; and in this way more awards may be won and even those firms that have been less successful in the award-winning competition may achieve success.

Conclusions

Based on the results of our research and the literature reviewed, the following conclusions can be drawn.

The overall result shows that commercially successful firms (Queen's Award winners) perceive design as the lifeline for staying in business. But to stay in business, firms have not only to fight competition, they have to maintain and sustain competitiveness. This is manifest in the special attention given to design and the factors that influence it. It was found, for example, that 85 per cent of those who managed design in the sample (ninety-two firms) investigated were directors and the remainder were chief engineers — suggestive of the great importance placed upon design by respondent firms. It also indicates the direct involvement in design activities by top management, and thus demonstrates their influence on new-product designs.

It was evident in our findings that innovation plays a significant role in influencing the design of new products. Commercially successful firms tend to see improvements (innovation leading to minor changes in or minor addition to existing products), major innovation (innovation leading to major changes in or major additions to existing products), and radical breakthroughs (innovation leading to entirely new technology or new scientific ideas) as powerful tools for breaking design into distinct customer requirements. They were also seen as an instrument for translating inventions into successful commercial realities, aimed at achieving international competitiveness.

Value engineering and quality circles are among the important manufacturing techniques used by management to improve the quality of new-product designs, thus supporting our hypothesized relationship between the introduction of new manufacturing techniques in commercially successful firms and their perceptions as competitive tools for improving the effectiveness of new-product designs.

Further, products with better aesthetic and ergonomic features were found to have a strong influence on the sale of new products against the rival offerings in international markets, thus lending weight to our hypothesis that commercially successful firms are reputed for the use of design, design-related, and design-influenced factors to achieve international competitiveness.

It was also clear from the survey that commercially successful firms pay great attention to customer/user needs. This is attested to by the high level of involvement of customers/users in the development of new products in the majority (96 per cent) of the cases. However, the result shows that the degree of involvement varies, probably depending on the type of product.

An investigation into the stages of customer/user involvement in the new-product development produced similar variations, which led us to develop a practical model of customer involvement in new-product development, with an illustration of its impact on successful manufactured exports. The model warns innovators (particularly commercially successful/technically progressive firms) of the dangers of resting on their laurels simply because they have achieved success (or received awards). It stresses that firms should see design/innovation as a continuous process for achieving commercially successful manufactured products, and earning profits. Specifically, the model emphasizes that firms (be it winners or non-winners of awards) should attempt at all times to enhance their competitiveness in manufactured products by ensuring that customers/users are actively involved in the process of developing new products. This means that customer requirements must be fully ascertained and clearly understood before any prototype design goes into commercial production; and this can best be achieved by involving users themselves in the process. In the end, it is the customers' decision to buy that matters more than the manufacturers' decision to make.

Finally, the review of the literature indicates that timing is a necessary ingredient for the successful introduction of new products onto the market, the absence of which could bring about failure or cause a new product to be withheld or withdrawn from the market for modification. Bad timing could also cause a project to be suspended, aborted/abandoned, or a new one started.

The major implication of the overall finding for innovators, industrialists, and management is that to stay in business in this highly dynamic environment, firms must prepare themselves for a hard battle (i.e. intense competition) in the marketplace. To fight competition effectively, firms must consider design in all its ramifications (dimensions). This means that firms should give attention to all factors (technological, management, and marketing) that influence design, not just design alone, when developing new products or modifying the existing ones. Giving adequate attention to customer/user needs, incorporating aesthetic and ergonomic features into new products in line with customer needs, applying value-engineering and quality-circle techniques to improve the effectiveness of new-product designs, and timing accurately the introduction of these products, all combine to maximize value while minimizing cost. However, to ensure successful results, these activities must be efficiently planned, carefully organized and co-ordinated, and enthusiastically directed from the top. In the next chapter, we examine the meaning of design itself, and demonstrate how it should be applied in the business world to enhance competitiveness.

Chapter six

Design: meaning and application

Introduction

Design is the lifeline of modern commercially progressive industries and its role in international competitiveness is growing significantly. Yet confusion seems to exist within and among the design profession as to what design really means and/or the benefits designers can bring to industries. Therefore, the objective of this chapter is twofold. First, it attempts to throw more light on the meaning of design with a view to providing a framework for a better understanding of the design concept as well as to add to the existing knowledge in design studies. Second, it focuses on design as a major contributing factor to non-price competitiveness.

The material is divided into five parts. The first reviews the definitions of design and examines the different interpretations attached. Our empirical findings on the innovators' perception of design is reported.

Part two highlights the differences/conflict between two disciplines – engineering design and industrial design (probably introduced by the varying definitions of design). They include gaps in education, status, and product/project presentation. It is concluded that although both disciplines are often perceived as distinct, they are none the less dimensions of the same activity, which calls for a marriage of both disciplines to achieve good results.

Consequently, the third part devotes itself to exploring ways of integrating engineering design and industrial design. The contributions made by the government and its agents, and a number of individuals towards achieving this objective are critically examined. Our research findings on how the skills embodied within the practice of industrial design and engineering design can best be integrated are then reported.

It is felt that the technological impact that computers are having on design can facilitate the integration problem, particularly in the area of communication. This is spelled out in the fourth part, which covers discussions on the major computer aids for design (CAD). Next, the results of our survey on the significant changes made by the Queen's Award winners to their manufacturing processes are presented. It was found that computer-aided design and computer-aided manufacture (CAD/CAM) was the second most frequently used technique, after automation, for introducing change to manufacturing processes by commercially successful firms.

Finally, the chapter is concluded with special emphasis on the need for professionalism in design management so an organizational framework may evolve whereby engineering designers and industrial designers and indeed all allied professions may work in harmony as a team to produce competitive and commercially successful products.

What is design?

The complex nature of design has made it difficult for authors to agree on a tightly controlled definition. Almost everything we do in life appears to involve one form of design or another. For example, design manifests itself in the organization of:

1. Self — the way you dress yourself — style, appearance, attractiveness, etc.
2. Home — the way the home is decorated. For example, decoration with wallpapers, paintings, crafts and even the positioning of tables, chairs, television, turntable/video set, in order to save space and be attractive to look at.
3. Office — the arrangement of filing cabinets, tables, chairs, telephones, office equipment such as typewriters, computers, and matching people with such equipment to generate maximum comfort (ergonomics), to conserve energy, to save time and money, and to increase productivity.

Other forms of design, and perhaps, the most noticeable of all, can be found in the manufacture of:

4. Consumer goods (durable and non-durable), such as cookers, washing machines, vacuum cleaners, hair driers, and packaged foods.

5. Industrial goods, such as, agricultural machines (tractors, earth diggers), X-ray equipment, safety-helmets, chemical products (drugs), and so on.

It is therefore no wonder that there is little uniformity in the definition of design. The apparent lack of uniformity can be attributed to differences in the interpretation and application of design, as shown by the following definitions:

1. Decision-making, in the face of uncertainty, with high penalty for error (Asimow, 1962).
2. Finding the right physical components of a physical structure (Alexander, 1963).
3. Engineering design is the use of scientific principles, technical information and imagination in the definition of a mechanical structure, machine, or system to perform specific functions with the maximum economy and efficiency (Fielden Committee, 1963).
4. Simulating what we want to make (or do) before we make (or do) it as many times as may be necessary to feel confident in the final result (Booker, 1964).
5. A creative activity — it involves bringing into being something new and useful that has not existed previously (Reswick, 1965).
6. The optimum solution to the sum of the true needs of a particular set of circumstances (Matchett, 1965; 1968).
7. A good directed problem-solving activity (Archer, 1965).
8. The conditioning factor for those parts of the product which come into contact with people (Farr, 1966).
9. Relating product with situation to give satisfaction (Gregory, 1966).
10. The performing of a very complicated act of faith (Jones, 1966).
11. The imaginative jump from present facts to future possibilities (Page, 1966).
12. An iterating decision-making activity to produce plans by which resources are converted, preferably optimally, into systems or devices to meet human needs (Wooderson, 1966).
13. Design starts and ends with the customer, and designers are in the main stream of translating ideas into hardware for money (Turner, 1968).
14. The refinement of a converging multivariable system operating within constraints (Britch, 1968).

15. A conscious plan (Cain, 1969).
16. Design is a significant way of augmenting the product (Levitt, 1969).
17. Fitness of purpose (Evan-Vaughan, 1970).
18. Design is a process whereby solution is found in principle for a new or improved product, process or material (or a new application of an existing product, process or material (Holt, 1977).
19. Product design is defined to include both engineering design and industrial (aesthetic) design (Corfield, 1979).
20. Engineering design is however, concerned with technical factors to a high degree, as well as with economic and human ones. Thus, the art of designing may be described as the ability to reconcile factors, be they marketing or production, social or political (McTavish, 1974).
21. Design is a means of adding value to products. In some cases one may identify design as a sophisticated packaging — in others it is the product (Watt, 1980).
22. Design is regarded as a means of making a product function technically (Moody, 1980a,b).
23. The initiation of change in man-made things (Jones, 1981).
24. Design is the very core of innovation, the moment when a new object is imagined, devised and shaped in prototype form (OECD, 1982).
25. The process of converting an idea into information from which a product can be made (SERC Report, 1983). This is an adaptation of Lord Caldecote's definition of design (1979).
26. The Open University Design Innovation Group, UK (1983), shows the percentages of how eight design-conscious firms out of the representative sample (forty-one firms) have defined 'design'. Of the design-conscious sample, 25 per cent defined design as 'safety'; 25 per cent as 'durability'; 25 per cent as 'fashion co-ordination with range'; 50 per cent as 'efficiency in production or use of materials'; 63 per cent as 'increased value, making products that sell or make a profit'; 63 per cent as 'shape, visual appearance'; and finally, 74 per cent as 'ergonomics, fitness for use'.
27. Design therefore is the activity in which ideas and needs are given physical form, initially as solution concepts, and then as a specific configuration or arrangement of elements, materials and components (The Open University Design Innovation Group, UK, 1984).

28. Design may be described as the 'realisation of' ingenious contrivances, conceived to serve a specific purpose' or more simply, it is the process of establishing, from a given need, a conceptual solution leading to a physical solution (Schwarz, 1984).
29. The definition of design as presented in a Report on 'Managing Design' (1984), jointly sponsored by the Council for National Academic Awards (CNAA), the Department of Trade and Industry and The Design Council, reads thus: 'designing is understood as the preparation of solutions to problems concerning creating, producing and marketing products'.

These definitions represent different ways in which twenty-nine different sources have perceived and defined the word 'design'. Although these definitions appear to convey similar meaning, there is no single standard definition of design. Design is clearly a complex concept and this often makes the understanding and possibly, the interpretation of design difficult. The recent findings of Schwarz (1984) and the CNAA Report (1984) confirm this view. Whereas Schwarz argues that: 'Design itself has no unique definition', the CNAA Report maintains that misunderstanding appears to prevail within the rank and file of design experts as to what design really means. The report states: 'Judging by the discussions (from seminars aimed at soliciting views and comments from design and management experts on what do we mean by design?) it seems that confusion even exists within and among the design professions.'

The report further argues that the differences in 'these definitions show the difficulty of attempting to encompass in a single sentence the scope and complexity of designing. This may explain the absence from recent reports about design issues of any really informative explanation of what designing may be' (CNAA, 1984).

However, the conveyance of similar meanings in definitions of design, may be explained by the fact that seemingly all definitions, by implication, alluded to design as a 'problem-solving activity' through which various customer problems, that is, customer/market needs and wants can be conceptualized and solved (satisfied), via development of acceptable new and/or improved products. Hence it is reasoned that the differences in definition of design notwithstanding, the design concept appears to remain one and a coherent concept.

A coherent concept is important, not only in understanding the meaning and interpretation of design, but it is also necessary in the application of the concept itself. To this end we offer our own model of 'design' in Chapter 8.

It is perhaps significant that although the New Collins Concise and Oxford dictionaries have both referred to 'drawing and sketching' in their various definitions of design, none of the definitions reviewed above appears to have made any reference to either sketching or drawing, which is usually associated with the design process. This may be explained by the fact that nowadays, more emphasis appears to be laid on model building, often aided by a computer, rather than on paperwork such as drawing, which seems to waste time and often tends to create problems of presentation of projects. For example, the presentation of a new car concept to a board of directors with a drawing rather than a model may not only be boring, but may no longer be convincing enough to explain the practicality or workability of the project. In other words, the directors tend to prefer visual demonstration (with a model) of how the product (car) looks (possibly with its precise costs), rather than a sketch or descriptive version of how it will look.

Although Clayton (1984) maintains that 'it will be some time before the draughtsman disappears altogether, from the drawing office', Creighton (1985) has predicted that 'by 1990 all components will be designed by computer and paper prints will no longer be available'. This suggests that by the 1990s, design by drawing or sketching will have become a thing of the past because of the impact of the computer on modern designs. This aspect of computer aids for design (CAD) will be discussed later.

Against this background of a multiplicity of definitions of design, we decided to solicit the views of our Queen's Award winners on the subject. Our findings are reported in the next section.

Factors encapsulating the concept of design: survey results

As noted, confusion still appears to exist between and among innovators as to the meaning of design. Accordingly, it was decided to find out how the Queen's Award winners (innovators) perceive design. Table 6.1 presents the answers to the question that asked innovators to rank from a list of variables, up to four dimensions that they feel best encapsulates the concept of design. The limitation to four dimensions was based on two main reasons. First, in the light of the complex nature of the design concept, it was considered that up to four variables would help to reduce the choice to a more manageable proportion that would form the basis

Table 6.1A Factors encapsulating the concept of design

	Rank in importance							
	1	*2*	*3*	*4*	*Base*	*n*	*N*	$\bar{x}$
	%	%	%	%	%			
Making products that make profit	51[a]	22	15	12	100	68	93	1.86
Making products that sell	35	40	15	10	100	52	93	2.00
Superior overall performance	36	15	23	26	100	39	93	2.38
Ergonomics/fitness for use	13	40	33	13	100	30	93	2.46
Making products that function technically	24	26	24	26	100	38	93	2.52
Fashion	—	33	67	—	100	3	93	2.66
Safety	13	38	13	38	100	8	93	2.75
Durability	25	—	50	25	100	4	93	2.75
Increasing value in a product	12	24	41	24	100	17	93	2.76
Reliability	6	16	42	35	100	31	93	3.06
Visual appearance/shape/style	12	24	6	59	100	17	93	3.11
Efficiency in production or use of materials	6	16	38	41	100	32	93	3.12
Drawing/sketching/draughtsmanship	—	—	—	100	100	1	93	4.00
Others[b]								

Note: a. To be read: 51% of the 68 responding firms ranked 'making products that make profit' as the factor that best encapsulates the concept of design.
n = the number of firms who ranked each factor.
N = the number of firms who responded to this question.
$\bar{x}$ = the mean value given to each factor.
b. Others are: Satisfying needs in a better way (ranked first by one firm).
N.B. All percentages are rounded up to whole numbers. This applies to all tables throughout this analysis.

for comparison against other factors. Second, all the variables were compiled from two major components of design, namely, design factors, and design-influenced factors; thus it was considered interesting to determine whether respondents perceive only design factors, design-influenced factors, or a combination of both dimensions as best encapsulating the concept of design.

However, it is important, at this juncture, to make the following points to enhance the interpretation of the analysis of the rank-order questions.

1. The interpretation of the rank-order questions, where appropriate, is based on the mean rank of each variable, supported by the number of respondents who make up the

Table 6.1B Reasons given by respondents for ticking instead of ranking the answers

1. 'Product must meet customer requirements, be reliable and of good appearance. These are of equal importance.'
2. 'Unable to rank — all important.'
3. 'Cannot rank on single basis because of different nature of various businesses in our company.'
4. 'All equal.'
5. 'You need balance.'
6. 'Can't differentiate in ranking order.'
7. 'Difficult to rank.'
8. 'Ranking is not sensible.'
9. 'All rank equal (add up to the same result anyway).'

Note: The reasons presented in this table represent all the reasons given by respondents throughout this questionnaire for ticking (✓) instead of ranking the answers (1st, 2nd, 3rd) as the question requires. (All questions were design orientated.)

mean. This is particularly important as it reflects the nature of the respondent's business. For example, only three firms ranked 'Fashion' out of ninety-three who answered the question as a whole. Clearly if you are Laura Ashley or Habitat, it is important. If you are British Aerospace, it is unlikely to figure significantly in your concept of design.

2. Rank 1 on the scale indicates 'most important' followed by rank 2, 3, 4, etc. (in descending order of magnitude), with the last rank-order number indicating the least important variable of those ranked.
3. The overall purpose of the analysis of the rank order is to determine the relative importance attached to each factor under investigation by respondent firms, with a view to drawing both specific and general conclusions from the results as to the factor(s) that best encapsulate the concept of design.

Overall, the pattern and distribution of the results confirms the conclusion drawn from our literature review — there is considerable divergence in the way in which people define design. However, care must be taken in interpreting the table given the variable response to each question. Thus, although 'reliability', 'visual appearance', and 'efficient use of materials' have a lower mean than 'fashion', many more respondents included these factors in their 'top four', indicating that a mixed sample of firms serving a variety of markets attaches considerable attention to these variables. It follows that the respondents' perception is likely to be situation-specific and it would be interesting to pose the question to a more homogeneous group of respondents, i.e. serving the same market, and to establish the degree of consistency between them.

To add to the disagreement amongst innovators as to the exact meaning of design, fifteen firms ticked in this question, instead of ranking the answers. Whereas some argued that they did so because they felt that all the factors were equal in importance, others maintained that they could not differentiate the factors in rank order and that it was difficult to rank them. In fact, as Table 6.1B indicates, one firm (innovator) defended his reason for not ranking thus: 'Product must meet customer requirements, be reliable and of good appearance. These are of equal importance', yet some firms were able to distinguish between 'reliability' and 'visual appearance' (see Table 6.1A).

However, it is clear from these findings that no one factor explains clearly the concept of design, rather, the design concept seems to be best explained by a combination of design and design-influenced dimensions.

It is likely that the differences in the definitions of design might introduce some differences/conflict between engineering and industrial design. For instance, some of the definitions of design appear to be engineering-orientated whilst others have industrial-design connotations. In the following section, we shall examine these differences and why they exist between the engineering- and industrial-design professions.

Differences/conflict between engineering- and industrial-design components

This section seeks to outline the major differences/conflicts that exist between engineering- and industrial-design components. The discussion starts with the definition of the two concepts, followed by the historical background of the conflict. Evidence is presented for the differences between the two disciplines and includes: gaps in education, status, and product/project presentation.

The definition of both engineering and industrial design will, it is hoped, provide a basis for a better understanding of the differences between the two components. The Fielden Committee (1963) defines engineering (design) as 'the use of scientific principles, technical information and imagination in the definition of a mechanical structure, machine or system to perform specific functions with maximum economy and efficiency'. On the other hand, Maldonado (1964) defines industrial design as 'a creative activity whose aim is to determine the formal qualities of objects (products) produced by industry. These formal qualities include the external features but are principally those structural and functional relationships which convert a system into a coherent unity both from the point of view of the producer and the user. Industrial design extends to embrace all aspects of human environment which are conditioned by industrial production.' This appears to conform with Roger Jones' (1980) view that 'industrial design determines the physical form of a product, its performance and cost'.

Maldonado's (1964) idea of quality has been confirmed by Flurscheim (1983), who indicates that 'the term quality is used to include a number of important attributes dependent on design: appearance and visual coherence, which are significant factors in achieving and demonstrating quality; compatibility of machines with their human and physical environment; reliability, especially as affected by maintenance facilities; safety, as influenced by control considerations; and the less specific quality of 'wholeness', which is dependent on the designer's ability to optimise every detail in its relationship to the unit as a whole, and in which the synthesising approach of the industrial designer makes a major contribution'. This further explanation of the attributes of design is necessary because it appears to throw more light on the important functions of industrial design and engineering.

Judging from both definitions, therefore, there appears to be a fundamental difference between the functions of engineer and industrial designer. Both seem to have separate roles to play in the process of new-product design and development (NPDD). For example, whereas the engineer appears to be mainly concerned with technical and functional performance of the product, the industrial designer is largely concerned with human factors (ergonomics, aesthetics, graphics, human environment, etc.). This appears to agree with Carter's (1977) view that 'initially a young engineering designer will be primarily concerned with technical performance while the industrial designer will be concerned with aesthetics and human values'.

In practice, though, the functions of engineering and industrial design are not mutually exclusive, the extent of involvement by each team depending, of course, on the type of product/project. In other words, both functions appear to overlap in most projects. For instance, Carter (1977) reports that 'on some projects it is almost impossible to distinguish between the work of the industrial and engineering designers'. He maintains that 'the skills embodied within the practice of industrial design frequently overlap those of related disciplines — not only architecture and graphic design, but also ergonomics and engineering design'. For instance, an agricultural harvester machine, while expected to perform mechanically well, is also expected to be easy to control, and symbols, diagrams, words of instruction, etc., all easy to follow by the operator, safe to use, and easy to maintain. All these combine to provide comfort to the user. The absence of one is likely to cause deficiency in the other. This suggests the indispensable nature of both dimensions of design in a project, which in turn, suggests that both skills should be combined in the industrial-design team in order to achieve an acceptable and competitive product.

Figure 6.1 demonstrates the mutual relationship between industrial and engineering design disciplines and the overlap of their functions; and supports the proposition that both dimensions of design are essential in the development of new products. On the other hand, the 'design spectrum' gives the impression that the more complex the product (computer terminal, machine tool), the more the engineer is involved and the less the industrial designer contributes. Conversely, the less complex the product (cutlery, lighting fitting), the more the industrial designer contributes, and vice versa. While it is not intended here to refute this contention, it is pertinent to note that Flurscheim (1983), Evans (1983), and Black (1984) hold the view that the more sophisticated the product or project, the more contribution the industrial designer makes in terms of aesthetics, ergonomics, graphics, and human environment, which plays a major part in the purchase descision aside from functional performance for mechanically driven products, and, of course, price.

Aside from the conceptual difference as discussed previously, the difference/conflict between industrial design and engineering design appears to date back to 1944 when the Council of Industrial Design was set up by government to create awareness of industrial design and to improve industrial products. Unfortunately, the Council appeared to be mainly interested in consumer products designed by industrial designers to the neglect of engineering products made by engineers. As a consequence, the Council appeared not to be seriously involved in improving the engineering products, such as agricultural machines. Gradually, the quality of such products faded and became increasingly uncompetitive when compared to their foreign counterparts. This became a cause for concern in engineering industries.

In 1972, the Council of Engineering Institutions Committee, on which the Council of Industrial Design was represented, met and recommended that the Council of Industrial Design be renamed

Figure 6.1 The design spectrum

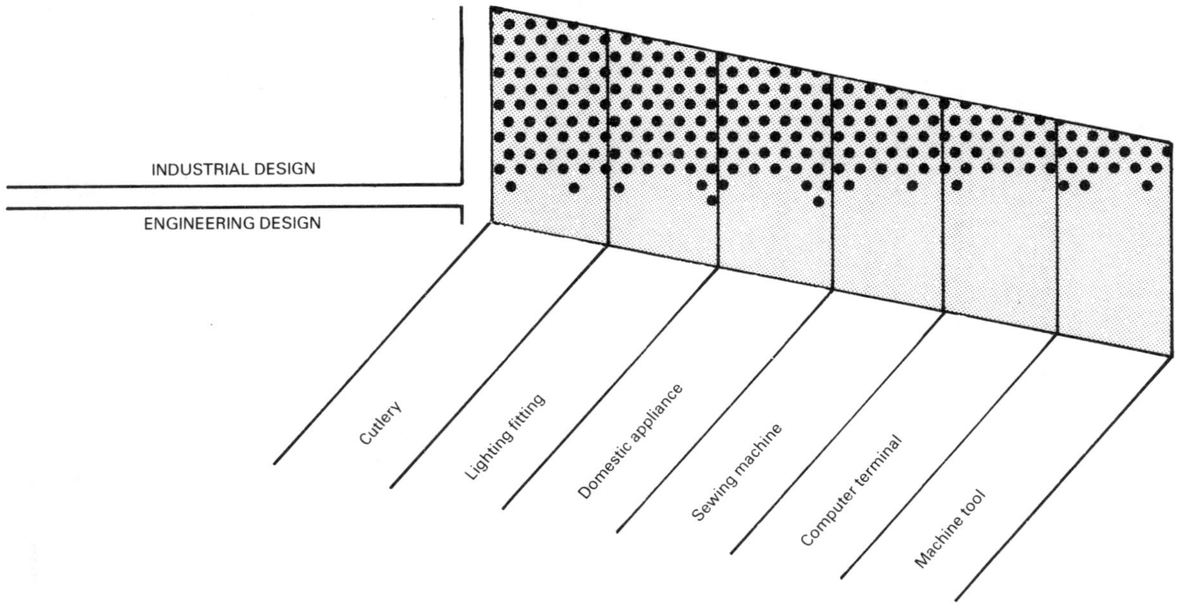

Source: Carter, D. (1977) 'Industrial design education in the United Kingdom'. A report to the Design Council's Education Study Group. Design Council, London, p. 13

the Design Council and its task be enlarged to embrace engineering products. As Black (1984) puts it, 'as the realization of the weakness of much British design of heavy capital and engineering products spread, the Council (i.e. Council of Industrial Design) was renamed simply Design Council and its work extended to cover the whole range of designed and manufactured goods'.

This change of name appears to have brought about a change in emphasis or direction. First, distinguished engineers appear to have dominated the Chair of the Design Council since its inception, to the neglect of industrial designers. Second, the Design Council seems to be mainly interested in engineering characteristics of a product, especially when making awards for new-product designs, to the neglect of industrial-design features (aesthetics, ergonomics, graphics, etc.) — almost the opposite of what happened when the Council of Industrial Design was in existence. The Design Council, however, has been criticized on this ground (i.e. for being technologically biased), as will be seen later in this chapter.

Thus, since the creation of the Design Council in the UK, the relationship between engineering and industrial-design professions appears to have weakened in some way. This tends to create a gap between the two components, and the evidence seems to suggest that the gap is widening. For the purpose of this study, 'gap' simply means the division that exists between engineering and industrial design. This is the gap created by the educational background and training given to the two professions. Cain (1969), the Moulton Report (1976), Carter (1977), the Finniston Report (1980), the SERC Report (1983), Gibbs and Flurscheim (1983), and Black (1984), all hold the opinion that there is a difference in education and training of the industrial designer and the engineer.

Cain states that 'the (basic) principles of industrial design, design for manufacturing expression, ... are not generally taught within the orthodox education of engineers or technicians many of whom become responsible for the design of engineering products' (Cain, 1969). This has been confirmed by the recent Science and Engineering Research Council (SERC) Report which reveals that (industrial) design is allocated only 'a few weeks' on the timetable of the overall engineering course, instead of being regarded as 'an essential part of the (engineering) course or the integrating theme of the various specialisations taught to the engineer' (SERC, 1983).

Emphasizing specifically the division between industrial and engineering design, the SERC Report (1983) comments: 'Although it is not within our terms of reference, we have been impressed by the weight of opinion deploring the present *schism* between engineering design with its tradition growing out of Mechanics Institutes, and industrial design, wholly rooted in art colleges.'

'The *divide*', the Report continues, 'between art and science and engineering which begins so early in schools is damaging to the society and ensures that individuals with artistic talents will almost never take up an engineering career.' Following this finding, the Report recommends the creation within the academic community of 'a greater awareness of (industrial) design throughout the teaching of engineering'.

Whereas the Moulton Report (1976) recommends that 'engineering should be taught in the context of design, so that design is a continuous thread running through the teaching of undergraduate engineering', the Carter Report (1977) on 'Industrial Design Education in the UK' recommends that qualified engineers undertake postgraduate courses in industrial design in order 'to provide an opportunity for the study of a specialised area of industrial design in greater depth than is possible during the first degree course'. This view has been confirmed by the Corfield Report (1979). One thing is certain. Whether or not industrial design should be taught at undergraduate or postgraduate level, the underlying fact as revealed by the aforementioned reports is that there is an educational gap between engineer and industrial designer that needs to be closed if British industries are to be as competitive as their foreign counterparts.

The difference in education and training appears to contribute largely to a difference in status — each claiming to be superior to the other.

Cain (1969), Moody (1980a,b), and Black (1984) share the belief that engineers seem to feel superior to industrial designers because they tend to feel confident in 'technological knowledge' acquired during the course of their engineering training. For example, Cain (1969) indicates that 'an enormous volume of poor design in the engineering industries' results from engineers' confidence in technology as well as 'a general lack of understanding of design principles'.

In his work, Black (1984) maintains that 'mutual suspicion' grows between engineers and industrial designers. As a result, he argues that 'the highly analytical or mathematically ingenious engineering designer, exploiting the latest advances in technology, regarded the industrial designer *merely as a stylist*, brought in at the last stages of design to wrap a more attractive "envelope" round his complex engineering product to help to sell it'. This appears to confirm an earlier work done by Moody (1980a,b).

Moody (1980a,b) found that engineers oppose the employment

of industrial designers — an evidence of conflict between the two components. He asserted: 'Even in this small design-conscious sample, there is enough evidence to suggest that opposition by engineers to the employment of industrial designers is fairly wide spread.' He revealed that engineers have higher status than industrial designers when they (industrial designers) are employed in-house. On the other hand, industrial designers have higher status than engineers when practising as consultants. He reasoned that 'the superior status of the consultant industrial designer in a firm is bound to cause envy if not chagrin especially since functionally he is dependent upon engineers'. This has been confirmed by Lorenz (1984) in his recent survey which states: 'over the past two years industrial designers and their consultancies have been propelled into the limelight after decades of languishing at the beck-and-call of the nether reaches of British marketing and engineering'.

All of these assertions suggest differences in status in one way or another, and this has led to a war of words between engineers and industrial designers.

Moody's (1980a,b) finding shows that the work of industrial designers has been variously described by engineers as 'cosmetic treatment, "tarting up" the product, giving the product a "face lift", or "styling" the product'. Other terms used to describe industrial designers include: 'technician, draughtsman, parasite'; an arty shiftless bunch; and Humpty-Dumpty.

Davies (1977) appears to deplore any conflict between engineer and industrial designer, caused by difference in status. This manifests itself in his opening address to the Engineering Specialists Exhibition when he declared that 'I abhor the gulf that has developed between "aesthetic" and "engineering" design: we have to some extent been confronted with a "two cultures" situation ...' He then said that "acceptance of the caricature of "aesthetic" design as the *epitome of glamour* and "engineering" design as in some way a second rate was "alarmingly wide spread".' In turn, industrial designers describe engineers as analytical, mathematical, and numerical thinkers. In the words of Jack Levy (Director of Engineering Profession at the Engineering Council) (1984) 'the profession (i.e. engineering) lacks glamour. It also suffers from a number of popular misconceptions. It is thought by many to be non-intellectual, an applied science, a blue-collar job. Many people still think of an engineer as *a man in a boiler-suit with a spanner.*'

All of these assertions buttress the argument that there is a difference in status between an engineer and an industrial designer.

Product/project presentation gap

Product presentation is referred to in this study as a method of introducing a new project/product to the board of directors and to the market, first, to gain approval for development and second, to gain the market acceptance of the end product.

Whereas industrial designers appear to be practical and model-orientated, engineers seem to be impractical, analytical, numerical/mathematical, and too detailed in presentation of new projects to the board members, so that the chief executives are often bored with what Moody (1980a,b) described in his finding as 'abstract argument' often put forward by engineers when presenting new projects. This may be interpreted to mean that engineers appear to be theory-led and less model-orientated in the explanation or presentation of new projects to directors.

According to the Research Director at Moore and Wright (one of the companies investigated by Moody, (1980a,b)), 'particular skills are required to adequately communicate research proposals to the boards of directors. Accountants, financial directors, and even technical directors are unfamiliar with the *minutiae of R & D*. They are not easily convinced by abstract argument. ... Models which demonstrate ideas simply and indicate the likely form of a finished product have proved to be an effective means of gaining the approval of those directors who have difficulty in visualizing the outcome of partially resolved projects.' This seems to suggest that the directors are in most cases not so much interested in the verbose and/or mathematical or numerical description of the new project/product as they are in the model itself or what may be called the practicality or workability of the project. In other words, the directors tend to prefer visual demonstration (with a model) of how the product looks (possibly with its precise cost) rather than a descriptive version of how it will look, as referred to earlier.

Thus, unlike the industrial designer, the engineer appears to be impractical in communicating the 'core of the matter' to chief executives. This strongly supports a comment made by Black (1984) in his review of Springer's book entitled *Industrial design in engineering: a marriage of techniques*, which states: 'The first major problem faced by this book is that engineering designers tend to think analytically and numerically, and are wary of long-winded descriptive passages with subjective statements about for example, surface finishes, with only vague comparisons of costs.'

Myerson (1983) quoted the Royal Society study group as rapping the knuckles of the National Institute of Agricultural Engineering (NIAE) when it says that 'R & D organisations do not always go far enough in developing a machine to the point where *practicality* of operation and manufacture has been fully illustrated'. Myerson said that the report further criticizes the NIAE's inability to present findings (of projects) in such a way that British industrialists can readily appreciate them. This strongly corroborates the argument that the engineer appears impractical in communicating the core of the matter (project findings) to chief executives.

Other well-known reports have also criticized the impractical nature of the engineering profession. The findings of the Finniston Report (1980) show that, 'the move to an all-graduate profession in recent years has not been accompanied by satisfactory training for graduates in crucial elements of engineering *practice and applications*'.

In its own criticism, the Corfield Report (1979) attributes the 'analytical' approach to the inability of engineering disciplines to provide an adequate scheme whereby engineering graduates should 'be able to design to specification, to appreciate the importance of these and to learn to use them to maximum advantage; which implies creativity on their part and this is what many engineering courses fail to develop'. The Report maintains that, 'Too many courses in this country teach engineering as applied science and are *so analytical in approach*, cramming so many facts, as actually to repress creative propensity in the average student.'

The presentation of 'finished' products to customers appears to create similar division between industrial designer and engineer. For example, while industrial designers believe in visual presentation of a product and, in fact, it is one of his/her main functions (i.e. aesthetics appeals), some engineers do not. For instance, Moody (1980a,b) points out that many engineers in the product manufacturing industry 'hold that engineered hardware has a visual characteristic which is intrinsic to the technology'. He explains that engineers appear to believe in 'if engineering is right, it (product) looks right' whenever they want to assess the visual merit of a product. Consequently, these engineers 'regard deliberate visual manipulation as disingenuous, and (product) presentation as amateurish'. According to Moody (1980a,b) 'because its elements (presumably aesthetics) are essentially non-scientific, product presentation is seen by many firms and their engineers as a task that can be done by anybody with a modicum of common sense'.

However, Moody (1980a,b) has made it clear that 'technology is the means of making a product function technically but technology alone does not make a saleable product'. This has recently (1984) been confirmed by the Prime Minister, Margaret Thatcher, who categorically stated: 'In recent years, management in this country has been greatly influenced by technology. But technical innovation is not an end in itself, and *an innovative product does not necessarily sell*'. This suggests that although an innovative product may be reliable or functionally performing well, if it is not presented in the right form to the customer, it may not sell, and this is where engineer and industrial designer appear to disagree.

Some engineers, however, 'acknowledge that appearance of a product has some bearing on its commercial acceptability ... but they regard it as a vagary of the marketplace, to which some concessions must be made'.

It is therefore evident from the foregoing argument that both engineer and industrial designer have conflicting views about product/project presentation that apparently reveal a gap or division between them.

There are other reasons to suggest that differences/conflict exist between the engineering and industrial-design professions. Fraser (1984) recognizes that there is a gap between industrial design and engineering practices that needs to be closed. Conscious of this division, she, soon after her appointment as the new Head of Industrial Design and Editor-in-Chief of *Design Magazine*, observed that, 'bridging the gap between industrial and engineering design is an important priority'. She declared: 'I am aware that in the past Design Council policies have been criticised by some designers, but I am convinced that the Profession should not be in the business of fighting its friends. I've always maintained that we are both fighting for the same cause — the recognition of design as an essential part of the industrial and commercial life in Britain.'

Another reason to suggest that there is a difference between the engineering and industrial design teams can be found in their professional affiliations. Professionally-qualified engineers belong to the appropriate engineering institutions, and according to Harris (1984) there are fifty-one of them in the UK, whereas industrial designers appear to be members of one body — the Society of Industrial Artists and Designers.

Last but not least, Black (1984) reveals that Flurscheim's lifetime ambition seems to be aimed at encouraging a marriage of industrial and engineering design 'into a single organisational framework within which all types of design can flourish'. He points out that 'Flurscheim's lifetime concern to integrate industrial and engineering design has led to his membership of Councils of the Institution of Electrical Engineers, the Royal College of Art, and

Committees of the Design Council.' Obviously, if no difference/ conflict, or gap exists between engineering and industrial-design professions, there could not have been any need to attempt to unite them.

The overall review suggests that although there is a gap between industrial design and engineering practices, both professions appear to be fighting for the same cause — the recognition of design as a powerful tool for achieving commercially successful manufactured products, and thus an essential part of the industry and commercial life in our society. Put another way, although industrial design and engineering disciplines are perceived as distinct, they are none the less dimensions of the same activity, which calls for a marriage of both disciplines to achieve commercial success.

The following section will examine the possible ways of marrying both disciplines.

Improving design effectiveness through better linkages

Differences and conflicts between engineering and industrial design have already been identified. The objective of this section, therefore, is to draw attention to better linkages for improving design effectiveness so that engineering and industrial design can be integrated, to borrow from Flurscheim (1983), 'into a single organisational framework within which all types of design can flourish'.

As has already been argued, 'design is the (customer's) perceived value of the end product'. Thus, whether a product is designed by an engineer or industrial designer, the term 'design' remains a common factor to both professions. This is crystallized in the Corfield Report (1979), which states that every product contains elements of engineering and industrial design.

The differences and conflicts between engineering and industrial design components, as discussed earlier, have, to some extent, contributed to the need to integrate industrial and engineering design. Corfield (1979) argues, as briefly mentioned previously, that 'since all products contain a varying proportion of both disciplines (engineering design and what is commonly called industrial design), every designer must have a knowledge of both'. Since both the industrial designer and engineer require each other's knowledge in order to perform their duties effectively in the process of NPDD, an integration of both disciplines appears to provide a solid base upon which such a knowledge can be acquired to enable them to communicate effectively with each other and to produce successful products (products that pass the test of market acceptance).

Second, the SERC Report (1983) maintains that 'the complexities of modern life prevent any but the most gifted of individuals to be expert in both fields (i.e. engineering and industrial design)'. By implication, only a few people can combine the qualities of industrial design and engineering. Thus, to meet the complexities of modern life (i.e. differences in tastes/preferences), both engineer and industrial designer should have to work together to achieve the level of competitiveness currently required of most British industries to match their foreign counterparts. Therefore, each of them has a vital contribution to make in the process of NPDD.

Third, the industrial designer appears to play a middle-of-the-road role in the process of NPDD. That is, he/she interfaces with machines and people. This is an attempt to reconcile the external qualities of a machine (appearance, colour, style, etc.) with its internal qualities — functional performance (an area where the engineer's interest is deeply rooted) on the one hand, with people (customers/users) on the other. In other words, the industrial designer tends to ensure that while the machine is in a presentable and saleable form, it is easy to manufacture, safe to use, and easy to use, control, and maintain. This is another way of saying that as an 'interfacer', the industrial designer appears to play the role of an arbiter in a dispute. He/she (the arbiter) does not go away until he hears from both sides, (the manufacturer and the customers/users) of the dispute. Although the 'arbiter's' decision may not be final, he/she does not abandon the clients until there is some feedback. In parallel, the industrial designer who interfaces with machine hardware (the product) and people (the consumers) should continuously collate and co-ordinate feedback from consumers/users of this product and then use his/her creative ability to create a kind of product that fits the people.

This reasoning strongly supports Ashby's (1958) study that maintains that 'technology is an essential aspect of culture in an industrial society; the link between science and people. Industrial design is at the interface of technological hardware and people.' This confirms an earlier work by de Ferranti (1956) (an industrialist) who argued that 'technological innovation is the function of transferring the progress made in science into new technical advances, and the purpose of industrial design is to relate these to the needs of human beings'. It is important to mention that the 'needs of human beings' appear to be the most difficult thing to satisfy. First, the behaviour of human beings is generally very unpredictable, particularly buying behaviour. For example, because customers/users have various needs, due to changes in tastes or preferences, it appears difficult to predict what the next need will be, especially when one level of need is satisfied. According to Rothwell *et al.* (1983), 'it is the function of the designer working for that manufacturer to translate this set of user specifications into a working product or process that best satisfies these needs'. The task of interfacing and/or co-ordination in the process of NPDD is, therefore, very significant.

However, this task may not be feasible should industrial designer and engineer continue to pursue their separate interests, as referred to earlier. There is, therefore, the need to combine both interests — the amalgam of which it is hoped, will produce a successful product. As an old adage has it, 'united we stand, divided we fall'.

Fourth, Moody (1980a,b) reasons that 'a product that has been designed only in respect of its technical functions is deficient in a number of counts. The product has not been designed to be a continuous extension of the user and is therefore psychologically incompatible. As a result, a potential customer may dislike the product for reasons he is unable to define, but which nevertheless could cause him to demur at purchasing the product. The product may not take into account all the users' related technical needs; these have to be met by many separate products. The user may have to adapt ergonomically to the product. This may induce discomfort, stress or fatigue in the user, causing the task to be less than enjoyable. It may lower the efficiency with which the user performs the task. By default, maintenance of the product may be irksome and consequently neglected by the user. All of these are said to detract from the commercial success of the product, and *this is the point at which the aims of technological innovation and industrial design merge.* Both aim to achieve success in the marketplace (a point made earlier by Fraser); the former by offering potential users a new technical facility at an economic price; the latter by presenting that technical facility to users in a form which appeals to them.'

It follows that industrial design and engineering should not be allowed to work separately. For example, Gibbs and Flurscheim (1983) suggest that 'industrial design capability should be made available locally in the design office, whether provided in-house or by external sources, throughout the design period, so that the problems requiring attention can be identified and dealt with as they arise and before design is too far advanced'. This strongly supports Black's (1984) finding, which states that, 'he (the industrial designer) often found that even when an enlightened engineering manufacturer called him to advise, it was usually so late that engineering of the product was already at such an advanced stage that, although it might operate satisfactorily as a machine, it could not be made attractive, and safe and pleasant to use by the operator or customer'.

Finally, there is evidence to suggest that some engineers are still strongly opposed to the employment of industrial designers, as alluded to earlier. Moody (1980a,b), for example, found that the management of Clement Clarke International Ltd and Racal Amplivox Ltd (two of the companies investigated) were strongly opposed to the employment of industrial designers. Whereas Clement Clarke International objected on the ground that industrial designers are 'stylists', Racal Amplivox rejected them on the basis that they 'impede progress in development'. Racal Amplivox further claimed that their engineers were capable of handling both human (ergonomics, aesthetics, etc.) and engineering factors in their new-product designs. This sharply contrasts with Gibbs and Flurscheim's (1983) work which showed that 'the training of engineering is rarely, if ever, sufficient for them to carry out industrial design in its major role ...' Consequently, industrial design appears to be neglected.

One can then reason that since industrial design and engineering processes are 'interwoven', as has been so far demonstrated; and since the industrial designer serves as an 'interfacer' between engineering hardware (the product) and people (the consumers — in the end, all products are made for consumption by customers, users, and/or the public at large), it appears reasonable to suggest that any continued opposition to the employment of industrial designers, is likely not only to hinder the progress of NPPD process, but will also yield at the end of the process, an unsuccessful product — a product unacceptable to the market. In the words of Rothwell *et al.* (1983) 'it is clearly nothing but a recipe for disaster to produce a product that customers do not want'.

Overall, one can conclude that the need to integrate industrial design and engineering appears to be consequent upon:

1. the increasing absence of aesthetics, ergonomics, and graphics in engineering;
2. the interlinked nature of industrial design and engineering — the separation of which may cause commercial and competitive disadvantage to the firm;
3. the invaluable interfacing or conciliatory role of the industrial designer, which involves the reconciliation of

engineering characteristics of a product (functional performance, reliability, etc.) with human characteristics (human environment);
4. the opposition to the employment of industrial designers by some engineers, the continuation of which may well impair the process of NPDD.

Contributions towards integration

Several contributions have been, and are still being made to unify industrial design and engineering. The government has played a leading role in an attempt to amalgamate industrial and engineering design. Jones (1980) discovered that one of the government objectives for the creation of polytechnics in Britain was to encourage the merger of components of engineering and industrial design so that they both have a common module. His finding shows that at Leicester Polytechnic, where the project appears to have been put in practice, 'an engineering design option is taught by staff from both the engineering and industrial design faculties. And there are other examples of shared facilities there, equipment and a joint technical information library.'

However, Jones (1980a, b) pointed out that 'in reality though they have developed little beyond the integration of the administrative structure with many art and design faculties otherwise leading an entirely separate existence'. Regretting this unfortunate failure, Jones declares: 'it would be ironic if the very economic climate that makes more integration desirable to improve engineering performance was used as an excuse to postpone it'. This appears to suggest some problems on the way to the successful implementation of the integration programme.

The government has also offered contributions in many other ways. They include: commissioning of surveys, such as the NEDO, Fielden (1963), Moulton (1976), Carter (1977), Corfield (1979), CNAA (1984) reports, and so on; organizing seminars, such as Number 10 Seminar (1983); and the appointment of such agents as the Design Council.

The Design Council is one of the government's principal agents. One of its main objectives is to organize 'action programmes' aimed at creating design awareness as well as integrating engineering and industrial design. This was made clear by Patrick Jenkin (1983) (former Secretary of State for industry), when he stated that 'the Design Council in this country is in the forefront of promoting the government's new total design concept, both industrial and engineering design'.

The Design Council organizes its action programmes in various forms. They include: awards for design excellence, exhibitions, publications (journals, magazines, newspapers, and reports), and the initiation of a 'funded consultancy scheme'. The latest in the series of the Design Council's efforts to unify industrial design and the engineering profession was the appointment of an industrial designer, June Fraser, as the new Head of Industrial Design as well as Editor-in-Chief of *Design* magazine. She has been principally charged with the responsibility for the management and development of the Design Centre Selection and, most of all, to contribute towards the development and promotion of industrial design in British industry.

The Design Council, however, has been severely criticized for being technologically biased when making awards for outstanding designs. Karen (1980) blamed the Design Council for making the integration of engineering and industrial-design disciplines difficult. He asserted: 'I might add that although it neither draws up nor assesses design courses, the Design Council may itself be partly to blame for the technology bias in design training. One only has to look at the Design Council awards to see the dangers of setting too high a store on functional or innovatory features and too low a value on industrial design.' Similarly, Lorenz (1984) noted that 'the Design Council should ensure that more of its activities span the whole breath of design, thereby pre-empting the now widespread criticism from industrial designers that it has become over-biased towards engineering'.

While it is not intended here to dwell on the causes of such 'bias', it has to be made clear that 'bias' does not seem to be desirable in the present attempt to unify industrial design and engineering in order to make the most of design. For example, neither too much attention to engineering nor too much devotion to industrial design can solve the problem of non-competitiveness currently being experienced by some British industries, particularly in electronics and in the manufacture of agricultural machinery. A mix of both approaches seems to provide a better solution.

Given that the difference in 'status' and 'education' are the causes of a lack of integration between the engineer and industrial designer, one can argue that since the industrial designer has a higher status than the engineer while practising as a consultant; and since the engineer, in turn, has a higher status than the industrial designer when he/she (the industrial designer) is employed in-house (industrial designers are treated as 'technicians' when employed in-house), the status differential may not be seen as the major cause of the rift between the two professions. For one thing, the question of status appears to be a matter of organizational structure. For example, the Fielden Committee (1963) found that although 'almost all the committee's witnesses ... agreed that the economic and organisational status accorded to engineers engaged in design was generally low by comparison with the status accorded to other staff employed in their professional capacities (such as industrial designers), ... in the aircraft industry, according to Sir George Edwards of the British Aircraft Corporation Ltd, the status of design staff in relation to other specialist staff is generally good ...'.

Similarly Moody (1980a,b) found that whereas the industrial designer plays the central role in product development at Gowlands Ltd, at Racal Amplivox Ltd and Clement Clarke International Ltd, the reverse appears to be the case. At the same time, Moody found that Perkin Elmer Ltd and Moore & Wright Ltd appear to accord equal recognition to both industrial design and engineering components. Moreover, in some organizations, status appears to be accorded in line with the level of achievement by the industrial designer or the engineer. For example, Watt (1980) found that 'where respondents felt that their company competed on design and was successful in the marketplace, the head of design was regarded as first amongst equals, or design has "a friend at court" in the shape of a chief executive or other director, who ensured that the importance of (industrial) design was recognized by other company divisions.'

On the other hand, Gruber Poensgen and Prakke (1973) and Watt (1980) found that 'in companies where both the standard of design and commercial success were believed to be above average, design was integrated with sales/marketing and production functions'. This appears to suggest that industrial designers are accorded higher status than their counterparts (engineers) when their achievement is higher, and lower status when their achievement is lower, and vice versa.

On the international level, the difference in status between industrial designers and engineers appears to be even more marked. For example, whereas Watt (1980) found that industrial designers are given high status, appointed directors, and become 'heavily involved in design management' in Italy, in Britain, Lorenz (1984) found that industrial designers are still under the 'beck-and-call of engineers ...', as referred to earlier.

In that 'status' is a function of organizational structure and this reflects managerial attitudes or 'culture' one may hypothesize that the root problem lies, not in the actual status, but in the educational background given to both professions. The merger of engineering and industrial-design syllabuses appears to form the basis upon which, it is hoped, the two disciplines can be effectively integrated. This calls for the standardization of engineering and industrial-design courses. In their respective reports, Moulton (1976), Carter (1977), and Corfield (1979) have all made useful contributions in this respect.

Whereas the Moulton Report (1976) was of the opinion that 'engineering should be taught in the context of design, so that design is a continuous thread running through the teaching of undergraduate engineering', the Carter Report (1977) held the view that it would be more appropriate for engineering graduates to undertake postgraduate courses instead, in order to 'provide an opportunity for the study of a specialized area of industrial design in greater depth than is possible during the first degree', as already mentioned.

Moulton's (1976) view seems to imply that engineering students should first qualify as professional engineers from which industrial designers could be selected, probably through special interest developed towards industrial designing in the course of their engineering study. This appears to conflict with Carter's (1977) Report, which saw 'design' not as a means, but as an end in itself (an admission that engineering is as important as industrial design) that should enable the industrial designer to communicate effectively with the engineer in the process of NPDD, hence it advocated separate study of each course at first-degree level, with specialization in each other's course at the postgraduate level.

In his report, Corfield (1979) appears to concur with Carter (1977) on postgraduate programmes, but seems to disagree with Moulton (1976) on the question of teaching engineering in the context of design. He asserted: 'unfortunately, concentration on design as a separate subject at too early a stage may raise many aspirations that cannot be fulfilled and leave us with a stratum of "failed" designers whose technological skills are inadequate for them to be absorbed into "design related" activities'. He went on to add: 'I would wish to see the early stages of teaching concentrate on science and technology and to offer those with the aptitude and aspiration the opportunity of postgraduate and post-experience education in design.' Corfield further explained, 'thus the maturing student who has taken postgraduate studies in design would take up higher studies in engineering or technology. The engineer or technologist, physicist or chemist who had pursued his or her higher degree in a specialised branch of technology would be well equipped to take the specialised advanced course in design.'

Postgraduate courses appear to be particularly helpful when

they are continued in the same discipline. The problem with the postgraduate approach, as suggested by Corfield, is that it tends to leave the fundamental problem of integration only partially solved. Corfield, for instance, suggested that engineering students should first 'concentrate on science and technology at first degree', and then do 'a higher degree in a specialised branch' of engineering after which, those with aptitude could pursue 'a specialised advanced course in (industrial) design'. This idea is excellent if the engineer and industrial designer have had a thorough background knowledge of each other's subjects. However, evidence suggests that this is not often the case. For example, Gibbs and Flurscheim (1983) found that 'the extent of theoretical sciences needed to cover the principal engineering-based technologies inevitably restricts the extent to which peripheral subjects (industrial-design subjects) can be studied within the limited time available in degree courses'.

They argued that 'because of the small amount of technical training engineers receive in the field of industrial design, their knowledge of this subject tends to be *very restricted*' (Gibbs and Flurscheim, 1983). A 'restricted knowledge' implies, in this context, that not enough ground has been covered in the field of industrial design by the engineer to enable him to pursue, say, a PhD degree course in industrial design. It is assumed, based on Corfield's recommendation above, that the engineer has undertaken both his bachelor's and master's degrees in engineering. The same 'restricted knowledge' is expected to pose similar problems to the industrial designer who may wish to undertake a postgraduate study in engineering. Thus appropriate postgraduate courses may offer only a partial solution to the problems identified.

Other experts however, have made positive suggestions towards the integration of industrial design and engineering. Gibbs and Flurscheim (1983) argue that 'the overriding principle in selecting a method must be that, the greater the difficulties in separation of man/machine interface from engineering-based components, the greater the need for closer integration between engineering and industrial design'. Based on this hypothesis, they put forward the following methods of achieving successful integration.

1. (A) Use an engineer with experience in industrial design techniques; or
   (B) Co-opt a professional industrial designer for the period of the project (ad hoc arrangement) as the head of an industrial design team.

(Gibbs and Flurscheim 1983)

They contend that if the engineer is selected as the team leader, he should possess the industrial-design qualities to enable him to influence all the industrial-design aspects of the project. On the other hand, they maintain, that if the industrial designer is made the head of the project, 'he should in turn possess an adequate engineering background', apparently for the same reason.

The problem with suggestion 'A' is that the engineer may not possess all the qualities of a professional industrial designer, as the former appears not to be adequately educated in this discipline. There may be problems of communication and status that may slow down the progress of the team work. Although the engineer's experience in the field of industrial design (due to long years of practice) may be an asset to the team, it is nevertheless a substitute for the qualities of a professionally trained industrial designer. As Gibbs and Flurscheim (1983) point out, 'certainly there are engineering teams who can achieve excellent industrial design in their own right, especially in the field of ergonomics ... It must always be remembered, however, that it can be very expensive in terms of time and uncertainty to employ a person (engineer) who does not thoroughly understand his discipline. It is always better to get advice from an expert (industrial designer) rather than muddle through.'

It can be argued that the use of a person (engineer), who is not professionally trained in industrial design, as the head of an industrial-design project, on the basis of 'qualification by experience', may be likened to the use of a medical general practitioner (GP) in place of a medical specialist on the basis of 'qualification by practice'." The GP, for example, appears to be trained with a background knowledge of general medicine in the course of his/her medical study. In the absence of a medical specialist, the GP can provide a basic first aid in most fields of medicine, when patients are under critical health conditions.

However the ethics of conventional medicine appears to preclude GPs, without a specialist practising licence, from giving specialist attention to patients (which some of them can, with years of medical practice, just as some engineers can, in industrial design, with their wealth of experience, as illustrated previously by Gibbs and Flurscheim, 1983). Thus, when it comes to special cases, the GPs tend to refer patients to specialist doctors. On the same grounds, when it relates to industrial design projects, it may be necessary to appoint a professionally trained industrial designer as the head of the project and not to use people with 'general experience'.

In the present competitive milieu where tastes and preferences change fast, 'speciality' not 'generality' counts in new-product designs. In other words, to be a market leader, it is 'best', not 'second best' that counts; and best can hardly be achieved by a person with an inadequate (restricted) knowledge in the discipline concerned.

This argument seems to show the reason suggestion 'B' p. 269 appears to be better, except that using a professional industrial designer in an *ad hoc* fashion may not auger well for the continuity of the overall design process. By this is meant that although the project may be short-term, the design process is a continuous task. At the end of a project, and after introduction, for example, the continuous nature of the design process tends to compel the industrial designer to interact with customers/users to get feedback upon which improvement of the existing product can be made or new ideas developed.

Thus in practice, the industrial designer's work does not and should not stop at the end of the product-innovation process.

A second approach to achieving successful integration between engineering and industrial design amongst the four put forward by Gibbs and Flurscheim (1983) proposes that:

2. (A) Employ a visiting industrial designer on a part-time consulting basis; or
   (B) Use an external consultant in an educational manner by spreading his advice more thinly over the whole engineering activity.

This seems a good method in the sense that the services of a consultant is being used. However, the problem with suggestion 'A' is that the project may be delayed as the industrial designer consults on a part-time basis. Consequently, the firm may not meet the delivery date or the product may be introduced late. One can reason that in a technology-driven and competitive industry, such as electronics and computing, the longer it takes to introduce a new product, the higher the risk of product obsolescence. That is, in the end the product may come out perfect, but may come out too late to compete.

Suggestion 'B' has some weakness, too. For example, 'spreading advice more thinly …', may be tantamount to providing a 'skeleton' advice that may provide an insufficient information input for taking adequate decisions on design problems. As a result, the project may be delayed, which in turn may involve extra cost and waste of time and human resources. In our judgement, success is rarely associated with spreading advice more thinly over the whole engineering and industrial-design tasks.

Moreover, what is the guarantee that the external consultant's advice will be strictly followed? That is, how are we sure that the management and development engineers will not object to the external consultant's recommendation? One example, is the often-noted management's 'subjective hunch' attitude of sticking to their traditional ways of doing things, say, designing and developing new products, rather than adjusting to changes in the market.

In his research, Ames (1970) found that this sort of attitude results from what he described as the management's concentration on 'the trappings of marketing rather than the substance' or management's failure to 'face up to critical deficiencies in product, price, or service'. He cited an instance where the marketing (consultant) 'recommended a major redesign of the product line', but the management rejected it following the advice of their manufacturing and engineering executives, which made them (management) believe that their 'current product design and cost structure were still superior to any competitor's, and all that was needed was a better selling effort'.

He then declared: 'Faced with these conflicting points of view, top management decided to stick with the original product concept and put pressure on the marketing group for a more aggressive selling effort. It was not until the company lost substantial market share and its entire profit structure was threatened that the president could bring himself to fly in the face of the expert opinion of his engineering and manufacturing executives and force the redesign through' (Ames, 1970).

Ames pointed out that, 'management has a natural tendency to view its products through rose-coloured glasses, and to conclude that any advantages claimed for competitors' products are exaggerated or insignificant or that competition is "giving the business away" or that it has a "cheaply engineered" or "shoddily manufactured line" if it is sold at a lower price'. He then concluded: 'What it takes is total company involvement — from the top down — in the marketing effort and management willingness to depart from traditional practices if this is what is required to be responsive to the market.'

These excerpts from Ames' (1970) study are illustrative of our argument that there is no warranty that both the management and their engineers would implement the recommendation or advice of the external consultant. Obviously, any company with this kind of management philosophy may not be expected to observe or carry out the terms of the recommendation or advice (even if thin) of an external consultant. As Ames (1970) put it, 'trying to convince these executives (i.e. general management executives with technical background) that a technically inferior product is what the market wants is next to impossible'. If such an 'expert' conviction is next to impossible; what is then possible? How can the consultants advice be obeyed?

According to Rothwell *et al.* (1983), one of the reasons 'many attempts at technology transfer have failed' was because of the 'inability of the firm's managers to understand the commercial implications of technical alternatives being offered by the consultant'. Consequently, the advice of the consultant, no matter how 'thinly spread' may not be adhered to. As a result, the product may be badly designed, badly developed, and badly presented. Consequently, the product becomes uncompetitive and unacceptable to the customer.

The third approach to achieving successful integration between engineering and industrial design amongst the four put forward by Gibbs and Flurscheim (1983) proposes:

> 3. Employ external consultant with basic work being carried out in their own design offices, thus permitting the maximum concentration of their experience and facilities on the design problems involved.
>
> (Gibbs and Flurscheim 1983)

The problem with this method of integration is that the two professionals are still working separately instead of working together as a team — the problem it is intended to solve. It is true that both the engineers and consultants will maximize 'concentration of their experience and use of facilities' while initially working separately. However, it seems true also that when it comes to the 'crunch of the matter' — the integration of the two ideas basically developed in their separate offices, problems tend to emerge. The problem of communication, presentation, and status, as already pointed out, are typical examples.

Besides, working separately, whether 'basic' or 'main' work does not appear to permit the right atmosphere for brain storming or 'cross-fertilisation of ideas', which tends to stimulate creativity.

The fourth approach to achieving successful integration between engineering and industrial design amongst the four put forward by Gibbs and Flurscheim (1983) proposes:

> 4. Organise an in-house industrial design department to serve all the design offices.
>
> (Gibbs and Flurscheim 1983)

This appears to be the most useful of the four methods. This is confirmed by Parkinson (1980b) and Rothwell *et al.* (1983). Parkinson found that one of the reasons the West German Machine Tools Industry was ahead of their British counterparts was because the West German manufacturers 'had greater involvement between in-house design and development and customer organisations'.

Emphasizing the importance of in-house skills, Rothwell *et al.* (1983) contended that although many companies (small and medium) would like to use expert knowledge from outside, it was not a substitute for in-house skills. They argued that 'while a company might employ design consultants to assist in the creation of its new product, it nevertheless needs some in-house design expertise to enable it to capitalise on the external advice'. This corroborates Rothwell's (1978) finding which showed that 'many attempts at technology transfer have failed because companies did not acquire the necessary in-house skills'. Rothwell *et al.* then went on to illustrate how 'the importance of in-house skills to the successful transfer and adoption of external technology is emphasised by the remarkable ability of Japanese companies both to adopt and adapt foreign technology: this has owed a great deal to the very high level of technical and design skills within Japanese corporations'.

Nevertheless, the successful organization of an in-house industrial design department appears to depend on a number of factors:

1. *Management attitude.* The corporate attitude towards the integration itself counts. For example, Moody (1980a) found that 'all the firms which employ an industrial designer prefer a consultant, even if they had enough projects to warrant employing an in-house designer . . .' This appears to be another way of encouraging both the engineer and industrial designer to continue separate practice.
2. *Availability of resources.* If a firm has insufficient resources, it may be unwise to embark on an in-house design. On the other hand, if a company has enough resources and fails to organize an in-house design department, it may not be able to 'tap' fully the potentials of technological transfer.
3. *Extent of co-operation and co-ordination.* Finally, the successful organization of an in-house design department appears to depend on the extent to which the engineering and industrial design teams co-operate to work in harmony. It also depends on the extent to which the contributions of both professionals can be co-ordinated and maximized

during the process of NPDD to create and manufacture an acceptable and competitive product — a product with value for money.

Summary

Combining all the approaches so far discussed, one could argue that although the Government and Design Council efforts, the postgraduate approach, and Gibbs and Flurscheim's four alternative methods have their advantages and are necessary contributions towards integration, they have, nevertheless provided only part of the solution to the integration problem. For example, the government's introduction of polytechnics in the UK, aimed at integrating engineering and industrial design 'developed little beyond the integration of the administrative structure of both disciplines in some Polytechnics'. Second, the organization of various awards by the Design Council has been met with scepticism from some industrial designers, on the grounds that the Design Council does not seem to take into account certain industrial design characteristics when considering the awards for technological achievements. Third, the postgraduate approach does not appear to provide the basis for a complete integration. The solution to the integration problem seems to lie, not at the postgraduate level of education but at the lower level, such as first-degree courses, where it is hoped, students are adequately prepared for future higher degree courses. At this stage, one hopes, the graduates will have developed the aptitude for the course he/she may wish to pursue at a postgraduate level. Finally, benefits of integration may not be maximized through Gibbs and Flurscheim's four alternative methods. However, a modest approach to suggestion 'A' of the first method (p. 269) appears to provide an answer to the problem.

Since the engineer should have to possess the industrial design qualities, as already mentioned, in order to be selected as the head of the industrial-design team, and since the industrial designer 'should in turn possess an adequate engineering background' to be chosen as the head of the engineering-design team, the integration of the engineering- and industrial-design disciplines seems to offer the best alternative. This will involve a reorganization of both courses that will permit the integration of both syllabuses. This approach, one hopes, will allow for equal weighting to be given to both course contents. This will enable the so-called 'peripheral subjects' in each discipline to be made part of the core subjects, which will count as a partial fulfilment for the award of a bachelor's degree or its equivalent in either engineering or industrial design. It is hoped that under this arrangement, the postgraduate approach can be adopted with some measure of success. For example, the engineers who have had a thorough background knowledge in industrial design, can now confidently pursue a postgraduate course in any of the industrial-design core subjects they have studied in their first degree, and vice versa.

It is clear from the overall review that opinions seem to differ as to the level of education at which the integration of engineering and industrial design skills would be most effective. Whereas some suggest undergraduate level, others propose postgraduate level. Below are the views of commercially successful firms (Queen's Award winners) on this issue, as revealed by our research.

It is evident from Table 6.2 that 43 per cent of the sample believed that the integration of the engineering and industrial-design skills would be most effective at the undergraduate level, suggesting that the integration at this level will create a better organizational framework within which all types of design can flourish. Twelve (15 per cent) of the sample supported the integration at the postgraduate level I (e.g. MSc, MA, MBA, etc.) while four (5 per cent) of the respondents said that the integration would be most effective at the postgraduate level II, such as PhD, DBA. Twenty per cent (16 out of 82) of the sample did not know the level at which the integration of both skills would be most effective.

However, fifteen (18 per cent) of the sample believed that the integration of both skills would not be effective at any level. A

Table 6.2 At what level of education will the integration of engineering and industrial-design skills be most effective?

	No. of firms	%
Undergraduate level (e.g BSc, BA, etc.)	35	43[a]
Postgraduate level I (e.g MSc, MA, MBA, DMS, etc.)	12	15
Postgraduate level II (e.g PhD, DPhil, DBA, DSc, etc.)	4	5
At none of the above levels	15	18
Don't know	16	20
	82	100

Note: a. To be read: 43% of the responding firms stated that the level of education at which the integration of engineering and industrial design skills would be most effective was at the undergraduate level.

number of reasons were put forward in favour of this belief, including:

(1) The level is not important, the ability to *use* the knowledge is the important factor, and is often inversely proportional to the level of 'education'.
(2) Training linked to projects in industry, at any level is the best way to effectively integrate both skills.
(3) At all levels: first degree, HNC, OND, etc.
(4) May be any or may be none, one of our best designers was a farm labourer.
(5) Substantial site experience of graduate engineers applied to the design office will enhance the integration of both skills.

These comments are particularly important because 'Training linked to projects in industry' (item 2 in the list), is one of the strong competitive advantages that France and Japan have over the UK as far as the effective management of design is concerned. Similarly, 'substantial site experience of graduate engineers applied to the design office' (item 5 in the list) featured strongly as one of the suggested ways in which the management of design may be improved (see Chapter 7). Overall, however, the view seems to be the sooner one can introduce joint consideration of the two dimensions of design into the curriculum, the greater the likelihood one will be able to inculcate positive attitudes to their synthesis in practice. It is also possible that the growing use of computers in design may help overcome this interface problem.

The impact of computers on design

The technological impact that computers are having on design today can enhance the integration of industrial design and engineering skills. With computers able to present design in three dimensions, a common ground may be found whereby some of the communication problems facing engineering and industrial designers could be eased. This section examines the impact of computers on the design of new products, and is aimed at covering the major computer aids for design or simply, computer-aided design (CAD).

It is pertinent to point out right from the outset that CAD is often used in parallel with computer-aided manufacture (CAM) in recognition of the fact that a new design may be unsuccessful in today's competitive environment unless it is contrived for 'ease of manufacture'. There is, therefore, a close correlation between design and manufacture on the one hand, and CAD/CAM on the other, hence both terms are often used as if they were synonymous. Nevertheless, the central focus here is on CAD, which Trafford (1984) (of the Cranfield Institute of Technology) has defined as 'a class of information system which supports the design task by satisfying some of the designer's information requirements'.

Engineering and industrial designers are increasingly becoming included in decision-making about product configuration, choice of materials/components, methods of manufacture, etc., all of which are crucial to the successful design, development, and manufacture of new products/processes. Often these decisions are taken amidst uncertainty as to what the outcome will be. For example, uncertainties about the following questions. How can a new product perform or conform to specifications or requirements? How reliable is the new product? How well will the body panels fit together before the tools have actually been made? How can the strength of the body of a product, such as a car, be analyzed and at the same time, reduce the weight of the components, and so forth?

In most cases, the answers to the these questions are not known until the product is designed, developed, prototyped, and manufactured. Yet these are the answers that will ultimately decide the fate of the product in the marketplace. In other words, if any defect results from these problems, it is likely to be the customer, instead of the designer, who will detect the defect. In some products, say, cars, accidents might occur and lives might be lost before the fault is known. This may lead to a recall of some or all of the cars concerned. Figure 6.2 depicts the percentage of total recalls during 1978–82 due to defects found in cars ranging from Subaru through General Motors (GM), Chrysler to Mazda. Although Subaru recalled over 90 per cent of its cars, Ford's figure would have been much higher if the US government had insisted on a recall of about twenty-three million Ford cars and trucks on the ground that their automatic transmission could slip from park into reverse. Through political connections, Ford succeeded in forestalling the recall and instead, 'sent cautionary notices to owners'. In a competitive car market, any such large-scale recall is likely to destroy the company's image, at least in the eyes of buyers. For instance, the existing customers could switch loyalty. Repeat buys could be discouraged. Finally, news may go round and prospective customers may not purchase the brand at all.

While the designers' experience will be paramount in anticipating such problems, the car-recall data points up the difficulties,

Figure 6.2 Automakers' recalls 1978–82 (as a percentage of sales)

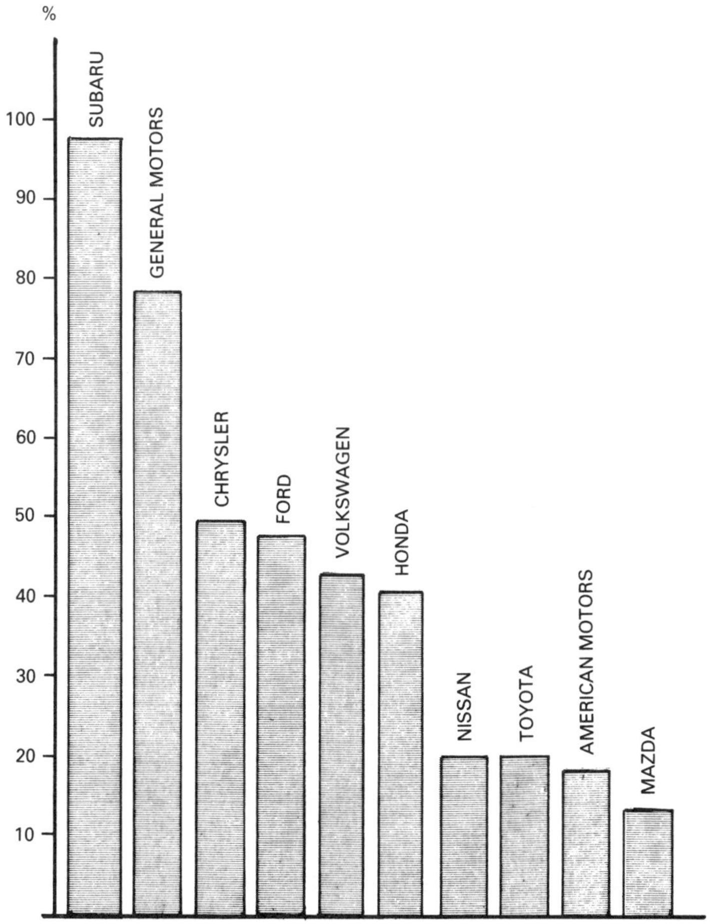

Source: Burck (1983) 'Will success spoil General Motors', *Fortune*, August 22, p. 98

particularly during periods of rapid technological change. It is this need for timely, accurate, and sufficient information to enable the designer to predict the probable outcome of a design decision, well before a project is embarked upon and thereby to minimize uncertainty, which appears to have prompted the introduction of CAD. Trafford (1984) takes this view when he argues that 'to reduce uncertainty of design decisions, the designer needs access to information on such things as customer's requirements, available materials and components, suitable (and available) manufacturing processes, relevant engineering science, configuration compatibility, product behaviour and design constraints'. He continues to classify CAD systems in terms of two main functions:

(1) Information-retrieval systems;
(2) Product-modelling systems.

Information-retrieval systems

These are design tools that provide direct information to the designer consequent upon interrogating the computer terminal. Alternatively, such information could be obtained through an information broker. Examples include: information about previous works in the field of engineering science, work of art, available manufacturing processes, materials/components, national and international patents, national and international standards, etc. A typical example of international standard can be illustrated with the case of Jaguar. In 1984, Jaguar — British Leyland's luxury car paid a fine of $5.5 million 'for failing to meet the US government's car fuel economy regulations' Gooding (1984). Obviously, if the information-retrieval system had been used, perhaps designers of Jaguar cars meant for the US market would have known about this US standard well in advance of such design. Consequently $5.5 million could have been saved and possibly ploughed back into business.

Thus, CAD provides a data bank that enables designers both to speed up design decisions and to spend more time on creativity (thinking about new ways of doing things such as improving the existing product/process or designing entirely new ones). This corroborates the view expressed by Naoto Sasaki (1984) who says that 'the manual components of managers' jobs is extremely simple — it is the thinking which is complicated ... production and clerical workers would work more with their brains than with their hands. If the job is simple they are able to think about ways of improving it'.

Product-modelling systems

A product-modelling system is a computer aid designed 'to model products' and/or components. They comprise:

(1) *Configuration modelling.* This model describes the shape or arrangement (configuration) of components or elements that make up the product, and is mainly used to forecast the behaviour of products. Control-circuit simulation, digital electronic-circuit simulation, stress analysis, and vibrational analysis, are a few of the examples. The aspect of the CAD system that does this configuration modelling is commonly known as design analysis, computational or simulation packages. Austin Maestro is currently using a design-analysis CAD package to describe the configuration of elements that make up the Maestro car.

(2) *Layout modelling.* This model shows the layout of components of a product such as, office and factory layouts; it is also used for checking the location of components.

(3) *Geometric modelling.* This is another useful model currently used by many car-manufacturing companies – Austin Montego, Maestro, Ford, Bedford, Austin Rover, Honda, General Motors, etc. A geometric model describes the physical outline or shape of products and components and is largely used for visualization and shape definition of products and components. In the Austin Rover Group (ARG) plant, for example, the operator uses a geometric modelling CAD package to define the surface geometry (properties and relations of lines, angles, solids, etc.) of the Montego car. Geometric models have three main characteristics: dimensions such as 2D, 2½D or 3D; representation such as solids, edge, and surface; and visualization such as projection, wire frame, and shaded.

Geometric modelling is particularly useful because aside from serving as a model for defining the properties of a product, it can also form the basis for future product assembly and/or component manufacture. For example, Clayton (1984) found that a full size model of a new Ford Sierra car in 'clay' form, used for a practical demonstration before a group of European car buyers, was later used as a basis upon which the actual Sierra car was developed and manufactured.

The origin of CAD

The CAD system was first invented, in the USA by the Bell Laboratories of General Motors (GM) in the 1970s. Central to its introduction was an attempt by GM to recapture its market share lost to Datsun Maxima car of Japan and Mercedes of West

Germany, especially the former due to the Japanese cost advantage. According to Charles Burck (1983) 'Since GM seeks not mere parity but world leadership, it is banking on technology to leap-frog competitors with more thoroughly automated plants and more sophisticated vehicles.'

In 1976, GM developed robots capable of handling 'sophisticated assembly work' but these appeared not to be as competitive as those of their counterparts due to lack of corporate interest. As Burck (1983) puts it 'Getting new technology into the cars will require persistent battling with one crippling habit embedded in the corporate culture, the tendency of divisions to reject innovations offered by the central staffs, and of the company as a whole to resist ideas from outside.' He points out that 'For too long, GM had taken too literally Sloan's dictum that the company's main purpose was to make money.' Thus, when GM realized they could not make money in a competitive market if they did not make competitive products, they hired Robert Frosch the former head of NASA (National Aeronautics and Space Administration (USA)).

This strategy led to further development of CAD. GM's CAD systems now handle the 'complicated jobs of designing engines, chassis and suspension components'. The Pontiac engineers, for instance, have successfully used GM's 3D (three-dimensional) CAD modelling system to redesign connecting rods in the 2.5 litre four-cylinder engine. This single CAD application resulted in the elimination of 25 per cent of the rods' weight, and consequently produced 'engines that run smoother and perform better'.

At Delco Electronics – a division of GM, more research projects are currently under way to improve upon the existing CAD Systems. Meanwhile, as Burck (1983) notes, 'Every gasoline-powered GM car has a Delco Computer that controls ignition, timing, fuel mixture, and idle speed.' It has been predicted that, in future, computers will control everything from transmissions to lights. Thus it was GM's corporate determination and competitive drive to seek 'not mere parity but world leadership' in the car industry that appeared to have given 'birth' to CAD.

CAD systems have been seen by researchers, writers, and practitioners to hold many great advantages for industries. Trafford (1984) argues that 'if correctly selected and utilised, computer-based tools for designers can result in considerable benefits, the principal of these being:

(a) Reduction in design uncertainty, which results in:
    (i) added value to the product

(ii) reduction in development costs
(iii) reduction in manufacturing costs

(b) Increase in design productivity, for example:
(i) reduction in time taken to conduct product design analysis studies (thus enabling more studies to be conducted).
(ii) reduction in draughting times.

(c) Increase in information flow between individual departments, etc.'

In its argument the SERC Report (1983) maintains that 'CAD will contibute to improving the status of the design profession in five ways:

(1) By its association with the high technology of computing.
(2) By reducing much of the drudgery of detailed repetitive drawing.
(3) By allowing much quicker changes in design it will facilitate better design optimisation.
(4) By reducing the ratio of draughtsmen to designers.
(5) With CAD/CAM the involvement and knowledge of the designer will become of increased importance in the manufacturing process.'

Johnson (1984) found that CAD/CAM:

— has improved the quality of printed-circuit-board design enormously (a typical example is Delco Electronics which designs a circuit board that is manufactured 'almost entirely without manual assembly');
— helps with the mechanical design of parts, which can be studied on the screen 'in the round';
— encourages flexibility in both design and manufacture.

Lord Nelson of Stafford (1984) asserts: 'New technology has, of course, invaded the design office as elsewhere, and methods of working and the equipment and capital employed have changed and are changing dramatically. Computers and computer-aided design (CAD) are today essential tools if designs are to be competitive in performance, in time-scales and in manufacture.'

He adds: 'With modern technology, much of the initiation of the manufacturing process starts in the design office. CAD can be linked to CAM and then in turn to the whole subsequent product manufacturing and service functions embraced by the terms 'factory automation system technology' (FAST) or 'flexible manufacturing systems' (FMS).' Robinson (1985) appears to agree with this view when he argues that 'The warehouse once considered in isolation and primarily from a technological point of view, must now be viewed in a context which starts at the end of the production line and ends on the customer's loading bay.'

Hilton (1983) points out that 'Advances in CAD:

(1) have brought computer representation of three-dimensional models within the range of economic feasibility
(2) can provide a three-dimensional picture of a man-model working in a machine and the surrounding visual environment, displayed on a graphic screen or plotted out as a drawing
(3) can make evaluation of changes in position and dimension both simple and rapid
(4) permit a variety of evaluations to be carrried out including a check on the ergonomic capability of the man-model to operate controls, or adjust to the working space and posture required
(5) permit an evaluation of the field view of the man-model in terms of instrumentation and the environment.

Jones (1981) believes that CAD 'uses graphical interfaces to speed up man-computer exchanges to the pace of thinking and conversation: ... and permits access to all published information and to all available sub-routines for automatic designing'.

In their study, Roy and Bruce (1984) contend that 'The output of the designer has traditionally been drawings and specifications, but with the spread of CAD/CAM these to a large extent have been replaced by magnetic discs or tapes from which the product can be made directly by computer-controlled equipment.' They further argue that 'CAD enables the engineering designer to select a particular design configuration by trying out different arrangements on the CAD terminal; perform the necessary stress calculation; and select components and materials according to their performance and cost.'

The awareness of the benefits of CAD has been so widespread that many companies have become increasingly interested. In the words of Lord Nelson (1984) 'Management must see that everyone in the design office is trained to these new (CAD) techniques and have appropriate facilities available to them. Any firm which neglects this need cannot exist today in the high-technology business.'

According to Finkelstein (1984); 'We understand what needs to be done for design and innovation formation; we now need the will and resources.' Basically, designers appear to have the will and often know what to create and how to create it. However, they often seem to lack the techniques to translate their thoughts into action. For example, Clayton (1984) states that the designers of Ford Sierra basically knew 'the parameters to which they would have to work', and that to be competitive, the overall length of the Sierra had to match the rival cars. Because Ford wanted Sierra to carry five passengers, it was agreed that the existing Cortina engines should be used. Thus the designers also knew the interior dimensions. What they probably did not know was the technology that would enable them to create and test this new Sierra model 'without welding a single joint'. CAD solved this problem.

According to Clayton (1984), all of the aforementioned characteristics were formulated into a concept, the result of which was 'a full size clay model of the new (Sierra) car, which is measured by a computer-controlled device that takes spot measurements all over the body. The computer translates the measurements into a mathematical representation of the shape of the car, and then smooths out the imperfections in the clay to produce the final shape.' He asserts: 'This is a perfect example of the way the car makers are using computers (i.e. CAD systems) to speed up the development programmes. In the past, it took a team of men several days to take all the measurements, which then had to be converted into drawings. The computer can do all this in a matter of hours.'

Austin Rover has modified its CAD system to the extent that it can now 'create and effectively drive a car that exists only in series of numbers within a computer, ... analyse the strength of the body and at the same time, minimise the weight of the components.' As Clayton (1984) describes it, 'In the laboratory electrohydraulic rigs can simulate anything from a British motorway to the infamous Belgian pave.' He continues, 'The sight and sound of a car mounted on four hydraulic rams one to each wheel being pounded over the equivalent of a normal road surface is awe inspiring.' Designers appear to believe that this type of simulation has the major advantage of producing precisely the same conditions in the laboratory as can be experienced on the motorway.

Collaborative design

The intensity of international competition, the astronomic costs of development and manufacturing, the increasing rate of failure of new-product designs, the search for ideas, and the increasing consumer expectations of new products mean that many companies can no longer afford to do it all alone. Consequently, companies tend to seek collaborative design arrangements whereby two manufacturing firms, making a similar range of products, for instance, can pool resources to maximize new design ideas, share design projects while operating from their respective locations, and be able to design and develop competitive products to serve various segments of international markets. By doing so, the overall company resources are optimized. Examples of collaborative arrangements include: Austin Rover and Honda, Bedford and Pontiac, and so on.

Who would have imagined some 20 years ago the possibility of a British and a Japanese car-making company, separated by thousands of miles, exchanging between them, drawings of a new-product design, information, and data on-line? Today Austin Rover (UK) and Honda (Japan) are jointly designing and developing cars. One such example is the development of 'a new executive car code named Project XX' in 1985 with the two design teams working simultaneously on the same project, but based in their respective countries. With a CAD system, both teams of designers are able to exchange between them information, data, drawings about the new car, on a VDU (visual display unit).

Similarly, Bedford, a truck maker based in Luton, UK, now designs new cars in collaboration with Pontiac, near Detroit in the US. Like Austin Rover and Honda, Bedford and Pontiac have two teams of designers working on the same project — one team is based in the UK, the other in the US. Again with CAD, both teams are able to communicate instantly, exchange information, data, drawings/sketches about their current design projects.

Some implications

The introduction of CAD technology implies that both engineering and industrial designers should be educated, trained, and retrained in the effective use of CAD Systems. The introduction of CAD, therefore, in the suggested 'integrated syllabus' means that it will now be much more feasible to instruct engineers and industrial designers in design skills, and thus ease the potential communication problems between engineers and industrial designers. That is to say, the CAD technique will make it possible for both professionals to communicate on design projects on the same wavelength. For example, Charlish (1984) found that 'Prodabas' a CAD system developed by Prosys Technology Company has a particular advantage of 'the removal of paper and human links between what were distinct design activities. As a result the data is consistent for all users. This avoids the anomalies that can arise when different engineers (and industrial designers) have private versions of particular items and when a design change produces more changes elsewhere.'

The second implication is that CAD will go a long way to close the gaps between industrial designers and engineers in terms of education, product presentation, and regard for individual status as discussed earlier. Thus, whilst industrial designers are model-orientated in project presentation, engineers are analytical, mathematical, or numerical as mentioned earlier. In the words of Schwarz (1984) 'It is axiomatic that engineering statements must be objective, i.e. numerate.' Since CAD techniques now take spot measurements of components, translate these measurements into mathematical representation of the shape of, say, a car, calculates and analyses, then this major difference appears to disappear and both may now regard themselves as 'first among equals'. This is because engineers and industrial designers will now receive instructions on the effective use of CAD for new-product designs. Consequently, both are knowledgeable in the use of CAD systems — an ideal tool for integrating industrial design and engineering components.

The third implication of CAD appears to lie in the relationship between design and manufacture. In the past, the relationship between design and manufacture has been relatively weak, with industrial designers and engineers (development and manufacturing) working independently. This means that the task of an industrial designer seems to end the moment his product model is handed over to the development or manufacturing engineer. He may be needed at the last stage of manufacture to stylize the product before it rolls off the production line. But while 'design for ease of manufacture' has been mentioned as one attribute of good design, little notice appears to have been taken about the apparent weak relationship between designers and manufacturing engineers or simply between design and manufacturing.

The introduction of CAD seems to strengthen these links between design and manufacture. This appears to open up a new dimensional guide to management thinking about the formulation of new design procedures for reinforcing the links between 'design for ease of manufacture' and 'manufacture' itself. For example, the introduction of CAM means that both designers and manufacturing engineers may have to change their various methods of assessment or evaluation of new designs. This again calls for education and training of industrial designers and engineers. This, it is hoped, will increase each other's knowledge and involvement in each other's field — another integrating mechanism.

The above argument strongly supports the SERC Report (1983), which states that 'CAD/CAM itself will strengthen the links between design and manufacture: the design department will simply no longer be able to be treated or to act as a separate part of the manufacturing process.' Similarly, Lord Nelson (1984) has expressed the opinion that the 'design office and the designer no longer stand on their own; they are the starting point of a long process, the success of which will ultimately be proved in the marketplace'. He asserts: 'Designers today must be trained and equipped to initiate successfully this long train of events, and they must be part of the whole team dedicated to the commercial success of the product in the world marketplace.' The introduction of CAD will go a long way to integrate industrial design and engineering components.

One can conclude that the introduction of CAD technology is an attempt to provide correct and up-to-date information or data bank and effective modelling techniques to enable designers to reduce uncertainties associated with designs and be able to take the right decisions at the right time, in order to increase design productivity. It also enables designers to calculate, measure, analyse, see how well the body panels of a product fit together 'without welding a joint', describe the physical shape, and predict the product behaviour.

Other benefits accrue. For example, design, development, and manufacturing costs will be reduced. Manasian (1980) reveals that 'the engine modifications of the Metro car alone have cost £30 million, but this is a small fraction of the cost of developing a new engine'. Similarly, it is hoped, that CAD will reduce the gestation period (duration between design and development and the launch of a new product). It is also hoped that CAD systems will minimize the incidence of defects in new products via the use of various CAD packages. Consequently, value-added is increased and the rate of product recall reduced.

Finally, CAD techniques will make it much more feasible to instruct engineers and industrial designers in design skills. In turn, CAD will enable designers (industrial and engineering) to cross-fertilize design ideas within and between organizations and nations. Besides, it is hoped that CAD will go a long way to integrate the industrial design and engineering disciplines.

CAD technology has come to stay, and its benefits to designers, manufacturers, and consumers will continue to grow for years to come. Our research findings support this conclusion.

Manufacturing processes and changes made

The rapid increase in the use of manufacturing techniques to improve the design of new products was evident in the literature. Accordingly, it was decided to investigate the manufacturing techniques used by the Queen's award winners and why.

As can be seen, Table 6.3A shows that seventy (78 per cent) of the sample introduced changes to their manufacturing processes during the last five years. Only twenty firms (22 per cent) did not introduce any changes during the corresponding period. Table 6.3B presents these changes.

It is clear from Table 6.3B that more than half (55 per cent) of the sample introduced automated flexible manufacturing system (FMS), materials requirement planning (MRP), automatic testing equipment (ATE), computer numerically controlled (CNC)

Table 6.3A Has your firm introduced any significant changes to its manufacturing process in the last 5 years?

	No. of firms	%
Yes	70	78[a]
No	20	22
Don't know	—	—
	90	100

Note: a. To be read: 78% of the responding firms said that they made changes to their manufacturing processes during the last five years.

Table 6.3B Changes made to the manufacturing process

Change made	*No. of firms*	%[a]
Introduced CAD/CAM	19	28[b]
Introduced robotics	10	14
Introduced automation	38	55
Introduced statistical process control (SPC)	3	4
Introduced subcontracting	2	3
Introduced computer programming	3	4
Modified existing process	15	22
Replaced existing process	8	12
Total responses	98[c]	142

Note: a. Based on 69 firms.
b. To be read: 28% (19) firms who responded to this question introduced CAD/CAM during the last five years.
c. A number of firms introduced more than one change during the period.

machine, automatic printed circuit board (PCB) etc.). Computer-aided design (CAD)/computer-aided manufacture (CAM) is the second most frequently mentioned manufacturing technique introduced by 28 per cent of the respondents during the last five years, while 22 per cent (15 out of 69) of the sample modified their existing processes during the same period. Other significant changes made during the same period include the introduction of robotics (14 per cent) and the replacement of the existing process (12 per cent).

All of the above are important findings in the sense that these techniques are designed to contribute to the design and manufacture of good-quality products. For example, CAD aids and speeds up the design of new products, while CAM facilitates the manufacture of the product itself. In this way the gestation period, the manufacturing costs, etc., are reduced.

Similarly, because of the great importance placed by firms upon the 'choice of materials' in making product decisions as revealed in our study, the MRP technique ensures that the suitable materials are not only selected but also supplied on time to ensure that the right product is produced with the right materials and delivered at the right time. In a similar vein, the ATE technique provides the facilities for testing products for reliability, performance, stress, safety, and the like, to detect faults before products are introduced onto the market. This is a particularly significant breakthrough in the design and manufacture of new products. As discussed earlier, often it is the customer who detects the faults, say, in a car, and sometimes lives are lost before such faults are detected. On the other hand, the introduction of computer programming appears to enable firms to use standard application packages (software) for facilitating the design and manufacture of new products. The use of standard application packages also helps engineering designers and industrial designers to communicate on the same wavelength, and thus contributes to the integration of the two professional skills.

Furthermore, firms which did not introduce any changes to their manufacturing processes during the last five years were asked to state the manufacturing techniques they currently use. It was found that of the sixteen firms who responded to this question, nine (56 per cent) use computer aids to design (CAD), six (38 per cent) use value engineering, while four (25 per cent) operate quality circles, as reported earlier (see Chapter 5, Table 5.8, p. 235), once again reinforcing the earlier findings on the growing influence of computers on the design of new products.

Summary

A review of the concept of design suggests that *design* is complex and has a range of meanings. This manifests itself in the confusion that seems to exist within and among the design professions as to what design means. The differences in definition of design by different authors is a typical example. This may be explained by the fact that almost everything and anything we do in life involves one form of design or another. For example, design can be found in the organization of self, home, office, and most noticeably, in the manufacture of new products (industrial and consumer). Consequently, individuals tend to see and interpret design differently, which in turn, appears to explain the differences in definition. This shows the difficulty of attempting to find a tightly controlled definition of design.

However, although there are differences in the definition of design, almost all definitions by implication refer to 'design' as a 'problem-solving activity'. In effect, this suggests that the definition of the design concept appears to remain one and a coherent concept. A coherent concept is important, not only in understanding the meaning and interpretation of design, it is also necessary in the application of the concept itself.

There is evidence in the literature to suggest that differences exist between industrial designers and engineers. This is illustrated in gaps in education and training, status, and presentation of products or projects that exist among them. This calls for integration between industrial design and engineering, to enable firms and innovators to make the most of design. For instance, the fact that most products, particularly high-technology products, cannot compete successfully in the international market (the ultimate objective of any progressive firm), without the combined effect of engineering characteristics (technical/mechanical) and industrial-design factors (ergonomics/aesthetics) appears to demand that the two professions be integrated.

Thus the need to integrate industrial design and engineering appears to be consequent upon:

(1) the increasing absence of ergonomics and aesthetic appeals in engineering;
(2) the interlinked nature of industrial design and engineering — the separation of which may cause commercial and competitive disadvantage to the firm;
(3) the invaluable interfacing or conciliatory role of the industrial designer, which involves the reconciliation of engineering characteristics with human factors;

(4) the opposition to the employment of industrial designers by some engineers, the continuation of which may not yield any fruitful result in the process of NPDD.

Contributions towards integration indicate that the Government, the Design Council, and a number of individuals have played leading roles in an attempt to integrate industrial design and engineering. For example, the Government provides funds for mass propaganda for design awareness in Colleges, Universities, and industries, seminars, and surveys, and the Design Council organizes various awards for technological achievements and/or excellent designs.
excellent designs.

Overall, it appears that the capabilities of engineer and industrial designer may continue to be under-utilized in the process of NPD, especially in the electronics, computer, and agricultural-machinery industries, as long as industrial design and engineering techniques remain divorced; which is regarded here as a waste of human capital. As a result, firms/innovators may not make the most of design, the result of which can cause British industries to continue to be followers in the market place.

It is therefore, our considered view that merging the syllabuses of engineering and industrial design disciplines will provide a solid base for succcessful integration. One hopes that this approach will offer the following advantages:

(1) A possible status gap will be greatly reduced.
(2) Mutual respect and trust among both professional groups will be restored.
(3) Communication between the two experts can be made a lot easier, so that projects can be presented to the board of directors without misunderstanding.
(4) Goal congruence during the process of NPDD will be ensured, as their interests will no longer be in conflict with each other.

The discussion of the 'impact of computers on design' suggests that the introduction of CAD is an attempt to provide accurate and up-to-date information or data bank and effective modelling techniques to enable designers to minimize uncertainties associated with design, and to make the right decisions at the right time, in order to increase design productivity and added value.

Since 'users' of CAD will be able to see on the VDU, how well the body panels of, say, a car, fit together 'without welding a single joint', as well as to select adequate materials/components, manufacturing processes, etc., and calculate their costs, CAD techniques will enable manufacturers to know whether or not to commit funds to developments and manufactures right from the outset. Also the introduction of CAD/CAM implies that the links between design and manufacture will be strengthened. This should yield increases in productivity, reduction in development and manufacturing costs, reduction in the gestation period, and improvement in quality.

Finally, and more importantly, CAD systems will make it much more feasible to instruct engineers and industrial designers on design skills. Consequently, both professions will be able to communicate on the same wavelength. This avoids anomalies that can arise when engineers and industrial designers have private versions, as is often the case, of new-product designs. In short, with CAD, both engineers and industrial designers will be better able to cross-fertilize new design ideas between and within firms and countries. CAD will, therefore, largely help in the integration of industrial design and engineering components.

Conclusions

Design is a complex phenomenon that applies to almost everything and anything we do in life in the way human needs are first conceptualized, next prototyped, and finally produced and marketed (possibly with room for further improvements to meet changes in tastes and preferences). This means that design awareness should be created throughout our society. In particular, design awareness must be created throughout manufacturing organizations (where the impact of design on new products is more pronounced) — from top (board room) to bottom (shopfloor). In effect, this means that design should not be isolated, as is usually the case, in the technical department alone. Our research findings reveal that technical people alone cannot produce a 'balanced' product — a product that provides value for money to the customer and 'reasonable' profit to the manufacturer. Rather, a balanced product is more likely to be achieved through contributions from everyone in the organization, hence quality is the concern of all in an organization.

Although there still exists some disagreement among innovators as to the factor(s) that best encapsulate the concept of design (probably due to its complexity as alluded to above), our findings indicate that no one factor explains clearly the concept of design, rather, the design concept seems to be best explained by a combination of design factors (e.g. superior overall performance) and design-influenced factors (e.g. making products that sell and make profits).

The differences in the design concept *per se* appears to have introduced some elements of differences/conflict between engineering designers and industrial designers. Such differences as gaps in the background education and training, status, and product/project presentation diminish the effectiveness of our industrial society. Products are often ill-designed and/or ill-produced either because the engineering characteristics (e.g. reliability, performance in operations, etc.) are wanting, or there is the absence of flair in industrial design (e.g. attractive appearance/shape, style, colour, etc.). The reason is clear. In many firms, engineers and industrial designers tend to lead separate existences. Yet, there is abundant evidence in the literature to demonstrate that the skills embodied in the practice of industrial design and engineering interrelate and overlap. In other words, although industrial design and engineering disciplines are perceived as distinct, they are none the less dimensions of the same activity, which emphasizes a greater need for the integration of both disciplines. It therefore seems to be a waste of human capital to encourage both professions to continue to work as separate entities when, in fact, they should work as a team to produce a product in which the customer's satisfaction is maximized at a reasonable profit to the manufacturer.

A search for effective ways of integrating industrial design and engineering disciplines reveals two methods. The first is closing the educational gap that exists between them. This calls for the integration of engineering and industrial-design syllabuses whereby equal weighting should be given to both topics. In this way, the so-called 'peripheral' subjects in each other's discipline would be made part of the core subjects, which will count as a partial fulfilment for the award of a bachelor's degree or its equivalent in either engineering or industrial design. Our research findings suggest that the undergraduate course is the best level of education at which the integration of engineering and industrial-design skills would be most effective.

In the second method, the need to capitalize on the benefits that computers are having on the design of new products is evident. It is found that computers, particularly via CAD, are capable of breaking the communication barrier between engineering designers and industrial designers, especially during the product/project presentation, and more so when both have their own versions, as is often the case, of new-product designs or when a change in design produces more changes elsewhere. This means that engineers and industrial designers should have adequate knowledge in the use of computers. An adequate knowledge in the use of CAD systems, for example, will make it much more feasible to instruct engineers and industrial designers on design skills. Consequently, both professions will be able to communicate on the same wavelength. That is to say, with CAD, both engineers and industrial designers will be better able to cross-fertilize new design ideas between and within firms and nations. In sum, the use of computers (CAD) in the design of new products will not only enhance the integration of industrial design and engineering components, it will also assist in the better understanding of the design concept and indeed, its interpretation.

Chapter seven

Improving competitiveness through design management

Introduction

Today's intense competition among nations in international markets, prompted by commercial and social changes in our society, dictate that a shift in emphasis in design activities is inevitable. Available evidence suggests that design is neglected in many British industries but when effectively managed may lead to high-quality products and competitiveness. But, good design without adequate management commitment may detract from the competitiveness of firms, suggesting that design management is a critical success factor to which adequate attention must be given so a company's overall competitiveness may be enhanced. Therefore the specific concern of this chapter is to explore ways in which the management of design may be improved to enhance the competitiveness of firms in international markets.

The chapter contains five parts. The first reviews the definition of design management. The second highlights the need to manage design so a firm's overall competitiveness may be improved, as already stated. The results of our survey on how the Queen's Award winners manage design are reported in part three.

It is evident from our discussions in Chapter 5 that value analysis/value engineering, aesthetics, ergonomics, quality circles, timing, leadership, and attention to user needs, all influence the design of new products as well as the redesign of existing ones. This means that to be successful in the marketplace, a firm should be design-driven, particularly in design-orientated products. That is, a firm should attempt to create an organizational climate whereby everyone should be made aware of the importance of design to the success of the firm and the international competitiveness. This is spelled out in the fourth part entitled 'Thinking Design'.

Finally, conclusions are drawn, among which is the confirmation of our research hypothesis, that the effective management of design leads to better design, greater competitiveness, and higher profits.

Design management

The term 'design management' has been described by many authors in different contexts. Farr (1966), one of the pioneers in this area, drawing on his experience as a consultant, defines 'managing design' as a professional activity involving someone acting as a 'go-between', bringing together companies and freelance designers and enabling them to communicate with each other. Smith (1978), states that design management emphasizes the need for concern by companies about moral, social, and ethical issues. In his study, Topalian (1980), classifies design management into two parts. The first concerns top management whose principal functions are to formulate policy, take decisions about setting of design standards, and organize design activities (such as whether or not to set up in-house development or to hire design consultants). The second describes the management of individual projects as well as routine administration and control.

Willcock (1981), sees design management as the matching of design talent with marketing opportunities. In a wide-ranging survey, CNAA (1984) reports that managing design is used to refer to the totality of awareness of the significance of the product by managers and to the techniques through which that awareness is articulated. Finally, Olins (1985) argues that managing design has no mysteries, except for those illusory ones that people have created for themselves over the years. It is his contention that 'if you know how to manage finance or R & D or personnel, then you also know how to manage design ... it must cover product, environment and communications design ... if there is a lack of coherence between product, environment and communications design, failure will follow.'

It is not surprising, therefore, that 'design management' differs in definition between authors. This is more so because there appears to be no tightly-controlled definition of the word 'design' itself as demonstrated in Chapter 6 pp. 244–248. None the less, these varied definitions seem not only to indicate the broad nature of design management, but also underline the magnitude of tasks facing innovating firms and their managers in the process of managing design.

'Design management' is defined here as the effective control, review and monitoring of new products by managers, as well as the efficient and timely application of techniques by which a product/ process can be improved upon in order to achieve international competitiveness. Improvement here means ways and means of adding better value to an existing or new product to make it more presentable, saleable, and more competitive than rival product offerings.

The need for design management

Evidence exists from previous studies to suggest that British managers are failing to exploit fully the design potential and thereby contributing to the poorer industrial and economic performance of Great Britain when compared with their major foreign competitors. The evidence also suggests that improved design management will lead to better design and higher quality products. For example, Oakley (1984a) finds that too many British companies are failing to exploit good product design. He asserts: 'In many cases, foreign competitors are enjoying considerable market gains at the expense of British producers who often seem unable to drag their products out of the past.'

The Fielden Committee (1963) states that there is 'a widespread view among managements of engineering businesses that design is something separate from management, which can be carried out in a back-room'. The Committee adds that 'in far too many cases there is a lack of forward thinking on the part of management which results in lack of incentive to develop new designs'. The Committee then concludes: 'The neglect of or inattention to detail in design suggests that there is something wrong with the organisation of the activity.' This has been confirmed by Wally Olins who argues that 'in most companies design of product is so strictly segregated from design of environments and communications material that people involved in the different fields are effectively locked away from each other. Even more serious, not only is there no mechanism for managing design across the traditional corporate frontier, there isn't even any recognition that such management might in principle, be desirable' (Olins, 1985).

It is also his contention that 'design is normally regarded as a jobbing task to be commissioned and executed — or not as the case may be — for specific products or show-rooms or brochures on a job-by-job basis, as though each is a completely separate task, having no relationship with anything else that emanates from the company. So in most companies design is not actually managed at all.' Illustrating the haphazard manner in which product design is organized or managed he declares: 'Product design is run by engineers. Communications (advertising, literature, instruction manuals, etc.) are run by public relations and marketing people. And in many companies, environments (implying shops, retail outlets, showrooms) are looked after by the janitor.'

He continues: 'In most companies there are no rules about design, or such rules as exist are made up by individuals as they go along – as with information management before the computer.' Olins (1985) then cites Chrysler as a classical example of product failure resulting from improper management of design. He ascribes the major cause of Chrysler's failure to an apparent lack of coherence in its product policy. For instance, 'In Britain, where Rootes factories were sporadically producing Hillman Avengers, Imps and Humber Sceptres for a market turning away to Minis, Morris and Austin 1100s and Triumph Heralds, Chrysler's product policy seemed to consist solely in changing the badges on the front and back of the car.' In France, for example, 'where Simca had recently introduced an excellent new medium sized product, Chrysler boldly changed the badge at the front of the car to Chrysler; but became afflicted with nervousness and left the Simca badge on the back'. What impression will these two incompatible badges create in the minds of prospective buyers? Does the 'swapping' of badges imply innovation or improvement upon Chrysler's existing cars? In other words, does the change of badges make one car sell better than the other? To both questions the answer is negative. In fact, Olins's finding reveals that 'the Hillman Avenger, which sold badly, did no better as the Chrysler Avenger' after badges were changed.

He concludes: 'Much of what went wrong with Chrysler was failure to manage design.... In product policy, however, there was either no policy or there were three policies. In any event, it didn't work. And much more important, there was no overall design management and no overall design co-ordination. The design of products should have been associated with the design of environments and of communications. But it wasn't. Consequently, Chrysler's message was hollow. The Chrysler idea did not exist. Customers did not know what a Chrysler car was. That's why they didn't buy the products.'

In brief, Chrysler failed because of its apparent failure to recognize that it is value added, not badges, that sells a car, and that this can only come about through design and design-related factors. For example, the fact that Chrysler differentiated identical cars with badges when it should concern itself with improving the quality of or adding value to its cars via redesign or designing new ones, exposes its inability to manage design effectively.

In her study 'Managing design function in the UK textile industry', Watt (1980) found that 'the major cause of concern was that the actual management of design was carried out by committees which typically comprised sales, marketing, production heads, none of whom usually had training in design or design appreciation'; and that management of design in Italy is completely the opposite of what occurs in the UK. Opposite in the sense that Watt's finding showed that in the UK, 'employers who recruited designers put them straight in the studio and did not consider: (a) giving them initial training in other areas of the firm's activity as is usual with other types of recruit, or (b) giving them any form of general management training at any time in their career. In Italy ... having finished an academic training in design ... a designer on joining the company would spend at least two years on the sales force and in other commercial areas of the company before actually going into the design department.'

She commented: 'it is optimistic, to say the least to expect that a designer who spends 46 weeks a year in a studio tucked away in a corner of one of our 'dark, satanic mills' is going to be able to draw together the different threads of what is happening in the world and produce first class designs' (Watt, 1980).

Explaining 'training in design or design appreciation' as already stated, Watt drew an analogy with finance: 'it is doubtful if a company would employ a qualified accountant and then set up a committee of non-qualified staff from other areas to actually decide financial policy and rely on the possibility that one or two members might be good at arithmetic'. This implies that design is not 'something everyone intuitively understands'. In other words, managers must be able to appreciate the problems involved in design in order to manage design effectively. In Italy, for example, senior managers, although not formally trained designers, 'engage empathetically in the creative process' and tend to regard themselves as part of the design team. Similarly, in Japan, the CNAA (1984) found that 'the design function is normally managed by senior designers, but because Japanese managers participate in extensive discussions and share decision-making other managers have considerable influence over design, and designers tend to be involved in debates outside their own immediate areas'.

Altogether, the top-management involvement is perceived 'to be more intimate than simply setting standards from the top'. This tends to lead to 'strong commitment among top managers and a high level of concern that it (design) should be done well'.

Earlier, the Carter Report (1977) had expressed concern about the lack of awareness in many UK industries of the benefits designers can bring. The Report strongly stressed that 'management in British Industry had for many years lacked insight into the relationship between effective design on the one hand and industrial performance on the other'. First, this appears to suggest a lack of understanding of the importance of design management. Second, it seems to portray the inability of some UK industrial managers to manage design effectively.

Rothwell *et al.* (1983) argue that excellent design by itself is insufficient; adequate product strategy is equally vital. They maintain that to achieve a 'high' standard in the design of a new product is one thing; to achieve an 'appropriate' standard is another. For example, a comparison of a BMC/BL 1100 of 1962 with a Ford Cortina MK1 (1962), both of which were acclaimed by 'Motoring Which' as '"joint best" in terms of performance, handling and economics', shows that it was Ford's conventional design tailored to suit customer requirements/specifications that performed better than that of British Leyland (BL) in the marketplace. They however, regret that after 20 years BMC/BL made little or no changes in design while Ford has consistently and continually made changes on its model through redesign 'to meet the demands imposed by customers and competition'. Rothwell *et al.* (1983) declare: 'Ford continually evolves and reworks its car designs for each mark and number while BMC/BL, until recently, tended to lock up on its design. Generally speaking, Ford has produced robust designs while BL has had lean designs. [A robust design is one that permits changes (major or minor) to be made in the existing product, that is, a design that is capable of being 'stretched' or reworked (redesigned) to arrive at versions of similar or new products or what Gardiner and Rothwell (1985) describe as 'new variants' and/or 'design families' ... to cater better for existing or new markets.] [A lean design offers little room for improvement (i.e. for rework or redesign). This tends to limit the chances of creating new design variants or family products to satisfy the existing or new market demands: lean design provides limited choice to the customer.] For instance, the Mark 1 Cortina had three engines and four horse power levels while the Mark 1 1100 had only one engine and could only compete with the Cortina at the first level. BMC/BL's designers in comparison with Ford's, have consistently been under-financed and poorly managed.'

Finally, Rothwell *et al.* (1983) draw the conclusion that 'during the past two decades Ford has built up an impressive portfolio of mainly incremental potential design changes and, for the period

1970 to 1979, Ford registered 1,441 patents in the UK compared to British Leyland's 239'. Once again, the importance of the need to manage design is stressed.

It is evident from the Open University Study (1983) on 'Plastics products' that 'the successful management and practice of design, as reflected in firms winning design awards for their products, increases the chances of business success'. The study reveals that whereas the eight design-conscious firms investigated 'deliberately aimed to expand their business through a strategy of producing well-designed "up-market" products of high unit value and quality ... over a third of the representative sample (about 41 firms) believed that design was not worth much time, effort or money'. Also skilled designers were seen as an investment in the design-conscious firms; in marked contrast, many representative firms viewed spending on advanced plant and machinery, rather than on skilled designers, as an investment.

The findings of the Open University (1983) strongly support the conclusion of several other recent reports and studies that 'competitiveness is gained from increased investment, not just in more productive plant and equipment but in the human resources needed for the effective management and practice of product design and innovation'.

The wide-ranging survey carried out by CNAA (1984) on 'Managing design' has demonstrably produced supportive evidence to show that most UK managers are not sufficiently exploiting the design potential in UK industries. For example, the examination of 'six overseas (West Germany, Italy, France, Netherlands, US and Japan) approaches to managing design' by the CNAA strongly suggests that these countries pay more attention to design and design management than their UK counterpart. Further, the fact that some of the overseas industrialists interviewed by two CNAA study groups 'declined to divulge their design experience and practice' to them clearly indicates the degree of importance attached to design and its management in overseas countries. Apparently, these are the major UK competitors in international markets. In Italy, for instance, 'design is widely viewed by Italian industrialists as an added value factor .... Recognition that design is important leads to strong commitment among top managers and a high level concern that it should be done well.'

Based on these findings, the CNAA Report (1984) concludes: 'Attention to design follows directly from changing conditions in the business environment. In many firms and industries, there is now a uniformity of competence in aspects such as the management of labour, process, distribution and finances. What now determines success for individual companies is the ability to develop other distinctive competences, dealing with such issues as product design and other forms of design, product and marketing strategy, corporate identity and image, and the management of new products and services.' Unfortunately these competences are wanting in areas of design and design-related factors, thus, stressing the needs for the proper management of design. For example, the Report continues: 'In the USA, especially in the large corporations, design-related matters are receiving increasing attention and the appointment of managers with specific responsibility for design is becoming more common. In Japan, interest is being generated in design as an important ingredient in economic development.'

Finally, the Report warns that 'the UK could be left with only low cost, low added value manufacturing activities unless the trends now clearly evident are recognised and more design-led strategies are adopted'. This strongly suggests the greater need for an effective and efficient management of design.

Other writers have also produced evidence to suggest that design has been neglected and thus requires adequate attention in order to compete effectively in the world trade.

In his study, Barratt (1984d) discovers that buyers think that 'design is something that is stuck to the outside of a piece of work. It is thought to be expensive, and because it has something to do with 'art', buyers are wary of it — you can measure the crack in a casting or the strength of a plastic component, but how do you tell good design from bad?' He concludes: 'These two attitudes — that design is expensive and a superfluous add-on, and that its benefits are intangible — infect UK and the public sector in depth. One result is that those items which are given to professional designers are often badly done.' This appears to expose the inadequate attention being paid to design.

Similarly, Lorenz (1984) contends that 'it is more of a lingering belief on the part of many retailers and manufacturers — plus the City — that design is merely a promotional veneer that can be tacked onto whatever lies beneath, and used as a cure-all'. Obviously, if importance is attached to design, some UK manufacturers should not perceive design as 'a mere promotional veneer'; or should it be an isolated case? The answer seems to be negative. Moody (1980b) also found that even engineers, who are expected to have a better understanding of the role of design in international competitiveness, often perceive design as "tarting-up" the product, giving the product a "facelift", or styling the product'.

In sharp contrast, 'design work is seen as a natural and normal part of a company's activities in Italy; this recognition does not lessen consciousness of the need for careful preparation to ensure good results' (CNAA, 1984). Similarly, the CNAA (1984) reports that in Japan 'the role played by design in economic competitiveness is well-understood and widely acknowledged'.

Last, but not least, Macdonald (1984) contends that the 'management view of quality (design) is similar to people's view of China: they know it's there, they know it's important, but they have never paid it a visit. Out of sight, out of mind.' He continues: 'Management begins to pay attention to quality when market share falls, customer complaints grow or the cost of after-sales service escalates.' Once again, the management has been exposed due to its inability to manage design.

All of the aforementioned evidence tends to strengthen the argument that an apparent lack of sufficient exploitation of design in many British industries has largely contributed to the UK's poor industrial and economic performance in international markets when compared to those of its major foreign competitors.

Managing design: some evidence

Olins (1985) points out the main elements involved in the effective management of design. Among them are:

(1) Board commitment. One director championing design within the boardroom – probably with active co-operation of external consultants.
(2) A design manager centrally located, with opposite numbers in each of the operating companies, firmly linking up with and giving more scope to internal design people.
(3) There should be relevant task forces, in purchasing, marketing, production, engineering, advertising, and so on, to make it work.
(4) A manual on design guidelines should be made available.
(5) There should be a rolling (design) programme that can be checked, monitored, modified, and where appropriate, amended.
(6) There must be a commitment to all of these things by way of financial resource. There must also be 'appreciation of power, purpose and nature of the resource'.

Further, a study done by CNAA (1984) on how overseas countries manage design shows that different countries adopt different approaches. Below are some of the results of the study, demonstrating how design is managed in Italy, Japan, and West Germany.

Italy

1. It is generally believed in Italy that design cannot be effectively managed by sticking to old ideas or concepts. Consequently, emphasis is often laid on diversity, novelty, and innovation.
2. The relationship between managers and designers is reported to be close, promoted by a cultural tradition that tends to give rise to a confident familiarity with matters concerned with art, architecture, and design. According to the CNAA report, in the UK, such close relationship tends to be distanced.
3. Designers enjoy high status. As a consequence of this, design is perceived as a top-management responsibility, particularly in ensuring the setting and maintaining of in-house design standards. In addition, the concept of 'design patronage' by companies is not only important, but well established. Olivetti is often cited as an example. However, it is evident in studies previous to this that design does not enjoy high status in the UK, (Fielden Committee (1963), SERC report (1983)). It is also worthy of note that the maintenance of in-house design standards does not seem to receive the attention it deserves in many UK industries. For example, such design-conscious companies as TI Raleigh (manufacturer of Raleigh bicycles) and Norton Villiers Triumph (producer of mopeds) still import a substantial proportion of their components from France and Italy, respectively. In fact, Norton Villiers Triumph, as referred to in Chapter 1, p. 17 imports all of its manufacturing parts from Italy, yet an Italian firm Garelli, which also manufactures mopeds is currently one of the major competitors of Norton Villiers in the UK market. It is argued here that design may not be effectively managed when major foreign competitors still maintain the design initiative and the manufacture of major components/parts used in British industries.
4. Design work is seen as a natural and normal part of a company's activities. This recognition does not lessen consciousness of the need for careful preparation to ensure good results. For instance, the cost of design work is not only controlled but also incorporated in basic product

costing, not treated as an additional expense.

5. Steps are usually taken to ensure 'balanced' design by taking account both of the production system and of customer desires. This suggests that Italian designers have both customers/users and ease of manufacture in mind when designing new products.
6. In most cases, design is managed by designers not managers. Further, it is widely seen as essential that all managers have an appreciation of the problems involved in managing design. Consequently, senior executives, particularly in successful companies, although not formally trained designers, tend to engage empathetically in the creative process and regard themselves as part of the design team. In turn, this tends to create smooth relationships between designers and top management. This helps to make the management of design more effective.
7. Also contributing to the effective management of design in Italy is the frequent initiation of new-product proposals by designers and a willingness by companies seriously to consider such proposals and frequently to permit the designer's view to predominate in any decision-making concerning new products, rather than passing instructions/orders to them from the top. According to the CNAA Report (1984), this recognition that design is important tends to lead to strong commitment amongst top managers and a high level of concern that it should be well done. 'This concern', the Report argues, 'ensures that managers take steps to provide the correct environment necessary to encourage creativity, not just the provision of appropriate accommodation and equipment but also as important is the maintenance of an atmosphere of "creative tension".' The 'creative tension' is explained as a factor that 'demands an understanding of what motivates designers, which is often felt to include the need for at least partial insecurity (the free-lance rather than in-house system) together with other, non-financial rewards such as the opportunity to tackle many different projects and kinds of design, the chance to work with eminent designers and the enjoyment of personal recognition, respect and prestige'.

Japan

The role played by design in economic competitiveness is well understood and widely acknowledged in Japan. This is manifest in a series of design programmes organized by the Japanese Industrial Design Promotion Organisation (JIDPO), under the auspices of the Ministry of International Trade and Industry (MITI) to promote and exploit design. An instance is the Design Development and Promotion Project for Regional Industries (DDPPRI), which is set up to show companies how important design is to them. This body, in co-operation with the local government, also formulates systems to give practical guidance to companies about managing design. In the UK, the National Economic Development Office (NEDO) plays the role of MITI whilst the function of JIDPO appears to be performed by the Design Council.

Although NEDO and the Design Council appear to be doing well, they seem to be less competitive when their operations are compared with the extensive design activities organized by their Japanese counterparts. One reason for this drawback may be the unwillingness on the part of some British industries to change as quickly as NEDO and the Design Council would want. Another reason can be traced to an apparent lack of industry-linked projects as in France and Japan, as well as too much emphasis on theory rather than practice in design education. For example, the CNAA study (1984) shows that, in France, 'although there is an "educational resources" section which can give students theoretical or practical background, no pedagogical structure is imposed upon students in order to give free rein to creativity'. The Report quotes the staff of the French Institute for Industrial Design (Les Ateliers) as saying, 'Spoon-feeding by means of traditional instruction in design, business techniques or any other disciplines would be far less effective.' This argument strongly supports the conclusion drawn by the CNAA which states: 'It is interesting to speculate whether a similar approach, particularly the industry-linked project, might be appropriate when addressing design issues in UK management courses, which have some tendency to concentrate on mastering of techniques and concepts rather than developing the initiative of students' (CNAA, 1984).

Further, the links between design on the one hand and production and marketing on the other are often significantly emphasized in Japan. In addition, companies are shown how to design products for efficient manufacture. As stated in Chapter 1, one of the reasons for the UK's laggardness is the tendency, inherent in many UK industries, to design for the production to produce and marketing to sell. In other words the links between design, ease of manufacture, and marketing, typical of Japanese industries, appear to receive little attention in many UK industries.

In Japan, the management of design is not isolated to the design department alone. For example, it was evident in the company visited by the CNAA team that 'collaboration between personnel from different departments was of a high degree'. Besides, in setting up projects, very detailed pre-planning and analysis of alternative approaches is carried out. Since design guidelines on technical and styling matters have been devised and thoughtfully presented in a series of in-house publications, projects can proceed rapidly and efficiently once the pre-planning is complete.

Finally, clear objectives are apparent in Japanese industries. It was evident in the company investigated by the CNAA (1984) that there was a clear written design policy and everyone working in design knows what the company is trying to achieve. In addition, 'planned programmes for the incremental refinement of existing designs are given a high priority and new designs are required to support and enhance the corporate mission'.

West Germany

In West Germany, design is managed through the following methods:

1. Thorough research into project requirements before design work begins. This strategy ensures that new products rarely fail to find market acceptance.
2. Attention to the official standard.
3. Any teaching about managing design should stress co-ordination, co-operation between functions, and interdisciplinary approaches — reminiscent of the Japanese practice.
4. Analysis of target customer groups, and customer desires and wants.
5. Drawing the attention of managers about the nature of the design process and its interactions with other activities in the firm.
6. Analysis of existing product benefits and features and reconciliation of these with customer needs.
7. Evaluation of market characteristics, competitors' activities, legal and other constraints, needs of the company.
8. Listing characteristics of new or improved products.
9. Critical examination of individual components or sub-assemblies of products to determine the appropriate use of materials, colour, textures, and forms necessary to satisfy customers' psychological desires. To achieve this, product

managers need a grasp of designers' methods of addressing these problems.

10. Analysis of methods of manufacturing products, distribution, after-sales service, etc.
11. Review of effects and problems after the product launch as well as forward planning for next generation of products.
12. Analysis of steps necessary for further development of market and consequences for product development.
13. More training in management practices for designers.
14. The need for design to be practised within a long-term rather than a short-term context if consistently successful results are to be obtained.
15. Involving designers in projects from the beginning rather than simply to do styling work at the end.
16. There should be a project brief that strikes the right balance between being sufficiently informative and excessively doctrinaire, and which is flexible enough to enable any truly innovative leads arising to be pursued.
17. A recognition that design is not a linear process but a cyclical process in which an end point is never reached, some improvement or enhancement being always possible and desirable, and that the process is iterative, each step being affected by feedback of information from earlier and later steps.
18. Ability and willingness to evaluate design results and an expectation that designers will always offer alternative solutions from which a choice must be made.
19. It is important to start designing products — and consumer products in particular — with some idea in mind of the desired appearance and ergonomics, and then attend to technical functions of the product. In other words, to design from the outside in, rather than as many firms try to do, to start with the technical aspects and add styling as an afterthought, designing from the inside out.
20. Finally, West Germany blames excessive concentration upon analysis at the expense of subjects that encourage emotional development and creativity, for the widespread suppression of intuitive design skills.

Managing innovation: some evidence

The management of design, as stated earlier, implies the management of innovation and vice versa; the reason being that design and innovation are interlinked (see Figure 7.2, p. 337). Consequently, the following section examines how the management of innovation can be improved in order to enhance the management of design.

Addressing the twenty-third Annual Yorkshire Joint Management meeting on 'The management of innovation', David Sainsbury (1983) (Financial Director of Sainsbury Ltd) listed five key conditions for the proper management of innovation in industry. They include:

1. Total commitment of top management. By definition innovation requires initiative, doing things in unconventional ways, and risk taking. Top management must, therefore, be convinced that innovation is essential to the survival and prosperity of their business, and they must also make certain that everyone in the organisation is aware of that conviction.
2. Top management must be prepared to invest the necessary resources and where necessary to take a long-term view. If possible, it is obviously desirable to introduce change in an evolutionary way, but this cannot always be done.
3. There must be a good intelligence system which alerts management to technological and competitive changes. A large number of Sainsburys' directors visit the US or Europe each year to see what developments are occurring, and this leads to a mass of innovations in the way they run their business.
4. There must be an organisational structure which breaks down barriers between departments, and a good supply of managers with experience of more than one function. Innovation will only be successful if all the key areas of a business are involved in their development.
5. Companies will only adapt quickly to new technology and to changing world markets if people feel that they have a vital stake in the company for which they work.

Sainsbury (1983) then identifies three factors that, he argues, impede the successful management of innovation in many UK industries.

First, a number of British innovations have not come up to commercial expectations because they have been based too much on 'technology push' and have not involved any proper assessment of market needs.

Second, not enough attention has been paid to manufacturing feasibility in the design of new products, with the result that there have been delays in bringing the product to market and unexpected costs of manufacturing start-up.

Third, the deployment of technological resources in the UK appears to be inadequate. For example, although the proportion of the UK's GDP spent on research looks impressive, a very high proportion goes on defence research and a relatively low proportion on industrial R & D. According to Sainsbury (1983) 'it has been estimated that 60% of our electronic engineers, and an even higher proportion of the rest, are working on defence-related projects. This cannot be the best use of our limited resources.' He asserts: 'We need to think seriously of increasing the funds allocated to industrial R & D and of aiming them more specifically at the commercial exploitation of new technology, the development of prototypes and the availability of new technology for small enterprises.'

It is pertinent to note that one of the reasons for the UK's non-competitiveness (see Chapter 3) is the concentration of R & D expenditure on defence as opposed to civilian industry.

Other ways of managing innovation abound. Harris (1984) argues that, 'although innovation may differ from one industry to another and from one company to another, the basic preconditions for its successful management remain the same'. Some of these conditions were spelt out by a number of speakers at the British Institute of Management's (BIM) Conference (1984) on 'The management of innovation'.

Rothwell (1984) list eleven factors that enhance the management of innovation:

1. Top management must be committed.
2. Innovation must play a role in the company's long-term corporate strategy.
3. There must be long-term commitment to major projects, based on future market penetration.
4. There must be corporate flexibility.
5. Top management must accept risk.
6. An environment must be created in which entrepreneurship can flourish.
7. User needs must be understood and met — failure to do this is the most frequent cause of company failure.
8. There must be good coordination between all in-house functions.
9. There must be effective coupling with the market place and with external sources of expertise.
10. Companies must retain gifted and committed entrepreneurs.

11. Efficient manufacturing procedures must be coupled with efficient after-sales service and education.

The following are comments by practitioners who also addressed the BIM Conference (1984), briefly describing how innovation has been and/or can be successfully managed.
Richard Cutting (1984) (Managing Director of Sinclair Research Metalab):

> In Sinclair, cost is treated as an integral part of product design and, to some extent dictates it. Working on the principle that 'the lower the price, the more you can sell', it had been decided that the price of our first personal computer must be kept under £100 (the Ford principle); this decision had been a major factor in design.

(Henry Ford believed in first, reduce the price; second, extend the operations; third, improve the article. It was his philosophy that naming a price so low will force everybody in the workplace to the highest point of efficiency. The low price makes everybody dig for profit. Thus, Ford started with reducing the price to the point where he believed more sales would result.)

David Jack (1984) (Research Director of Glaxo):

> You must have at least one fellow who knows what ought to be done and how to do it — and preferably he should be in charge. People can be divided into two classes, simplifiers and complicators — in research you need simplifiers.

Barrie Marson (1984) (Chairman of The Oxford Instruments Group):

> . . . new or innovative products are rarely initiated within the organisation itself; virtually every significantly new product has grown from initial work in a university or research organisation. . . . The management of an innovation from start to finish calls for three different types of people. At stage one there is the inventor, at stage two comes the developer and at stage three the manager, maybe of a separate company, who is going to handle the product through production and marketing. [He concludes:] The UK and Europe are not good places to introduce new products, the people are conservative and there is not much money. The best plan is to develop in the UK, launch in the US, and if it goes there it should go in Europe.

Ronald Denny (1984) (Managing Director of Rediffusion):

> Ideas come from people not organisations. The task is not finding ideas but spotting potential winners. Therefore the criteria for successful product innovation include:
> 1. The validity of the market must be tested as well as the feasibility of the product and its development cost.
> 2. Market timing is also a vital factor.
> 3. Finally, one has to be sure that the product will fit the group's strategy and, if so, whether it can be handled by the existing business unit or whether a new one will have to be called into existence.
>
> He concludes: innovation is a creative activity; instant profit is not the be-all and end-all.

Finally, writing on 'Management for profit — by product design', Turnbull (1984) propounds five steps, essential to the successful management of design. They are:

1. Define the product range in accordance with company strategy.
2. Make sure before proceeding, that a product champion is appointed. In many cases, this can be the key to the success of the new product; if no-one is enthusiastic about it, it is unlikely to succeed.
3. Look at the best ways of implementing new techniques where appropriate.
4. Set up a new multidisciplinary team for each new product under the product champion.
5. Conduct a general review before being hindered with detailed specifications. This concept phase is intense, and distraction must be avoided.

All of the aforementioned are ways of managing design effectively as documented in studies previous to this. In the following section we shall report our findings on how the Queen's Award winners manage design. This will be presented in two phases. The first highlights some points that suggest inadequate design management or the neglect of design in the firms investigated. The second presents the different methods of managing design in the award-winning firms, to enhance their competitiveness in international markets.

Survey results

Some elements of inadequate design management

Design is the lifeline of modern progressive industries, and its role in international competitiveness is growing significantly. However, a 'balanced' design, particularly in consumer products, may not be successfully accomplished without an effective integration of the skills embodied within the practice of industrial design and engineering design.

An investigation into how the Queen's Award winners for export and technology managed design revealed that significant differences existed in terms of the rate of employment of industrial designers and engineering designers for the purpose of designing and developing new products, satisfying industrial design needs, managing design itself, and indeed, educational status.

Of the ninety-three respondent firms, 77 per cent employed engineers for the purpose of design and development of their new products, compared with 34 per cent who made use of the services of industrial designers. Similarly, the question that sought to determine how the firms' industrial design needs were satisfied showed that 50 per cent of the sample (fourteen firms) satisfied their industrial-design needs by drawing on the knowledge of an engineer member of the design team who is experienced in industrial-design techniques, or by employing engineers with knowledge of industrial design, rather than employing professional industrial designers themselves. Only 29 per cent of the sample made use of professional industrial designers. As argued in Chapter 6, knowledge gained through an engineer's self-practice of industrial design should not be confused with that gained through pure professional industrial-design training; the reason being that both professionals have different educational backgrounds as will be attested to later in these findings.

Further, the question which sought to establish how many engineering designers and industrial designers were employed in-house revealed that seventy-eight respondent firms employed a total of 19,904 engineering designers and industrial designers. Of this number, 19,449 (97.7 per cent) were engineering designers. In marked contrast, only 455 (2.3 per cent) were industrial designers. This calculates to a ratio of 43:1, suggesting that for every forty-three engineers employed in-house in this sample, only one industrial designer was employed — indicative of the apparent neglect of the role of industrial designers in the design and development of manufactured products.

Overall, 71 per cent (55 out of 78) of the sample did not

employ industrial designers in-house, compared to only 8 per cent (six firms) who did not employ engineering designers (see Table 7.1). Yet as demonstrated in Chapter 6 (Figure 6.1, p. 254), the skills of engineering designers and industrial designers overlap in many product areas — electronics, computers, machine tools, domestic appliances, to name a few. Regrettably, these are the product areas where many UK companies do not compete effectively in international markets (NEDO, 1983).

It is also clear from Table 7.1 that those who default in the employment of industrial designers are mostly medium and large companies. Again, these are companies who mostly compete internationally and thus are vulnerable to 'attacks' in international markets. These are companies who mostly lag behind their major foreign competitors.

Next, respondents were asked: How many engineering and industrial designers are college/university graduates (or equivalent)? Of the 19,449 engineering designers employed in-house, 85.6 per cent were university or college graduates. In sharp contrast only 16.4 per cent of 455 industrial designers employed in-house were college or university graduates (Figure 7.1).

Although industrial design and engineering disciplines are often perceived as different, they are none the less dimensions of the same activity. This suggests that any 'divide' between these two overlapping functions are likely to do damage to British industrial society and thus calls for adequate management of design to minimize and where possible, eliminate such problems.

Finally, the question that sought to establish who was responsible for managing design in the award-winning firms produced the following results. Of those who managed design in a sample of ninety-two firms, 45 per cent were technical directors, 26 per cent and 15 per cent were R & D directors and production directors, respectively; whilst the 'chief engineer' managed design in 25 per cent of the cases.

The least-mentioned personnel assigned the responsibility for managing design in the responding firms were: industrial-design director (2 per cent), purchasing director (1 per cent), and financial director (1 per cent). Although these are the least frequently named personnel for the purpose of managing design, they nevertheless present interesting results. As expected, fewer firms assigned the management of design to the industrial-design director, suggesting the apparent neglect of the role of industrial designers in the development of new products in this sample. It is, however, surprising that the financial director was given the responsibility to manage design, yet both functions appear to be

Table 7.1 Engineering designers and industrial designers employed in-house

	No. of firms employing engineering designers in-house	*Actual no. of engineering designers employed in-house*	*No. of firms employing industrial designers in-house*	*Actual no. of industrial designers employed in-house*
None employed	6 (8%)		55 (71%)	
1–10	29 (37%)[a]	121	20 (26%)[b]	60
11–20	9 (12%)	142	1 (1%)	20
21–30	8 (10%)	209	—	—
31–40	4 (5%)	150	—	—
41–50				
51–60	1 (1%)	58	—	—
61–70	1 (1%)	70	—	—
71–80	2 (3%)	155	1 (1%)	75
91–100				
81–90	1 (1%)	90	—	—
101–200	7 (9%)	1,210	—	—
201–300	2 (3%)	550	1 (1%)	300
301–400	1 (1%)	350	—	—
401–500	1 (1%)	500	—	—
601–700	2 (3%)	1,294	—	—
701–800	2 (3%)	1,550	—	—
901–10,000	2 (3%)	13,000	—	—
Total	78 (100%)	19,449	78 (100%)	455

Note: a. To be read: 29 (37%) of the responding firms stated that they employed between 1–10 engineering designers in-house
b. To be read: 20 (26%) of the responding firms stated that they employed between 1–10 industrial designers in-house.

Figure 7.1 Bar charts of engineers/industrial designers employed in-house and level of education. *Source:* Ughanwa 1986

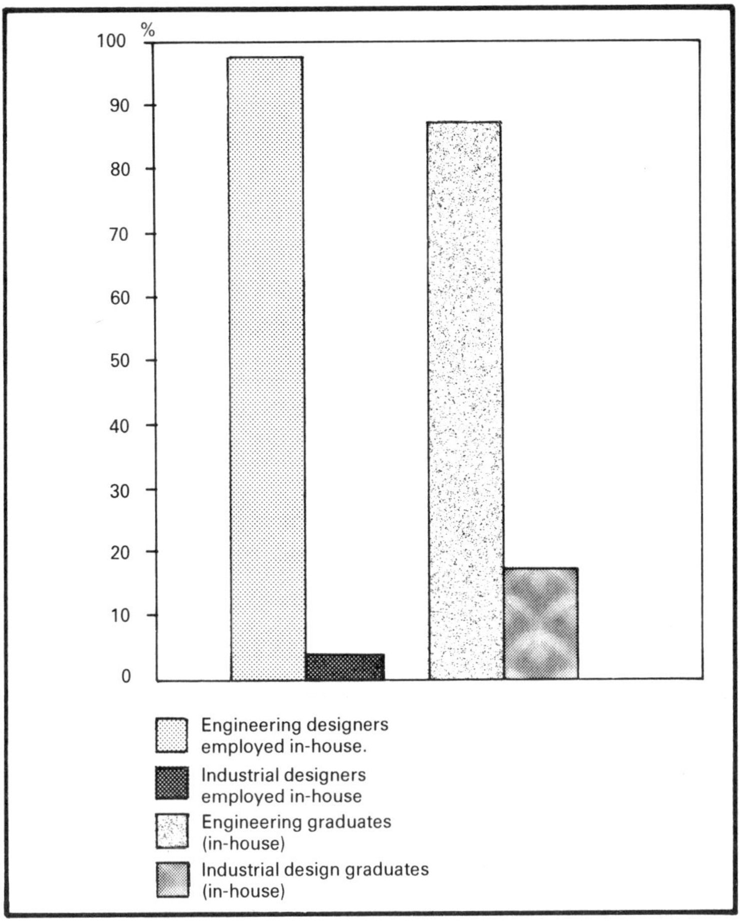

unrelated. However, since only one firm was involved, it may well be that this firm was heavily involved in cost-reduction exercise as a competitive strategy, and found that competence in their financial director, hence he was asked to manage design. Similarly, the purchasing director managed design in one firm, yet 'choice of materials' (a factor that is within the portfolio of the purchasing function) was considered by respondents as the most important factor when taking product decisions. Perhaps, this may be explained by the fact that the 'choice of materials' is but one aspect of design activities.

Indeed, in some firms design was not managed at all. According to one respondent (a marketing controller), 'We are not a manufacturing organisation. We are only a retail organisation.' This may be associated with the still-lingering belief among innovators that the management of design is only for manufacturing firms and manufacturing firms alone. In our opinion this is wrong. Design can also be managed by non-manufacturing organizations. There are many dimensions of design activity that vitally require management to enhance the overall good management of design. For instance, the management of materials that featured prominently under automation (see Chapter 6) and the management of design-related factors such as delivery and after-sales services, or design-influenced factors such as value engineering/value analysis are all implicit in design management. In the service/overhaul industry, for example, the adequate management of service equipment implies cleaner car engines, which in turn, increases the engine efficiency, reduces fuel consumption and maintenance cost, and enhances the durability of the car. Since all of these factors are largely influenced by design, it follows that their proper management implies the adequate management of design itself. Even salespersons, customers/users, retailers, etc., can contribute to the effective management of design through suggesting ways (e.g. via specifications) of improving an existing design or leading to the design of new products, hence it is hypothesized that design can be managed by organizations other than manufacturing firms. As we have frequently observed, confusion as to the nature of design is all pervasive.

While it is not argued that engineers and industrial designers should be given a 'fifty-fifty' treatment in all cases, the attention given to engineering *vis-à-vis* industrial design appears to be so disproportionate in terms of education, employment for the purpose of design, and development of new product, or indeed, for satisfying industrial design needs that it does not seem to create a conducive atmosphere necessary for formulating an organizational framework whereby the two disciplines can mutually work together to create a balanced product — a product good enough not only to compete on equal footing but also to excel those of the major foreign competitors in international markets.

These elements of inadequate design management are corroborated by the following selected excerpts from respondents' comments:

1. The key to successful, effective design is the integration of design, development, production, manufacturing, sales and product support engineers with minimal boundaries. But the established UK practice is to divide engineering into quite distinct groups and particularly into product and production engineers. This must give way to a coherent engineering force who together see a product through from 'cradle to grave'.
2. Too many design groups have performed badly because they have reported to the marketing director in companies where the skills are concentrated on commercial aspects. It is imperative that the leader is skilled in both technical and commercial functions and has the capability to plan for the future.
3. I can only state that in my view if management concentrated on the simpler points of life in getting the product right and gradual improvement will be of in-house experienced engineers producing the desired moulds, we may end up by selling more. Regrettably, large slices of UK management that we read about are those involved in satisfying the shareholders with growth by acquisition, shutting of factories rather than steady change. The bulk of UK factories are working and designing to traditional basic concepts; unfortunately, the 'British' method of being different, leads to widespread importation even if UK goods are available to correct prices and design.
4. Design has been neglected in this company .... We intend to strengthen our design team by recruitment of a degree qualified engineer who can co-ordinate our activities more effectively. We believe that new technologies must be introduced in the form of CAD, Robotics, and the use of Automatic Assembly Equipment .... We will concentrate on design for ease of assembly and test at low component cost and high reliability.
5. We do not design hardware for sale. We design only process plants. (Consequently, this respondent did not suggest any method of managing design as he felt that the management of design was unnecessary in his company.)

Although all the firms investigated were winners of the Queen's award and thereby judged to be commercially successful, it is apparent from the above findings that design is ill-managed in some of the firms. This may probably be explained by the fact that design management is only one aspect of a company's overall commercial activities necessary to achieve international competitiveness.

The following section will present ways suggested by the Queen's Award winners in which the management of design may be improved, to enhance the competitiveness of firms in manufactured products/exports.

Better ways of managing design: the innovators' view

The following ways in which the management of design may be improved were suggested:

1. Better definition of design.
   This is essential because design is of vital importance even when engineers are not involved (e.g. in cosmetics, it is the chemists and marketing men who rule; in the catering/food processing business, the food technologists are 'King'; and in the pharmaceutical industry pharmacists, 'medics' etc., are the principal designers).
2. Top management involvement and commitment.
   (a) Design must be accepted by top management — adequately funded and professionally managed and staffed, to maximize achievements with regard to time scale, technical, quality assurance, produceability, maintainability, and reliability.
   (b) All new design and innovation must be stimulated and initiated by top management.
   (c) Strive to continually improve product, reduce cost and weight.
   (d) Believe the criticism of own product by users and be prepared to change it.
   (e) Have a formal product plan for the future with input from marketing, manufacturing, etc.
   (f) Recognize that collaboration and adoption of outside ideas and licences have a major role in business expansion and successful design.
   (g) Be closely involved in, and committed to, R & D in the company, the requirements of the company's customers, and trends and developments throughout the industry, throughout the world.
   (h) Implement good product research.
   (i) Be more aware of, and committed to, good design.
   (j) Road blocks to successful design must be effectively addressed by senior management.

- (k) More financial backing for and stamina and drive to go through the design process.
- (l) Management must be committed to creating good design.

3. Effective co-operation and co-ordination.
   - (a) Ensure that a close liaison exists between sales, marketing, and R & D to assure the real market needs are known and understood.
   - (b) Ensure that a close relationship is fostered between quality assurance, production, and R & D at the development stage of products.
   - (c) Ensure total co-ordination with marketing/sales, and accounting functions in order to define and monitor objectives.
   - (d) Ensure more co-ordination between R & D, engineering, and marketing.
   - (e) Ensure a very close co-operation between design and production engineering (manufacturing) departments.
4. Employ the best people.
   - (a) Utilize designers with practical experience, i.e. having spent some considerable time on various shop-floor departments before being allowed to design.
   - (b) Recruit graduate engineers/industrial designers, or equivalent to co-ordinate design activities more effectively.
   - (c) Hire young people, encouraging enthusiasts and champions.
   - (d) Find people who are capable of creativity; if you have these, management (of design) thereafter becomes relatively simple.
   - (e) Practising professional engineers must have a sound grounding in the understanding of engineering practices and materials.
   - (f) Employ innovative designers and pay them to design (not seek increased financial reward by becoming managers).
   - (g) Ensure better selection of those with talent.
   - (h) To succeed in the marketplace, the product must meet the customer requirements and this requires a good level of technical expertise/innovation.
   - (i) Use a team with interrelated skills who work as a group rather than being segmented.
5. Motivation.
   - (a) Create personal identity with design (e.g. in the way

architects have their names recorded on a plate attached to structures they design), i.e. recognition.
- (b) Create a better environment for team working/briefings, etc., and continual communication of project status and other related and important aspects.
- (c) Give better recognition to good design. (This tends to boost the morale of the designer(s).)
- (d) Ensure high status for engineers and manufacturing industry.
- (e) Give better pay for performance.
- (f) Give more status to R & D in an organization as it is the future of that organization.
- (g) Create a willingness among suppliers to design better products.
- (h) The management philosophy/style should create an environment suitable for engineers/scientists.
- (i) Keep R & D committed to the product to ensure success and market fit.
- (j) Creative ability should be given full encouragement by the Board.

6. Continuous design evaluation and review.
   - (a) Ensure more widespread adoption of design evaluation and critical review procedures with manufacturing participation at an early stage.
   - (b) Pay more attention to life-cycle aspects of design.
   - (c) Ensure regular design and progress reviews by senior management.
   - (d) Review of design concept during initial stages must be supported at a senior level by all departments that will eventually be involved after design release.
   - (e) Good management of design involves ongoing design and advanced development, as well as the integration of computer-aided equipment/computer and design systems (particularly in the 'high tech' industries).
   - (f) Continued participation throughout the product life may prove advantageous.
7. Effective communication.
   - (a) Improve communication between marketing, design, and production functions.
   - (b) Keep information freely available on products/markets/needs/progress.
   - (c) Provide better marketing information and specification on a worldwide basis.

(d) Shorten the communication link with marketing/market/customer.

8. Clear identification and understanding of the market/customer.
   (a) Positively identify the market.
   (b) Positively identify the product(s) for the market, including price and essential functions.
   (c) Gain a clear understanding of the product target market.
   (d) Gain a clear understanding of customer needs and customers' business in the target market.
   (e) Ensure product requirements are well defined. (The earlier stages of better marketing will provide a clearer definition of requirements.)
   (f) Design must not be an end in itself, but must be targeted at a current or forecast market need.
   (g) Focus on real market-driven needs.
   (h) Better market research, (i.e. in-depth market study).
   (i) Success or otherwise depends on a knowledge of the market.
   (j) Products have evolved to meet market needs over the years.

   This suggests that better management of design correlates with 'creative imitation' (improving existing products). In other words, design may be better managed through evolution rather than revolution.
9. Closer interaction with customer/user.
   (a) Designer involvement is necessary in customer meetings and trade shows.
   (b) Ensure more travel and contact with markets and users worldwide.
   (c) Ensure more and constant interface with the market.
10. Create overall design awareness.
   (a) Achieve by simply an awareness throughout the business of the role for design in helping the company meet its market needs.
   (b) Awareness by general management of the benefits of computer-aided equipment in the design of new products.
   (c) Introduce the concept of 'design quality'.
   (d) Clarify what is meant by 'quality', e.g. Is it the feel? Is it freedom from defects? Is it reliability? Is it meeting customer requirements? Is it features?
   (e) Quality of product is the concern of all and designs do

need support/comments at a stage that impact can be effected — customer support, commissioning, purchasing, manufacturing, etc.

(f) Create a design awareness among buyers and salespeople.

(g) Ensure that the design process involves the whole organization and is not just left to the 'technical' department.

(h) Design should be involved at the initial product concept, in order to improve the product quality.

11. Adopt a multidisciplinary team approach.
    (a) Ensure that there is a multidisciplinary approach to new-product development and commercialization.
    (b) Develop team commitment across departments.
12. Effective use of project management.
    (a) Use critical path analysis (CPA) in design projects. This among other things, enables projects to be finished in time and delivered in time.
    (b) Use 'matrix management' of different design disciplines.
    (c) Use formal working standards and methodologies, including critical reviews of original work by a peer group.
    (d) Divide innovation into manageable modules that can be designed, implemented, and tested independently.
13. Overall departmental involvement.
    (a) Involve other departments (i.e. non-R & D) at an early stage (investigation/definition), namely marketing, manufacturing, and quality assurance.
    (b) Involve industrial/production-engineering personnel at an early stage to ensure that the product designed can be made effectively.
    (c) Ensure clear and greater involvement of design with sales/marketing, to enable the design team to be freely aware of what the market/customer requires and of the problems that must be overcome.
14. Design with ease of manufacture and marketing in mind. During design the following factors must be more carefully considered:
    (a) Ease of manufacture/cost of manufacture.
    (b) Ease of assembly and test at low-component cost and high reliability.
    (c) Ease of marketing (performance in operation, reliability, durability, etc., i.e. competitive whole

product/unit cost).

(d) Emphasis on operability and efficiency of equipment

(e) Design for reliability, function, style, manufacturability, cost, and serviceability.

15. Introduce new technology.

(a) Use advanced design automation equipment (e.g. CNC machine, ATE, etc.) to aid design of new products.

(b) Use more computer-aided design and computer-aided draughting to reduce lead time, to ensure more accurate design of products, and to reduce the number of components (e.g. per set).

16. Adequate education and training.

(a) Better teaching of design at undergraduate levels using case-study material is needed.

(b) Better training of good people by good people is needed.

(c) Identify people with good technical grounding who have the necessary flair and charisma to be trained managerially rather than to expect a good designer to automatically make a good manager.

(d) To obtain a good technical grounding in industrial/engineering practices, materials etc., designers must undergo full training, some of which 'requires months of unglamorous hard work in poor environment' or 'cloistered surroundings'.

(e) Provide the motivation to make this happen, i.e. to make time and conditions acceptable to trainee designers. 'Often people's ambitions and expectations make the time and conditions unacceptable' to designers, implying that engineering/industrial designers are not trained correctly in some UK industries.

(f) Better education and training of designers is needed. In some industries, 'the basic engineering knowledge, ability and experience is lacking in young designers who in many cases have no idea how to make the results of their designs'.

(g) Production and commercial requirements are essential before any engineer is allowed to work on a design task.

17. Input of sufficient resources.

(a) Input adequate resources into design projects and in good time too. ('All too often input is too late and then cannot be effective'):

- More thinking/alternatives
- More time
- More money
- More manpower/expertise
- More machines (CAD/CAM, CNC, ATE, etc.)
- More materials
- More information
- More technical know-how, etc.

b) 'In cases of resource shortage industrial design is viewed as a luxury. This is intellectually incorrect but in many cases all too true in practice.' There must be a shift from this practice to give way to effective management of design.

(c) Company structure and investment decision-making processes should be attuned to the company's need to have 'good design'.

18. No 'nonsense' objectives.

(a) The objectives of the product design should not be compromised for expediency of early introduction, since once a new product is introduced it is then more expensive to reverse the design in production. In other words, 'do not launch immature product'.

(b) Ensure the 'exact definition of the product brief, including financial return'.

(c) Adhere to 'strict control of development process, including function and cost'.

(d) 'Management by objectives (MBO), agreed and shared by all members of a design team is the prime aim.'

(e) Set clear targets.

(f) Monitor at top level.

(g) Design to meet the need.

(h) Do *not* overdesign.

(i) Clarify goals/specifications and manage deviations.

(j) Do not commit to manufacture until the problems with design are solved.

(k) Ensure accurate definition of objectives.

19. Bridging the gap between design and other functions.

(a) The key to successful, effective design is the integration of design, development, production, manufacturing, sales, and product support engineers.

(b) Coherent engineering. The established UK practice is to divide engineering into quite distinct groups and particularly into product and production engineers. This must change to give way to a coherent engineering force who together see a product through from 'cradle to grave'.

   (c) Greater integration (of design) with marketing departments, but marketing emphasis should be strategic not commercial.
   (d) Marketing-design/R&D interface. The bringing together of these two departments is a simple but crucial operation.
   (e) Industrial designers (even when external consultants) must be integrated into the design, manufacturing, and marketing team.

20. Encourage in-house design.
   (a) A strong, well-qualified, experienced, competent, in-house team of designers/engineers is of the greatest importance to ensure technical commercial leadership, and to properly assess worth and potential of external proposals, as well as professionally introducing them.
   (b) In-house innovation must continue as part of the forward plan (to improve upon design management).
   (c) Management should first concentrate on the simpler points of life in getting the product right, then through gradual improvement in-house by experienced engineers the desired moulds can be achieved.
   (d) Basic design is carried out in-house, and only where aesthetic design is required do we employ outside consultants.
21. Change of attitude
   (a) If the individual charged with the general management function recognizes the essential need for both new and developing products, then it becomes part of the corporate objective.
   (b) The bulk of UK factories are working and designing to traditional basic concepts; unfortunately, the 'British' method of being different leads to widespread importation even if UK goods are available at correct price and design. Management should, therefore, attempt to fall in line with the conventional (modern) design concepts.
   (c) Large slices of UK management that we read about are those involved in satisfying the shareholders with growth by acquisition, the shutting of factories rather than steady change through gradual improvement (evolution). This suggests that management seem to be myopic about the great impact of good design on these

so-called company objectives/goals (e.g. growth by acquisition). This is so because the effective management of design can help to achieve all of these objectives, including opening new, rather than shutting the existing factories. For example, if design is right, *ceteris paribus*, sales will increase, profits grow, acquisition becomes more feasible, existing factories sustained and new ones built (if need be), would-be redundant employees retain their jobs, and shareholders get even 'fatter' dividends.

It is pertinent to note that the best acquisition seems to be that made with company profits, not by borrowing. This is because borrowing *per se* is a risk and borrowing money to acquire another company implies increasing business risk.

On the other hand, the best dividend appears to be that based on growing and sustained profit not profit made through cost savings from closed factories or rather profits made at the expense of victims of the shut-down factories.

22. High-quality leadership.
    (a) Establish an effective reporting system. 'Too many design groups have failed because they have reported to the Marketing Director where the skills are concentrated on commercial aspects.'
    (b) It is therefore imperative that the leader is skilled in both technical and commercial functions and has the capability to plan for the future.
    (c) Top management must be innovative.
    (d) Ensure effective delegation of responsibility so that individual team members are committed to the team's goals and don't feel creatively inhibited.
    (e) Sow design 'seeds' in fertile design 'fields', i.e. invest in design where there is:
        — enthusiastic, effective, and encouraging leadership;
        — open management style;
        — company pride;
        — job pride;
        — product pride.
    (f) The day-to-day management of design should be simplified from the need to struggle to convince your peers, to a position of working collectively to a common goal, with all benefiting.
23. Honour the marketing concept.

'Do not design for design's sake. If you can't sell it don't design it.'

This is another way of saying, design what people (customers/users) want to buy, not just what you like designing.

24. Collaborative arrangements.
    (a) Joint co-operation between industry and colleges/universities can provide more opportunity for undergraduates to become familiar with the design process through vacation work.
    (b) Greater co-operation with the end user of the product is necessary.
25. Create cost awareness.
    (a) To succeed in the marketplace, the product must be available at a lower cost than other solutions/competitors. To achieve this, 'it is necessasry to establish a targetprice (and then carry out cost-reducing exercises to arrive at the lowest (compliant) solution. ('It is in this area that much of UK industry is weakest'.)
    (b) 'The value engineer makes the designer cost-aware and then lets him use his imagination.'
    (c) Expertise must be developed in specialist areas and reliable components developed for use (against stringent criteria) for application to a whole product-line range to give the benefits of high volume and lower cost.
    (d) Education into aspects of total economy cost is needed.
26. Do not be discouraged by failure — try again!
    Essential to persist despite setbacks.
27. Government support.
    (a) Support (financially) the design-commitment drive throughout industry, to enable companies to promote the design message downwards and outwards by holding seminars for principal suppliers.
    (b) Create a 'National Design Awareness Campaign' emphasizing the role that designers can play in commercial success.
    (c) Introduce 'design management packs' for managers and designers in large as well as small companies, to aid the effective management of design.
    (d) Increase the 'Support for Design' — a Department of Trade and Industry-backed programme of subsidized consultancy for small and medium-sized companies

(administered through the Design Council).
    (e) Establish 'design-management quality standards' to assist companies more broadly in establishing better design-management procedures.
28. Introduce problem-tracking/management very early in design.
29. Avoid compartmentalization of the design team. Make them feel responsible for manufacture and service.
30. Learn how Britain's most successful foreign competitors manage design and apply the lessons.

In all, about 140 ways of managing design effectively were suggested. These have been analysed and summarized in Table 7.2.

Clear definition of objectives

In the survey, 43 per cent (26 out of 60) of the respondents considered the setting of clear and definite objectives as a better way in which the management of design may be improved. This is important because the achievement of success in an organization is, in most cases, not a one-person task, but a collection of individual efforts. This means that the objectives of the company must be clearly spelled out to all workers so they know at the outset what they are expected to achieve. Better definition of design, for example, is of particular importance in the light of different professionals — chemists, engineers, technologists, etc. — involved in the design of new products. Similarly, providing the exact definition of product briefs, including financial returns, setting clear and achievable targets, clarifying goals, specifications, and so on, will enable workers, particularly design teams, to organize their efforts towards the achievement of the overall goal of the company.

Further, because design is the concern of all in the organization as stated in Chapter 4, there is the need to create both design and cost-awareness throughout the organization, aimed at improving the quality of the product.

In the same vein, the continuous evaluation and review of the existing products must remain in the forefront of a company's objectives, particularly in this highly dynamic environment where tastes change rapidly, so the changes in tastes/preference may be updated. This is also vital in the light of the emergent new technologies that could render the existing products obsolescent if they are not continuously evaluated, reviewed, and updated.

Table 7.2 Suggested ways in which the management of design may be improved

		No. of responses	*Total responses*	*Total %*[c]
Clear definition of objectives	Better definition of design	1[a]		
	No-nonsense objectives	8		
	Create overall design awareness	7		
	Create cost awareness	4		
	Continuous design evaluation and review	6	26	43
Market and technological needs	Honour the marketing concept	1		
	Clear identification and understanding of the market/customer	8		
	Closer interaction with customers	4		
	Design for ease of manufacture and marketing	4		
	Introduce new technology	3	20	33
Effective co-operation, coordination and communication	Effective co-operation and co-ordination	4		
	Effective communication	4		
	Effective use of project management	2		
	Introduce problem management tracking	1		
	Adopt multidisciplinary approach	2		
	Bridge the gap between design and other functions	4		
	Overall departmental involvement	3	20	33
Management efforts	Top management involvement and commitment	11		
	Change of attitudes	2		
	Input sufficient resources	4		
	Employ the best people	10		
	High-quality leadership	5		
	Motivation	8		
	Adequate education and training	7		
	Collaborative arrangement	2		
	Government support	1		
	Encourage in-house design	3		
	Avoid compartmentalization of the design team	1		
	Learn lessons from foreign rivals and apply them	1		
	Keep trying despite failure	1	56	93
		122[b]		

Note: a. One firm suggested 'better definition of design' as a way in which the management of design may be improved.
b. A number of firms suggested more than one way in which the management of design may be improved.
c. Based on 60 firms who responded to this question.

Improving competitiveness

It seems clear from the previous discussions that setting clear objectives is a necessary ingredient for achieving commercially successful products.

Market and technological needs

In the survey, 33 per cent (20 out of 60) of the respondents were of the opinion that paying adequate attention to the market and technological needs will enhance the management of design. This is an interesting finding, because almost all the objectives discussed earlier appear to be geared towards achieving commercially acceptable products. To produce acceptable products, attention must be paid to the market needs and indeed to new technologies — CAD, CAM, automation, etc., with which these needs may be exploited effectively.

In the end, it is the customer who will decide whether or not the product will be accepted. There is, therefore, the need to clearly identify and understand the market/customer needs which, in turn, implies that firms should research the market in depth and interact closely with customers in order to design and develop new products in line with customer/user specifications/requirements as well as for ease of manufacture.

In sum, interacting closely with customers, identifying and understanding their needs/wants, and designing for ease of manufacture and marketing with the aid of new technologies are all necessary instruments for improving the effectiveness of new-product designs.

Effective co-operation, co-ordination and communication

Of the firms who suggested ways of managing design, twenty (33 per cent) believed that design management may be improved through effective co-operation, co-ordination, and communication between and among functional areas — R & D, design, production, marketing, etc. This is a particularly important finding. Since design is the concern of all in an organization as alluded to earlier, it follows that the effective co-operation and co-ordination of efforts as well as communications between departments or relevant teams involved in the design project is of paramount importance. In this way, rapport and team spirit necessary for working on design projects may be established. This is especially important in design projects where division was found in the literature to exist between engineers, industrial designers, and sometimes between R & D, production and marketing. Improving communication between the relevant functional departments is also vital, to ensure that the necessary information required for the execution of design projects is provided and that everyone involved in the project is fully informed of what needs to be done to achieve good results.

Similarly, the effective use of project-management techniques, tracking problems very early in design and controlling them, and adopting the multidisciplinary approach for design projects, all seem to require co-operation, co-ordination, and communication between and among the relevant functions to succeed. This suggests the need for an overall departmental involvement to improve the management of design. In turn, this seems to suggest the need for greater integration between design and other functions.

In sum, the management of design may be greatly improved through effective co-operation, co-ordination, and communication both between and within functional departments.

Management efforts

In the survey, fifty-six (93 per cent) of the respondents considered management efforts as the best way in which design may be improved. It is not surprising to see that a larger proportion of the respondents opted for management efforts as a way of improving design management. This is because defining the objectives of the company, encouraging the search for market and technological needs, and promoting co-operation, co-ordination, and communication between departments appear to be largely dependent on management efforts. In other words, the success of the activities in an organization appears to depend on the extent to which the top management get themselves involved and committed. This may involve some change in management attitudes (e.g. attitudes to investment). This demands that adequate resources be put into design projects and indeed other related activities, and in good time, to enhance the overall success of design management. Among such resources are: more technical know-how, time, money, manpower, machine, materials, information, etc.

In addition, the right people must be employed to use such resources optimally. They include people with practical industrial experience, leadership qualities, talent and enthusiasm, and product champions and the like. Further, these people should be adequately educated, trained, and motivated to achieve success in the management of design and thus in manufactured products.

It is also the responsibility of the management to arrange for joint co-operation between the firm and the end users of the product, colleges/universities, etc., to seek government support (e.g. financial), and to encourage in-house design as part of a forward plan to enhance technical and commercial leadership in manufactured exports. However, in-house design, does not necessarily mean locking up designers in the studios! As stated earlier, they have to go out to see things for themselves — interact with customers and production people, and monitor the progress of their new-product designs, with the overall objective of improving upon them or designing further new ones in accordance with the customer purpose.

Also, management should attempt to send people abroad (the design team, managers, etc.) to learn how Britain's most successful foreign competitors manage design with a view to applying the lessons to suit the British industrial environment.

Finally, it sometimes happens that management efforts to improve upon their new-product designs may fail which, in turn, may result in lack of competitiveness. In such circumstances, the management should not be discouraged, rather they should stand firm and resolute, learn lessons from their failure, and try again. Business is a risk and the biggest risk is not taking a risk at all!

The fact that some of the thirty ways suggested in Table 7.2 have been grouped or singled out as significantly important in the successful management of design does not mean to say that the other factors are not important. Far from it. Each suggested method is important and should be considered carefully in the light of its contribution to the overall management of design. It is our considered view that each method or a combination of methods when carefully and optimally applied is likely to result in a more efficient and effective management of design. This reasoning is based on the following parameters:

(1) All the firms (ninety-four) under investigation in this study are Queen's award winners.
(2) All the Queen's award winners in this study are judged as being commercially successful (see Chapter 6).
(3) All the suggested ways are assumed to be based on each individual respondent's experience. In other words, all the answers are based on the existing practice in the company.
(4) The majority (63%) of the questionnaires were completed by company directors, and the remainder by senior managers. Since top management support has emerged significantly as a major contributor to the successful design and development of commercially successful manufactured

products throughout this study, it seems logical to accept the validity of the answers presented in Table 7.2 and detailed previously.

If the foregoing argument carries conviction, then our hypothesized relationship between the effective management of design and commercially successful products is strongly confirmed. In sum, the management of design appears to be efficient and effective when top management are involved in, and commit themselves to design projects, employ the best people to manage the projects, set clear and achievable objectives, design to suit the market/customer needs, introduce new technologies to improve the effectiveness of new-product designs, and encourage effective co-operation, co-ordination, and communication between and among functional departments.

It is clear from the overall findings that design management plays an influential role in the competitiveness of firms in manufactured products. Both the strengths and weaknesses identified in the samples investigated, relative to design management, all suggest the greater need for effective design management, so acceptable, competitive, and commercially successful products may be achieved. This implies that innovators/firms must think positively towards the effective management of design. In other words, it is important that innovators 'think design' at all times to achieve good results. It is this dimension of design awareness that the next section will address.

Thinking design

It is evident from Chapter 5 that value engineering/value analysis, aesthetics, ergonomics, quality circles, timing, leadership, and attention to user needs (technical and economic), all influence the design of new products or redesign/modification of existing ones. It follows that to be commercially successful in the marketplace, a firm should be design-driven, particularly in design-orientated products. That is to say, a firm should attempt to create an organizational climate whereby everyone should be made aware of the importance of design in the success of the firm and/or international competitiveness.

More importantly, to compete effectively in today's world markets, a firm should get its design right first time. To do this means that the firm should first find out who the customers/users are. What are their needs and wants? When and where will they want and/or need them? The first two factors will enable the firm to design to customer specifications/requirements. The last question will help the firm to get the timing right for the design, development, production, launch, and delivery. This is particularly important, because it is no use launching a product prematurely or in haste only to withdraw; or introducing a product too late, only to discover that your competitors' products have already gained a wider market acceptance that may limit your share of the market and thus, lead to regret. This strongly supports Baily (1978) who argues that 'in practice forms (products) are often designed in haste, to be repented at leisure'.

Further, getting the timing right or getting it right first time appears to be paramount in creating confidence as well as the image of the brand in the minds of customers, hence it is postulated that 'first impressions count'.

It is, therefore, reasonable to conclude that to be commercially successful and/or internationally competitive, it is necessary that firms should think design by creating design awareness between and among all staff from top to bottom. This can be achieved through the proper management of design which, in turn, can be promoted, via adequate management of all technological, management, and social factors that influence the design of new products, such as value engineering, quality circles, aesthetics, ergonomics, and so on.

In the words of Sir John Harvey-Jones (1985) former Chairman, Imperial Chemical Industries, 'At the end of the day the marketplace is the king and the companies which will survive are those that understand best the needs of their customers — sometimes before the customers themselves understand them.' Understanding best the needs of customers implies effective management of design and design-related factors, once again stressing positive thinking about design.

It is, however, pertinent to note that 'thinking design' does not just stop at creating design awareness in the organization. It also involves the understanding of the relationship between design and innovation (both are often perceived as separate entities) or the 'mechanics' of design/innovation process (see Fig. 7.2). In turn, the understanding of the design/innovation concept tends to help innovators to set clear project objectives and definite parameters for effective project (design/innovation) management aimed at achieving commercially successful manufactured products (Fig. 7.3).

Figure 7.2 Design/innovation process and international competitiveness
Source: Ughanwa 1986

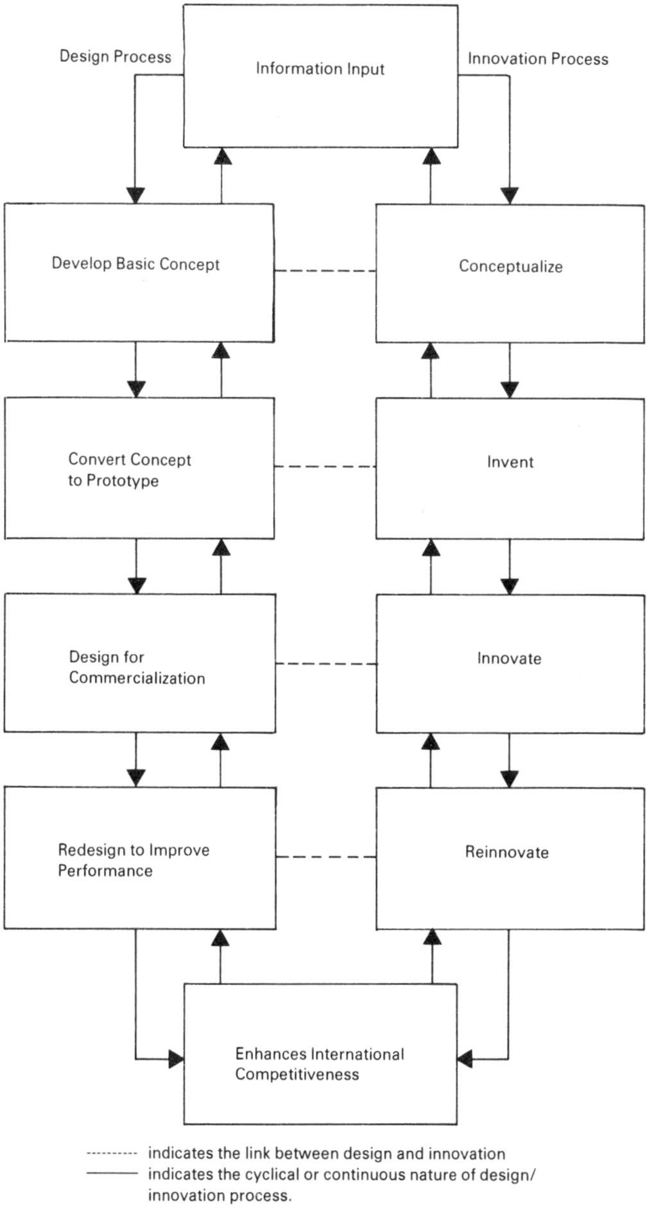

Figure 7.3 The impact of clear understanding of design/innovation concept on international competitiveness. *Source:* Ughanwa 1986

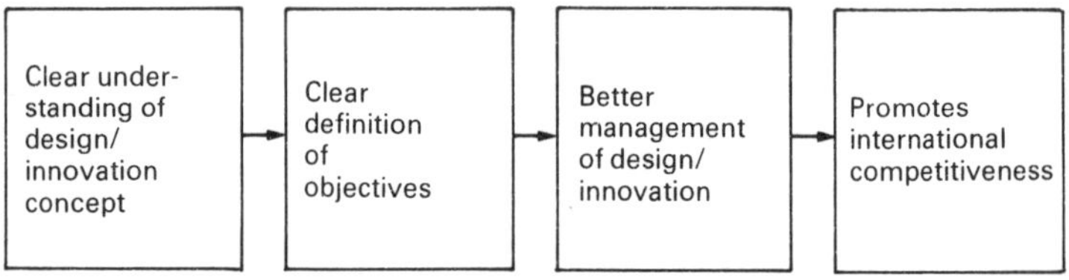

Design/innovation linkage and competitiveness

Traditionally, design is associated with drawing, draftsmanship, industrial designing, etc., while (technological) innovation is associated wih engineering. In other words, design and innovation seem to be perceived as two separate things. But as Moody (1980a) strongly maintains, 'strictly, engineering is design', suggesting that there is a link between design and innovation. Figure 7.2 demonstrates this relationship. It suggests that the development of the basic concept drawn from the available information relates to conceptualization while conversion to prototype correlates with invention. Similarly, designing for commercialization suggests innovation whilst redesigning to improve performance has a re-innovation connotation, suggesting the mutual 'co-existence' between design and innovation. In other words, the activities of design and innovation appear to run concurrently, hence we argued earlier: 'manage design and you manage innovation'.

As the basic idea is developed, converted to prototype, and commercialized (produced for sale in the market), the product performance is monitored. With the performance feedback obtained from customer/users, salespersons, dealers, agents, distributors, representatives, etc., the product is redesigned or re-innovated, or an entirely new one designed. This leads to a better or more acceptable product that in turn, enhances the competitiveness of firms in international markets. The process is then repeated to create products to satisfy the ever-changing tastes/preferences of customers/users as well as to keep up-to-date with new technologies. It is assumed that a firm should be commercially successful as long as it designs to customer/user specifications/requirements, matches good product designs with price, minimizes costs whilst maximizing the value content of a product, and last but not least, updates its technology in line with those of its competitors and/or in the light of the changing environment.

It is therefore argued that a clear understanding of the design concept will enable innovators to 'organize their own thinking' towards their overall approach to design and innovation. It is contended that the clear understanding of the design/innovation concept will help innovators to set clear objectives that, one hopes, will enhance the management of design and promote international competitiveness as shown in Figure 7.3.

Design/innovation checklist

As an aid for monitoring the design/innovation process, a flow chart has been developed as a checklist for designing and developing new products (Figure 7.4).

Figure 7.4 starts with the identification of the user needs upon which the basic concept is developed. This will be followed by the question, is it workable? That is to say, is the basic idea or concept practicable? If the answer is 'no', a new concept will have to be developed. On the other hand, if the answer is 'yes', then the choice of materials, colour, etc., will be made. As argued in Chapter 5, the type of materials and colour tend to influence the design of new products. It was also argued in Chapter 5, that the best alternative approach to design – particularly the design of consumer products, is 'to design from the outside in rather than from the inside out'. There is therefore, the need to have some idea in mind as to the desired appearance and ergonomics before developing the prototype.

Is it technically and economically viable? This question seeks to know the extent to which the developed prototype meets the customer specifications (technical and economic). Technical feasibility is a test of performance, safety, and flexibility in use, reliability and durability, and so on. Economic viability is mainly a test of price. Therefore, if the prototype does not meet the technical and economic needs of the customer/user, it should be referred to the development team for verification. However, if the answer is 'yes', then the product can now be developed in a commercial quantity. This will be followed by assorted technical and economic checks to ensure that the product does not fail market acceptance. These checks include: performance in operation, reliability, ergonomics/fitness for use, aesthetic and graphic features, and price. Is the quality of the product consistent with the price? Is the price within the affordable income of the customer?

Figure 7.4 Design/innovation flow-chart checklist. *Source:* Ughanwa 1986

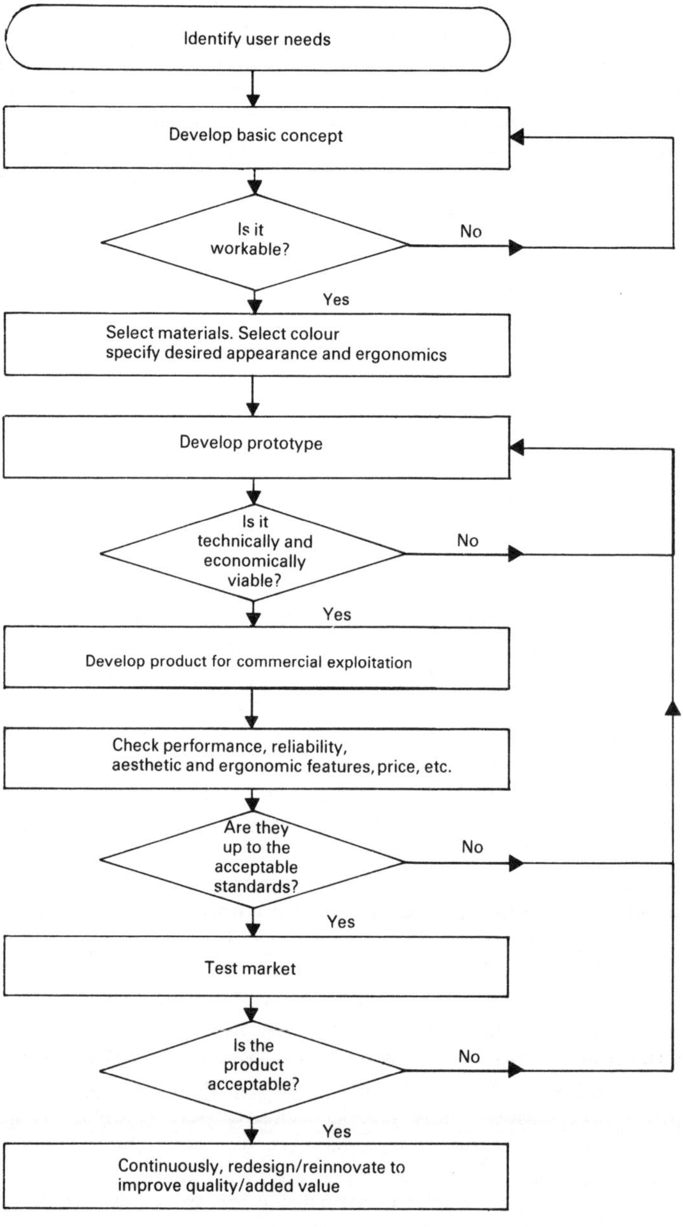

In other words, is the price consonant with the income of the target group?

However, because, in most cases, technical and economic factors have standards (national and international), it is essential to check these standards — safety, reliability, etc., hence the question, Are they up to the acceptable standards? Once again, if the answer is 'no', it should be referred to the development team. On the other hand, if the answer is 'yes', that is, if the firm is satisfied with the aforementioned 'checks and balances', then the product can go to the market for testing.

The final question in this checklist is, Is the product acceptable? This is important because as argued in Chapters 3 and 5, the marketplace is the final arbiter. In other words, in the end, it is the customer/user who has the final say. If the answer is 'yes', that is, if the product is accepted by the customer/user, then the firm should continuously redesign/reinnovate the existing product to improve quality. Aside from improving the quality of the product, the concept of redesigning or reinnovating also appears to derive from the fact that we live in a dynamic environment where tastes and preferences change fast. Consequently, customers/users often tend to be in want of new things to replace the old or to satisfy changes in tastes/preferences.

However, should the answer to the last question be 'no', which means that the product was rejected during the test marketing, then the firm should refer the product back to the development department for review in the light of the feedback from customers/users, sales-persons, distributors, agents, etc.

Conclusions

The overall review of the secondary and primary findings suggests that although the management of design differs from country to country, industry to industry and product to product, to introduce a satisfactory product to the customer/user, the entire organization must work together. It also reveals that the management of design in the UK is not as effective (successful) as it should be when compared with those of the major rival countries. Based on this weakness, a number of conclusions have been drawn, which, it is hoped, will enhance the management of design in British industries and elsewhere:

1. There must be a recognition by all managers 'that design is not a linear process but a cyclical process in which an end point is never reached, some improvement or enhancement

being always possible and desirable, and that the process is iterative, each step being affected by feedback of information from earlier and later steps'. Consequently, design should continuously be reviewed, monitored, modified and where appropriate aborted, and a new project started.

2. There is the need to lay emphasis on industry-linked projects to encourage creativity rather than concentrating on the mastering of techniques and concepts, in design education.
3. There must be close co-operation between design, production, and marketing. In other words, the management of design should not be isolated to the design department alone. There should be collaboration between and among personnel from all departments in the organization.
4. The successful management of design and thus, innovation requires that both technological and human resources be properly deployed. This should be accompanied by clear design objectives.
5. All factors influencing design (aesthetics, ergonomics, value engineering, etc.) should not be treated as separate activities but as part and parcel of the design process.
6. Companies should always aim at balanced design. This means that companies should bear both customer/users and ease of manufacture in mind, when designing new products.
7. Thought should be given to designing products, particularly consumer products, 'with some idea in mind of the desired appearance and ergonomics, and then attend to technical functions (operational functions) of the product'. In other words, to design from the outside in, rather than as many firms try to do, to start with technical aspects and add styling as an afterthought — designing from the inside out. In some cases, however, a mix of both approaches may offer a better alternative.
8. Generally, design should be managed by managers. However, where appropriate, designers should be given the opportunity to manage design. In addition, designers should be encouraged to see new-product designs not only through to, but throughout commercialization and beyond.
9. Design cannot be effectively managed by sticking to old ideas or concepts. Emphasis must be laid on 'diversity, novelty and innovation'.
10. For any of the above conditions to be successfully executed,

someone at the very top must be committed and provide direction. However, it is essential that all managers have an appreciation of the problems involved in managing design.

11. The overall results confirm our view that effective design management leads to better design, greater competitiveness, and higher profits.

In Chapter 8 we offer a model encapsulating all aspects of the design dimension that will enable innovators to improve their competitiveness in manufactured products/exports.

Chapter eight

A model of the role of design in international competitiveness

Introduction

The specific concern of this chapter is to develop a model of association between the various dimensions of design and international competitiveness that will enable innovators to 'model' their own thinking or 'think positive' towards their overall approach to product innovation.

The chapter contains three parts. The first summarizes the study as a whole. The second introduces the model *per se*, whilst the third outlines the objectives accomplished by the model. The implications of the model are discussed and conclusions drawn.

The dimensions of the study

The overall objective of this book as developed through the earlier chapters is to explore the role of design in international competitiveness, with special emphasis on its impact/effect on the decline of British industries in world trade. To establish such an impact it is important to have a background knowledge of the decline of British industries as documented in previous studies. To meet this objective we looked first at seven industries in which Britain has declined competitively when compared with those of the major foreign counterparts. It appears that the inability of much of UK industry to compete on non-price factors (technical superiority, reliability, efficient delivery/after-sales service, etc.) has accounted for the British decline in manufacturing exports.

Consequently Chapter 2 examined the role of price and non-price factors in competition. The result shows that commercially successful firms pay more attention to non-price factors such as performance in operation and reliability, than to sale price when designing and developing commercial products. However, it is evident in our findings that non-price factors alone cannot maximize the sale of a product without compromising with price.

Nevertheless to confirm the validity of the above findings we took stock of the current views of massive penetration of foreign manufactured imports into the UK. This was spelled out in Chapter 3. The rationale was mainly to find out the reason why domestic customers/users were increasingly opting to purchase foreign in preference to home-made goods. The results reveal that British customers are willing to buy from home sources if manufacturers can make products as good as or better than the foreign rival offerings. Regrettably, they tend to perceive home-made products as less appealing in terms of aesthetic and ergonomic characteristics, and overall performance in operation, once again stressing the importance of non-price factors, all of which are largely influenced by design and vice versa.

The characteristics of commercially successful firms provided the focus for Chapter 4. Our analysis shows that a number of design and design-related factors enhance competitiveness. Examples include product uniqueness/superiority, technical complexity, efficient after-sales service, delivery, etc., and led us to explore the factors influencing the design of new products in Chapter 5. Our findings are summarized in Figure 8.1.

The major components of innovation — radical, major, and incremental — were found to influence the design of new products or redesign of the existing ones. Value engineering minimizes costs and maximizes the value of a product. Each process involves changes being made, in one way or another, during the design of new products, hence the thesis that value engineering influences

Figure 8.1 Factors influencing the design of new products. *Source:* Ughanwa 1986

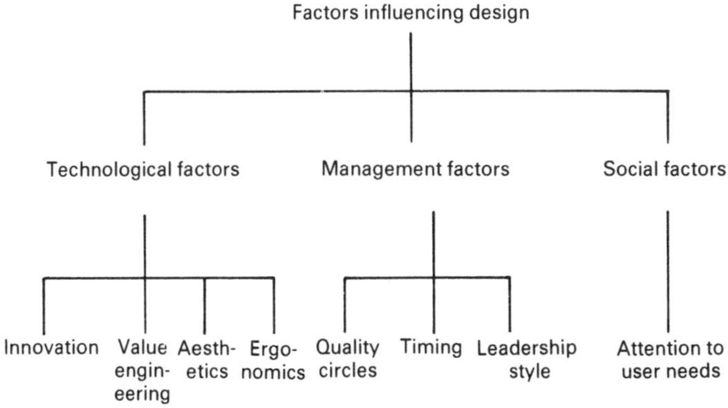

design. Further, it was established that aesthetic features (appearance/shape, colour, style) and ergonomic attributes (control and operator comfort, noise, and temperature) influence the design or redesign of new products. It was also evident in the review that such non-price factors as quality circles, timing, leadership style, and attention to user needs all influence the design or redesign of new products. Given adequate attention to all of these factors, a firm can achieve, maintain, and sustain competitiveness in manufactured products/services.

The concept of design is examined in Chapter 6, aimed at providing a framework to permit better understanding of a complex subject. The results indicate that design is a complex phenomenon that applies to everything and anything we do in life, which probably helps to explain why confusion still seems to exist among innovators as to what design really means. This suggests the need to evolve a model whereby all dimensions of design may be accommodated to help minimize confusion or misunderstanding about the design concept.

Since the review of the literature and our empirical findings reveal that most of the important factors stimulating international competitiveness are design-driven, it became necessary to review the management of design both in the UK and overseas. The overriding result shows that inadequate attention is paid to the management of design in some British industries when compared to foreign competitors. Further, all the identified ways of managing design are basically non-price factors, most of which are influenced by design, once again emphasizing the importance of design in international competitiveness.

From the overall review, the design dimensions contributing to success in manufactured exports can be identified. Figure 8.2 presents these dimensions.

Figure 8.2 suggests that success in manufactured exports is associated with design dimensions such as performance in operation, design-related dimensions such as quality of after-sales service, and design-influenced dimensions such as aesthetics. As alluded to earlier, a product is the lifeline of a company. In sum, it was established that attributes within each dimension contributed to the success of manufactured products.

There is, therefore, the need to organize and present aspects of design dimension in a sequential pattern that will enable innovators to organize their own thinking towards their overall approach to design/innovation. In other words, there is the need to understand the relationship between the various dimensions of design and international competitiveness — a model where all

Figure 8.2 Design dimensions contributing to success in manufactured exports. *Source:* Ughanwa 1986

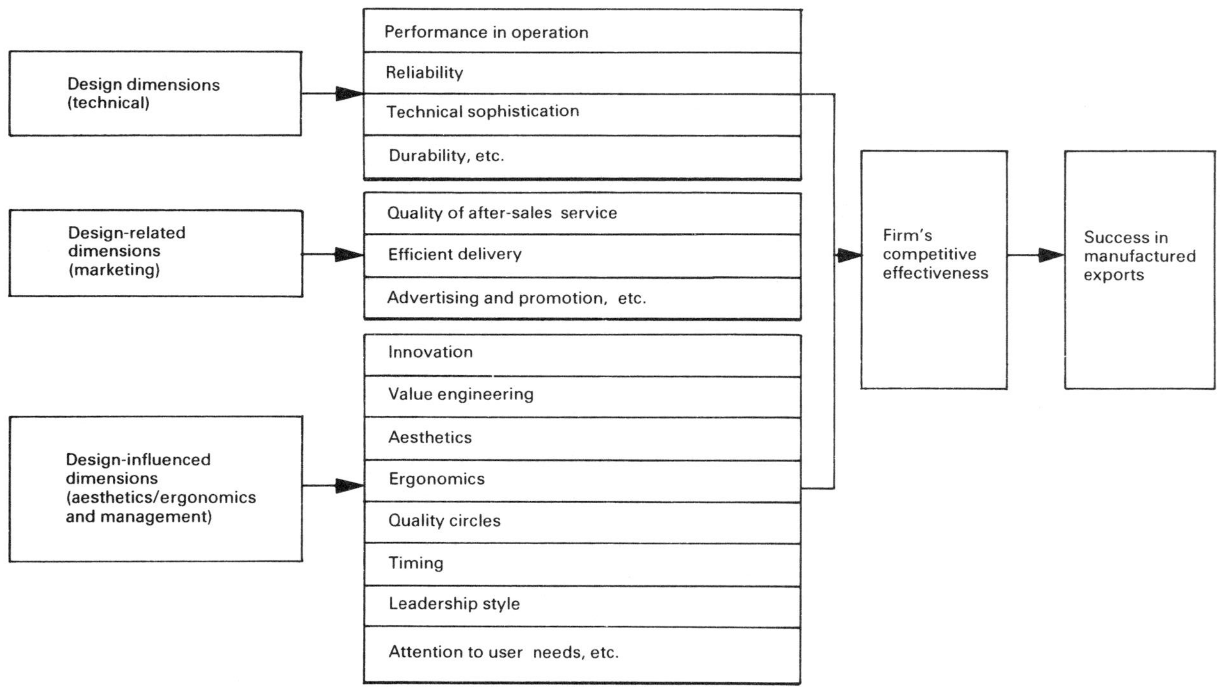

aspects of the design dimension can be conveniently accommodated, efficiently organized, co-ordinated, and adequately managed.

The theoretical model

Putting together all of the above factors, a model of the role of design dimensions in achieving international competitiveness has been developed and is presented in Figure 8.3.

This model stresses the importance of the various dimensions of design in achieving success in manufactured exports, and thus, success in international competitiveness. The model suggests four strategies to be adopted in order to increase the firm's ability to compete in manufactured exports. They are:

(1) Pay adequate attention to design factors — reliability, performance in operation, technical sophistication, product uniqueness, durability, etc. This takes account, to some extent, of some of the technical possibilities of the product.
(2) Ensure that these design factors are matched with design-related attributes such as prompt delivery, efficient after-sales service, etc. In other words, ensure that a reliable product is delivered in time and that efficient after-sales service is given, where the product requires such service. As argued in the earlier chapters, prompt delivery, efficient after-sales service, etc., are all part of the benefits (satisfaction) sought by customers/users in new products.
(3) Consider factors influencing the design of new products such as value engineering, innovation, aesthetics, quality circles, timing, etc., as part of the design activity. That is to say, these influencing factors should be considered in parallel with, not in isolation from, the other aspects of design activities. For instance, it was evident in the literature that aesthetics (e.g. styling) is often regarded as a 'superficial add-on' that can be called for at the last stage of the new-product design and development. There was also evidence in previous studies to prove this belief wrong.
(4) Organize and co-ordinate the activities of design, design-related and design-influenced factors so an effective result can be achieved.

The model suggests that in order to achieve, maintain, and sustain competitiveness in world markets, all the dimensions of design must not only be effectively and efficiently combined/matched, but also must be adequately managed. Therefore, the model

Figure 8.3 A model of association between the various dimensions of design and international competitiveness. *Source:* Ughanwa 1986

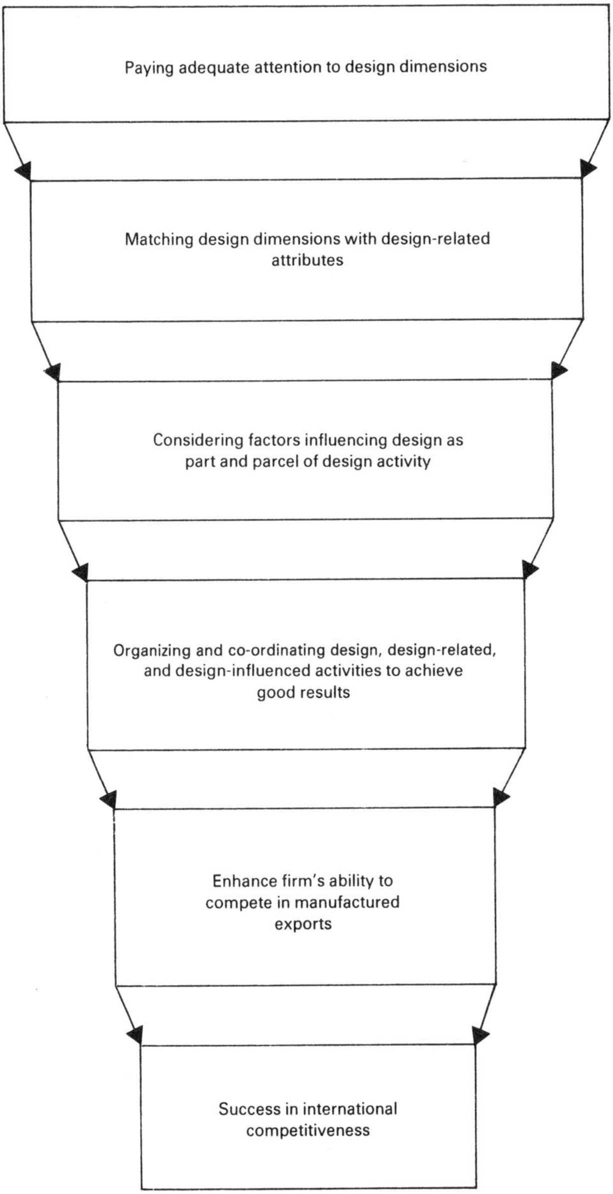

suggests the best fit to be made between and among the various dimensions of design, so a product with better value for money may be created. The model emphasizes the adequate management of design because it appears to make little sense or yield less favourable results if design, design-related, and design-influenced activities are not properly managed. The model therefore suggests that someone with experience, enthusiasm, power, and authority (it doesn't matter if it is delegated) must be at the head to direct these activities.

As argued in Chapter 5, the quality of a product is the concern of all in an organization. Consequently, this model proposes that the success of these design activities largely depends on the creation of design awareness throughout the organization, effective communication, and effective co-operation and co-ordination between and among the relevant departments, as well as close interaction with customers/users. The model stresses close interaction between firms and customers/users because success in manufactured exports principally depends on the acceptability of such manufactured products by customers/users in international markets. The model, therefore, stresses that these design dimensions (reliability, durability, aesthetics, delivery, etc.,) may contribute little to the success of a product if they are not in accordance with customer/user specification (i.e. if they do not meet the requirements of the customer or user). For example, whereas a customer may be concerned with the aesthetic appeal, say, appearance, of a product, another customer may concern him/herself with the durability of the same product.

In sum, the model emphasizes the relationship between these various dimensions of design and success in manufactured exports and thus, in international competitiveness.

It is pertinent to note, however, that in designing this model, the authors are aware of the 'dangers in attempting to present a typical theoretical or "cure-all" model of any human activity', and this model is not an exception.

In the words of Topalian (1980), 'because creation can be such an individual act, there are a multitude of ways of designing (a model) ... some processes may carry greater chances of resulting in successful outcomes than others, and some processes may be more efficient than others. 'But', he continues, 'no one process of designing carries with it a guarantee of success, irrespective of the problems and people involved', hence it was noted earlier that this model is not a 'cure-all'.

However, it is felt that the development of this model will help to explain the relationship between the efficient and effective combination or matching of the various dimensions of design on the one hand, and success in manufactured exports or international competitiveness on the other. This model fulfils the following objectives:

(1) The model focuses on the role of the various dimensions of design in achieving success in manufactured exports.
(2) It suggests that, all things being equal, it is those firms that efficiently and effectively combine the relevant dimension of design at any one time, in the formulation and implementation of the design strategies, that successfully compete and/or excel over their opponents in international markets.
(3) The model emphasizes that in addition to matching the design dimensions, clear objectives should be drawn as to how these dimensions should be effectively managed to yield good results.
(4) The model assumes the firm as a unit of enquiry.

All the evidence points to the situation-specific nature of competitiveness. It is essential to look at the firm in terms of its products/markets before jumping to premature conclusions based on the generalized kind of analysis developed here.

That said, it is hoped that the model will serve as a constant reminder that ultimately, success depends not only on doing the right things (effectiveness) but on doing them well (efficiency). We have attempted to diagnose some of the factors that have undermined UK competitiveness as well as those that have operated to the benefit of our more successful competitors. This diagnosis points to the importance of non-price factors as the basis upon which customers develop perceptions of value. In turn these non-price factors depend heavily upon effective design based upon a clear understanding of user needs and translated through integrated management into satisfying products and services. Understanding and implementing these insights truly constitutes 'A Design for Success'.

Bibliography and further reading

Abbot, L. (1955) Quality and competition: an essay in economic theory, Columbia University Press, New York.

Achilladelis, B., Jervis, P., and Robertson, A. (1971) 'Project SAPPHO: a study of success and failure in innovation', in Science Policy Research Unit (SPRU), University of Sussex.

Aitchison, D. (1978) 'Europe's international top people', *Marketing*, December, p. 41.

Albu, A. (1976) 'Causes of the decline in British merchant shipbuilding and marine engineering', *Omega*, vol. 4, no. 5.

Albu, A. (1980) 'British attitude to engineering education: a historical perspective', in K. Pavitt (ed.) *Technological Innovation and British Economic Performance*, Macmillan, London, pp. 67–87.

Alder, L. (1966) 'Time lag in new product development', *Journal of Marketing*, Vol. 30, January.

Alderson, W. (1965) *Dynamic Marketing Behaviour*, R.D. Irwin Inc., Illinois.

Alexander, C. (1963) 'The determination of components for an Indian village', in J.C. Jones and D.G. Thornley (eds) *Conference on Design Methods*, Pergamon Press, Oxford.

Ambridge, P. (1980) 'Metro: open up or shut down for BL', *Design*, November.

American Ordinance Association (AOA) (1964) 'Fringe effects of value engineering', The US Department of Defense, Washington DC, May.

American Society of Tool and Manufacturing Engineers (ASTME) (1967) *Value Engineering in Manufacturing*, Prentice-Hall Inc., Englewood Cliffs, New Jersey.

Ames, C.B. (1970) 'Trappings vs substance in industrial marketing', *Harvard Business Review*, July–August.

Archer, L.B. (1965) Systematic method for designers, Council of Industrial Design, London.

Ashby, E. (1958) *Technology and the Academics*, Macmillan, London.

Asimow, M. (1962) *Introduction to Design*, Prentice-Hall Inc., New Jersey.

Bache, D. (1980) In D. Manasian 'The man who made the Metro', *Design*,

November, pp. 40–1.
Baily, P.J.H. (1978) *Purchasing and Supply Management*, 4th edn, Chapman and Hall, London.
Baker, M.J. (1975) *Marketing New Industrial Products*, Macmillan, London.
Baker, M.J. (1979a) *Marketing, an Introductory Text*, 3rd edn, Macmillan, London.
Baker, M.J. (1979b) 'Industrial buying behaviour and the adoption of innovation', in M.J. Baker (ed.), *Industrial innovation, technology, policy, diffusion*, Macmillan, London.
Baker, M.J. (1979c) 'Export myopia', *Quarterly Review of Marketing*, Spring.
Baker, M.J. (1979d) 'Export myopia could lead to blindness', *Marketing*, March.
Baker, M.J. (1982) 'Innovation — key to success', *Quarterly Review of Marketing*, Winter.
Baker, M.J. (1983) *Market Development: a Comprehensive Survey*, Penguin Books Ltd, Harmondsworth.
Baker, M.J., Black, C.D. and Ughanwa, D.O. (1986) 'Profit by design', monograph prepared for the Design Council — Scotland and Scottish Development Agency, University of Strathclyde.
Barratt, M. (1984a) 'Facing the Japanese challenge', *Purchasing and Supply Management*, September.
Barratt, M. (1984b) 'Home truths for purchasing and supply', *Purchasing and Supply Management*, September.
Barratt, M. (1984c) 'Quality management: Japan's secret weapon', *Purchasing and Supply Management*, November.
Barratt, M. (1984d) 'How to buy print if you don't do it often', *Purchasing and Supply Management*, April.
Berkman, H.W. and Gibson, C. (1981) *Consumer Behaviour: Concepts and Strategies*, 2nd edn, Kent Publishing Company, Boston, Massachusetts.
BETRO Report (1976) 'Concentration on Key Markets', 2nd edn, Royal Society of Arts, London.
Black. J. (1984) Book Review, *Times Higher Education Supplement*, 25 May.
BIM (1978) 'Spur Initiative', Field Survey Report, Key Problem Areas in Engineering, October.
BIM (1984) (ed. S. Harris), BIM Report: 'The management of innovation', *Management Today*, April.
Binder, H.F. (1981) 'Executive guide: quality circles', Binder Hamlyn Fry and Company.
Bond, C. (1985) 'Food firms slammed for R & D failings', *Marketing*, 2 May.
Booker, P.J. (1964) Written contribution appended to Conference on the Teaching of Engineering Design, (Ed. P.J. Booker), Institution of Engineering Designers, London.
Booz, Allen and Hamilton (1965) *The Management of New Products*, 4th

edn, New York.
Boston Consulting Group (1975) 'Strategy alternatives for the British motor-cycle industry, the Boston Consulting Group to the Secretary of State for Industry', HMSO, London.
Boston Consulting Group, (1978) 'A comparative study of production economics in the UK, Japan, West Germany and South Korea', Report carried out for the Consumer Electronics Sector Working Party, London.
Bright, J. (1969) 'Some management lessons from technological innovation research' a paper read at a National Conference on Technological Innovation, University of Bradford Management Centre, 12–13 March 1969.
Britch, A.L. (1968) 'Building design aided by the computer', *The Computer Bulletin*, vol. 12, no. 4.
Brown, M. and Conrad, A.H. (1967) 'The influence of research and CES production relatives', in M. Brown (ed.), *The Theory of Empirical Analysis of Production*, NBER, Columbia University Press, New York.
Burck, C.G. (1983) 'Will success spoil General Motors?', *Fortune*, 22 August.
Bushell, T. (1985) 'IM's lesson for industry', *Marketing*, 23 May.
Cain, W.D. (1969) *Engineering Product Design*, Business Books Ltd, London.
Caldecote, Lord (1979) 'Investment in new product development', RSA lecture, April.
Carter, D. (1977) *Industrial Design Education in the UK*, The Design Council, London.
Carter, C.F. and Williams, B.R. (1957) *Industry and Technical Progress*, Oxford University Press, Oxford.
Carter, C.F. and Williams, B.R. (1958) *Investment in Innovation*, Oxford University Press, Oxford.
Carter, C.F. and Williams, B.R. (1959) *Science in Industry: Policy for Progress*, Oxford University Press, Oxford.
Catherwood, Sir Fredrick (1978) At the Product Quality Conference, British Overseas Trade Board, London, 27 November.
Chamberlain, E.H. (1957) *Towards a More General Theory of Value*, Oxford University Press, Oxford.
Chamberlain, E.H. (1962) *The Theory of Monopolistic Competition*, 8th edn, Harvard University Press, Cambridge, Mass.
Channon, D. (1985) 'Managing in tomorrow's world', British Institute of Management Report, *Management Today*, January.
Charlish, G. (1984) 'Computer aided design: Prosys links up complex processes', *Financial Times*, Thursday 31 May.
Chilton, R. (1985) 'IPC scorches the Holborn "gossip",' *Marketing*, 23 May, p. 16.
Clayton, K. (1984) 'High tech hits the road', *Marketing*, 5 July.
CNAA (1984) *Managing Design: an Initiative in Management Education*, CNAA, London.
Consumer Electronics Economic Development Committee (1983) 'Quality and reliability' *FACT*, Consumer Electronics EDC, London.

Cooper, A. (1980) 'US union says "ban Jap cars",' *Marketing*, 18 June.
Cooper, R.G. (1975) 'Why new industrial products fail', *Industrial Marketing Management*, vol. 4.
Cooper, R.G. (1979a) 'Identifying industrial new product success: Project NewProd', *Industrial Marketing Management*, vol. 8.
Cooper, R.G. (1979b) 'The dimensions of industrial new product success and failure', *Journal of Marketing*, vol. 43, Summer.
Cooper, R.G. (1984a) 'The strategic performance link in product innovation', *R & D Management*, vol. 14, no. 4, October.
Cooper, R.G. (1984b) 'The performance impact of product innovation strategies', *European Journal of Marketing*, vol. 18, no. 3, October.
Corfield, K.G. (1979) 'Product design', NEDO, London.
Cortazzi, Sir Hugh (1984) 'Why "Eyes and Ears" are necessary', *Financial Times*, Thursday, 9 February.
Cortazzi, Sir Hugh (1985) '14 questions from Japan', *Management Today*, June.
Cox, J. (1984) 'Purchasing and quality circle movement', *Purchasing and Supply Management*, December.
CPDM (1985) 'Survey of current practices', UK distribution management publication, Corby, Northants, January.
CPRS (1975) *Future of the British Car Industry*, HMSO, London.
Creighton, M. (1985) 'Towards best quality at lowest cost', *Purchasing and Supply Management*, February.
Crisp, J. (1984) 'ICL computer is first result of Fujitsu deal', *Financial Times*, Thursday 25, p. 1, 18 April.
Cutting, R. (1984) BIM Report: 'The management of innovation', *Management Today*, April.
Dale, B.G. (1984) 'Quality circles in the UK' *Journal of General Management*, vol. 9, no. 3, Spring.
Dale, B.G. and Ball, T.S. (1983) 'A study of quality circles, in UK manufacturing organisations', Department of Management Sciences occasional paper, No 8306, UMIST.
Dale, B.G. and Mortiboys, R.J. (1984) 'Management of quality and the quality of management: seven case histories', *Purchasing and Supply Management*, July.
Dastur, B.T. (1984) 'Management of quality: an Indian view', *Purchasing and Supply Management*, October.
Davies, D. (1977) 'Demolishing the caricature of the two design cultures', *Design* 347, November.
Davidson, H. (1976) 'Why most new consumer brands fail'. *Harvard Business Review*, 54, March–April, pp. 117–22.
Denny, R. (1984) BIM Report: 'The management of innovation', *Management Today*, April.
Department of Health and Social Security (1976) 'Exports and the public sector', *Design*, 333, September.
Department of Trade and Industry (1978) 'The agricultural engineering industry', HMSO, London.
Department of Trade and Industry (1978) 'Export performance: no room for complacency', *Trade and Industry*, September.

The Design Council (1984a) 'Car 2000', *Design Selection*, The Design Council, London.

The Design Council (1984b) 'Driving forward towards the year 2000', The Design Council, London.

Director of Fair Trading Report (1981) 'TI-Raleigh Industries Limited and TI-Raleigh Limited', Office of Fair Trading, London.

Drucker, P.F. (1978) 'Tasks, responsibilities, practices', in J. Wilmshurst (ed.) *Fundamentals and Practice of Marketing*, Heinemann, London, p. 3.

Drucker, P.F. (1985) *Innovation and Entrepreneurship*, Heinemann, London.

Economists Intelligence Unit Report (1980) 'Japanese face new car export policy', *Marketing*, October 15.

EDC (1968) 'Market — the world' *Mechanical Engineering*, EDC, HMSO, London.

Edwardes, J.U.V. (1976) 'No substitute for good workmanship', *The Times*, Thursday, 8 April.

Egan, John (1985) 'Improving management performance', Report of a BIM Panel, 2nd edn, p. 19.

Evans, C. (1983) 'The lie of the land', *Design*, January.

Evan-Vaughan, G.F. (1970) 'Design in industry', The Administrative Staff College, Henley-on-Thames.

Farr, M. (1966) *Design Management*, Hodder and Stoughton, London.

de Ferranti, B. (1956) 'Machine for to live with', Proc. Profit by Design, COID, London, October.

Fielden Committee (1963) 'Engineering design', Department of Scientific and Industrial Research, London.

Finkelstein, L. (1984) 'Education for design', *IEE Proceedings*, vol. 131, no. 8, November.

Finniston, Sir Monty (1980) 'Engineering our future', A Report of the Committee of Inquiry into the Engineering Profession, Cmnd. 7794.

Flurscheim, C.H. (1983) *Industrial Design and Engineering: A Marriage of Techniques*, The Design Council, London.

Fores, M. (1978) 'How Britain can compete', *Marketing*, December.

Foxall, G. (1980) 'Marketing models of buyer behaviour: a critical review', *European Research*, September.

Fraser, J. (1984) 'Fraser joins the Design Council' *Design 428*, August, p. 5.

Freeman, C. (1978) *Financial Times*, management page, 24 May.

Gabor, A. (1979) 'Pricing gets marketing-oriented at last', *Marketing*, November.

Gardiner, P. and Rothwell, R. (1985) 'Tough customers: good designs', *Design Studies*, vol. 6, no. 1, January.

Gerstenfeld, A. (1976) 'A study of successful projects, unsuccessful projects, and projects in progress in West Germany, IEEE Transactions on Engineering Management, 23 August, pp. 116–23.

Gibbs, A. and Flurscheim, C.H. (1983) *Industrial design in engineering: a*

marriage of techniques (ed. C.H. Flurscheim), The Design Council, London.
Gilligan, C. and Myerson, J. (1983) 'Ploughing ahead', *Design*, January, pp. 32-5.
Gloag, H.L. and Keyte, M.J. (1957) 'Rational aspects of colouring in building interiors', *Architects Journal*, Vol. 125.
Gofton, K. (1985) 'Biotech comes to life', *Marketing*, 23 May, p. 22.
Gooding, K. (1984) 'US car fuel rules will cost Jaguar £7m "fines",' *Financial Times*, Thursday, 21 June.
Gooding, K. (1985) 'Renault holds back medium truck launch', *Financial Times*, Tuesday, 28 May.
Gregory, S. (1966) *The Design Method*, Butterworths, London.
Griffiths, F. (1985) 'Is the customer sometimes right?', *Purchasing and Supply Management*, January, pp. 26–33.
Griffiths, J. (1985) 'Towards the glued lightweight car', *Financial Times*, Tuesday 19 March.
Griffiths, J.T. (1983) 'Electronics components and television manufacture', in *Standards Quality and Competitiveness*, NEDO, August.
Gruber, W.H., Poensgen, O.H. and Prakke, E. (1973) 'The isolation of R & D from corporate management', *Research Management*, November.
Haine, M. (1980) in Ian Murray's 'Four ways to launch successful new products', *Marketing Week*, 23 May.
Hall, M.A. (1984) 'A passive or active role for development', *IEE Proceedings*, vol. 131, no. 8, November.
Hamilton, M., Sullivan, V. and Ward, T. (1970) 'Launching a new product, a checklist for marketing management', Institute of Marketing.
Hardy, A.C. (ed.) (1966) *Colour in Architecture*, Leonard Hill Books, London.
Harris, D. (1984) 'Professional engineers: true unity: the most impossible: the body with fifty-one parts', *The Times*, Friday, 5 October.
Harris, S. (1984) (ed.) BIM Report: 'The management of innovation', *Management Today*, April.
Harvey-Jones, Sir John (1985) 'View from the Top', in *Improving Management Performance*, 2nd edn, Report of BIM Panel, BIM and Professional Publishing, London.
Hayward, A.L. and Sparkes, J.J. (1985) *The Concise English Dictionary*, Omega Books Ltd, Ware, Hertfordshire.
Hilton, J. (1983) 'Models as an aid to design', in C.H. Flurscheim (ed.) *Industrial and Engineering: A Marriage of Techniques*, The Design Council, London.
Holt, K. (1977) *Product Innovation*, Newnes-Butterworth & Co. (Publishers) Ltd, London.
Hopkins, D.S. (1981) 'New product winners and losers', *R & D Management*, May, 12–17.
Horrocks, R. (1985) 'View from the Top', in *Improving Management Performance*, 2nd edn, Report of BIM Panel, BIM and Professional Publishing, London.

Isaac, D. (1984) 'How Jaguar lost its spots', *Management Today*, April, pp. 39–45, 117, 120.
Jack, D. (1984) BIM Report: 'The management of innovation' *Management Today*, April.
Jenkin, Patrick (1983) An address to representatives of the industry and design profession at the Number 10 Seminar on Mass propaganda for design in the UK, *Design*, January, p. 52.
Johne, F.A. (1982) 'Innovation, organisation and marketing of high industrial products', PhD thesis, University of Strathclyde.
Johne, F.A. (1983) 'How to lead by innovation', *Management Today*, September.
Johne, F.A. (1984) 'Innovation in marketing of new technical products', *The Quarterly Review of Marketing*, Spring.
Johnson, D.L. (1984) 'Design — the requirement of maintenance', *IEE Proceedings*, Part A, vol. 131, no. 8, November.
Jones, J.C. (1966) 'Design methods reviewed', in S. Gregory (ed.) *The design method*, Butterworths, London.
Jones, J.C. (1981) *Design methods: seeds of human futures*, 1980 edn, John Wiley & Sons, New York.
Jones, R. (1980) '3-D design in a poly sandwich' *Design*, November, pp. 44–5.
Jordan, C. (1984) 'Drive forward', The Design Publication, London, June.
Karen, T. (1980) '3D design in a poly sandwich', *Design*, November, pp. 42–7.
Karen, T. (1984) 'Shape of the future', *Design Selection*, The Design Council, London, June.
Kotler, P. (1972) 'A generic concept of marketing', *Journal of Marketing*, vol. 36, April.
Kotler, P. and Fahey, L. (1982) 'The world's champion marketers: the Japanese', *Journal of Business Strategy*, vol. 3, no. 1, Summer.
Kraushar, P. (1969) 'Graveyards for new products', *Marketing*, November.
Kulvik, H. (1977) 'Factors underlying the success or failure of new products', Helsinki: University of Technology, Report no. 29, Finland.
Labouchere, C.M. (1984) 'The total cost of imperfection', *IEE Proceedings* vol. 131, no. 8, November.
Lancaster, G.A. and White, M. (1977) 'Industrial buying behaviour', *The Quarterly Review of Marketing*, Autumn.
Langrish, J. (1968) 'Industrial innovation: case studies of some recipients of the Queen's Award to Industry for Innovation', PhD thesis, University of Manchester, October.
Langrish, J., Gibbons, M., Evans, W.G. and Jevons, F.R. (1972) *Wealth From Knowledge*, Macmillan, London.
Leech, D.J. (1972) *Management of Engineering Design*, John Wiley & Sons, London.
Lester, T. (1985) 'Blowing the whistle', *Marketing*, 16 May.
Levacic, R. (1984) 'Do mercantilist industrial policies work? A comparison of British and French TV manufacturing', *National Westminster Bank Quarterly Review*, May.

Levitt, T. (1960) 'Marketing myopia', *Harvard Business Review*, July–August.
Levitt, T. (1965) 'Innovative imitation', *Harvard Business Review*.
Levitt, T. (1966) 'Imitative innovation', *Harvard Business Review*, September–October.
Levitt, T. (1969) *The Marketing Mode*, McGraw-Hill, New York.
Levitt, T. (1985) 'Return of the global guru', *Marketing*, 4 July.
Levy, J. (1984) 'A bridge too far for industry and education', *The Times*, Friday, 5 October.
Lorenz, C. (1984) 'A resurgence at last for UK designers', *Financial Times*, Wednesday, 23 May.
Lorenz, C. (1979) 'Investing in success: how to profit from design and innovation', *Anglo-German Foundation*, London.
(Keith) Lucas Report, (1980) 'Design education at secondary school level', The Design Council, September.
McConnel, D.J. (1970) 'An experimental examination of the price — quality relationship', *Journal of Business*, vol. 41, October.
Macdonald, J. (1984) 'Finding the right key', *Marketing*, 24 May.
MacDougall, D. (1952) 'British American exports', *Economic Journal*, December, 1951 and September.
MacFarlane, G.D. (1977) 'A comparative study of export marketing mixes of Scottish Queen's Award to Industry recipients', MSc dissertation, University of Strathclyde.
MacLaurin, W.R. (1949) '*The Massachusetts Institute of Technology Studies of Innovation: Invention and Innovation in the Radio Industry*'. Macmillan, New York.
McTavish, R. (1974) 'A study of selected problems of the evolutionary cycle of highly technical new capital goods', PhD thesis, University of Strathclyde.
Maldonado, T. (1964) UNESCO Seminar Report on Education of Industrial Designers, Bruges, March.
Mansfield, E. et al (1972) *Research and Innovation in the Modern Corporation*, Macmillan, London.
Mara, H. (1984) 'Hardware for the handicapped', *Financial Times*, Tuesday, 29 May.
The Marketing Team (1985) 'Making sense of the dollar', *Marketing*, 4 April.
Marson, B. (1984) BIM Report: 'The Management of innovation', *Management Today*, April.
Maslow, A.H. (1954) *Motivation and Personality*, Harper and Row, New York.
Mason, D. (1980) 'Why Britain's chips lag behind', *Marketing*, 18 June.
Matchett, E. (1965) 'The application of method study techniques in the design process, and 'The controlled evolution of engineering design', *The Engineering Designer*, IED, February. See also 'Control of thought in creative work', *Chartered Mechanical Engineer*, vol. 14, no. 4, 1968, pp. 4, 46, 190.
Meadows, D. (1969) 'Estimate accuracy and project selection models in

industrial research', *Industrial Management Review*, Spring.
Michell, P. (1979) 'Infrastructure and international marketing effectiveness', *Columbia Journal of World Business*, vol. 14, no. 4.
Miles, L. (1987) 'Sinclair suffers déjà vu with delays on lap top', *Marketing*, May 21, p. 2.
Miles, M.B. (1967) (ed.) *Innovation in Education*, Teachers College Press, Columbia University, USA.
Minasian, J. (1969) 'R & D, production functions and rates of return', *American Economic Review*, May.
Minasian, D. (1980) 'The man who made the Metro', *Design*, November.
Moody, S. (1980a) 'The role of industrial design in technological innovation', *Design Studies*, vol. 1, October.
Moody, S. (1980b) 'The role of industrial design in technological innovation', *Chartered Mechanical Engineer*, November.
Moser, C. and Kalton, G. (1977) *Survey Methods in Social Investigation*, Heinemann Educational Books, London.
Moulton Report, (1976) 'A report on engineering design education', The Design Council.
Murdoch, P. and Flurscheim, C.H. (1983) 'Form', C.H. Flurscheim (ed.) *Industrial Design in Engineering*, The Design Council, London.
Murray, I. (1980) 'Four ways to launch successful new products', *Marketing Week*, 23 May.
Myers, S. and Marquis, D.G. (1969) *Successful Industrial Innovation*, NSF, Washington.
Myerson, J. (1983) 'The lie of the land', *Design*, January.
Nagashima, A. (1970) 'A comparison of Japanese and US attitudes toward foreign products', *Journal of Marketing*, vol. 34, January.
Nagashima, A. (1977) 'A comparative "Made In" product image survey among Japanese businessmen', *Journal of Marketing*, vol. 34, July.
NEDO (1963) 'Export trends', NEDO, London.
NEDO (1965) 'Imported manufactures', NEDO, London.
NEDO (1977) 'International price competitiveness, non-price factors' *Export Performance*, NEDO, April.
NEDO (1981) 'Industrial performance: R & D and innovation', NEDO, London.
NEDO (1983) 'Standards, quality and competitiveness', NEDO, London, August.
NEDO (1983) 'The problem', *FACT*, NEDO, London, November.
Nelson of Stafford, Lord (1984) 'Export competitiveness: a factor of management and/or design', *IEE Proceedings*, vol. 131, no. 8, November.
Number 10 Seminar (1983), a seminar on 'Mass propaganda for design in the UK', held at 10 Downing Street when the Prime Minister met representatives of the industry and design profession, reported by John Thackara in *Design*, January, p. 52.
A.C. Nielsen Company Limited, 'How to strengthen your product plan', Nielsen House, Oxford.
NIESR (1966) 'Export competitiveness: British experience in Eastern

Europe', in David Stout 'International price competitiveness, non-price factors and export performance, 1977, NEDO, London, p. 37.

NIESR (1977) 'Export competitiveness: British experience in Eastern Europe', in D. Stout's 'International price competitiveness, non-price factors and export performance', NEDO, London.

Nott, J. (1980) in editorial opinions: 'Counting the costs of importing from Japan', *Marketing,* 18 June.

Oakley, M. (1984a) 'How to manage product design', BIM's *Management News,* No. 9, July.

Oakley, M. (1984b) *Managing Product Design,* Weidenfeld and Nicolson, London.

O'Donnell, M. and O'Donnell, R.J. (1984) 'Quality circles — The latest fad or a real winner?', *Business Horizons,* Indiana University Graduate School of Business, vol. 27, no. 3, May/June.

OECD, (1982) Innovation in small and medium firms: background reports, OECD, Paris, France.

Olins, W. (1985) 'Management by design', *Management Today,* February.

Onah, J.O. (1977) 'The concept of product differentiation revisited', *Quarterly Review of Marketing,* Summer.

Open University Design Innovation Group (1983) V. Walsh and R. Roy (eds), 'Plastic products: good design, innovation and business success', Open University, Milton Keynes, pp. 26–9.

Open University Design Innovation Group (1984) R. Roy and H. Bruce (eds), 'Product design, innovation and competition in British manufacturing: background aims and methods', Open University, Milton Keynes, p. 6.

Orr, Sir David and H.F. van den Hoven, (1981) 'Research and development in Unilever', in *Unilever 1981,* London.

Osola, J.V. (1984) 'Money' *IEE Proceedings,* vol. 131, no. 8, November.

Page, J.K. (1966) 'Contribution to building for people', Conference Report, 1965, Ministry of Public Building and Works, London.

Panic (1975) 'Why the UK's propensity to import is so high', in David Stout's 'International price competitiveness, non-price factors and export performance', NEDO, London, p. 39.

Parkinson, S.T. (1980a) 'New product development and international trading success (A comparison between West Germany and the UK), in M.J. Baker and M.A. Saren (eds) *Marketing into the Eighties,* March, Edinburgh.

Parkinson, S.T. (1980b) 'User-supplier interaction in new product development', PhD thesis, University of Strathclyde, vol. 1.

Pavitt, K. (1983) 'Characteristics of innovative activities in British industry, *Omega,* vol. 11, no. 2.

Peplow, M.E. (1960) 'Design acceptance', in S.A. Gregory (ed). *The Design Method,* Butterworth, London.

Peppal, E. (1984) 'Drive forward towards the year 2000', The Design Council publication.

Peters, T.J. and Waterman, R.H. (1982) *In Search of Excellence: Lessons from America's Best Run Companies,* Harper, New York.

Pilditch, J. (1978) 'How Britain can compete', *Marketing*, December.
Piper, J. (1978) 'Making the most of West Germany', *Marketing*, March.
Popham, P. (1982) 'Spaghetti Eastern', *Design*, February, pp. 32–5.
Project SAPPHO (1974) 'Project SAPPHO Phase II: Project SAPPHO Updated', a study undertaken by Rothwell, R., Freeman, C., Horsey, A., Jervis, V.T.P., Robertson, A.B. and Townsend, J., Research Policy, vol. 3.
Queen's Award Committee (1985) 'Press notice', The Queen's Award Office, London.
Ray, G.F. (1972) 'On defining diffusion of new techniques', *Business Economist*, vol. 4, Summer.
Reid, M.F. (1984) 'Maintainability — a marketing asset', *IEE Proceedings*, vol. 131, no. 8, November.
Reitsperger, W.D. (1982) 'Comparative management practice: the case of Japanese, US and British TV manufacture in the UK', PhD thesis, University of London.
Reswick, J.B. (1965) Prospectus for engineering design centre, case of Institute of Technology, Cleveland, Ohio.
Rhys, G. (1984) 'Heavy commercial vehicles: a decade of change', *National Westminster Bank Quarterly Review*, August.
Rines, M. (1980) 'Counting the cost of importing from Japan', *Marketing*, 18 June.
Robertson, A. (1970) 'The management of industrial innovation', Industrial Educational and Research Foundation, London.
Robertson, A. (1977) 'Technological innovation and the management of R & D, in D. Ashton (ed.) *Management Bibliography and Review*.
Robinson, J. (1932) *The Economics of Imperfect Competition*, Macmillan, London.
Robinson, R. (1985) 'Handling the recovery', *Purchasing and Supply Management*, February.
Rollason, B. (1971) 'What the successful exporter knows', *Marketing*, November.
Root, F.R. (1978) *International Trade and Investment*, 4th edn, South Western Publishing Company, Ohio.
Rothwell, R. (1972) 'Factors for success in industrial innovations', from 'Project SAPPHO — A comparative study of success and failure in industrial innovation', SPRU, University of Sussex, Brighton.
Rothwell, R. (1974) Hungarian SAPPHO: 'Some comments and comparison', Research Policy 3, pp. 30–8.
Rothwell, R. (1976a) 'The role of technical change in international competitiveness: the case of textile machinery industry', *Management Decision*, vol. 15, no. 6.
Rothwell, R. (1976b) 'Innovation in the UK textile machinery industry: the results of a postal questionnaire survey', *R & D Management*, vol. 6, no. 3. Note: Adapted from Rothwell's analysis of the classification of innovation.
Rothwell, R. (1977a) 'The characteristics of successful innovators and technically progressive firms (with some comments on innovation

research)', *R & D Management*, vol. 7, no. 3.

Rothwell, R. (1977b) 'The characteristics of successful innovators and technically progressive firms'. *R & D Management*, vol. 7, no. 3, 1977, pp. 199–200.

Rothwell, R. (1978) 'Some problems of technology transfer into industry: examples from the textile machinery sector', *IEE Transactions on Engineering Management*, February.

Rothwell, R. (1979) 'The relationship between technical change, and economic performance in mechanical engineering: some evidence', in M.J. Baker (ed.) *Industrial Innovation, Technology, Policy, Diffusion*, Macmillan, London.

Rothwell, R. (1981a) 'Non-price factors in export competitiveness of agricultural engineering products', *Research Policy*, vol. 10.

Rothwell, R. (1981b) 'Why a new product fails', *Marketing*, 29 July.

Rothwell, R. (1983) 'Design matters more than price', *Design*, January.

Rothwell, R. and Robertson, A.B. (1973) 'The role of communication in technological innovation', *Research Policy*, vol. 2.

Rothwell, R. (1984) 'The management of innovation'. BIM Report (ed. S. Harris), *Management Today*, April, pp. 105–8.

Rothwell, R., Freeman, C., Horsley, A., Jervis, V.T.P. Robertson, A.B. and Townend, J. (1974) 'SAPPHO Updated: Project SAPPHO Phase II,' *Research Policy*, 3.

Rothwell, R. and Zegveld, W. (1978) 'Small and medium-sized manufacturing firms: their role and problems in innovation — government policy in Europe, the USA, Japan, Canada and Israel', Report to the six countries programme on innovation, The Netherlands, June.

Rothwell, R. and Zegveld, W. (1981) *Industrial Innovation and Public Policy*, Frances Pinter (Publishers) Ltd, London, p. 31.

Rothwell, R., Schott, L., and Gardiner, P. (1983) *Design and Economy*, The Design Council, London.

Roy, R. (1984) 'Product design and innovation in a mature consumer industry', The Open University, UK.

Roy, R. and Bruce, M. (1984) 'Product design, innovation and competition in British manufacturing: background, aims and methods', Working Paper (WP-02), The Open University, UK, September.

Roy, R., Walker, D., and Walsh, V. (1983) 'Product design, innovation and competitiveness in British manufacturing industry', Faculty of Technology, The Open University, UK.

Sainsbury, D. (1983) 'The management of innovation', BIM's *Management Review and Digest*, vol. 10, no. 2, July.

Saren, M.A. (1980) 'Marketing and R & D — their organisational roles for successful new product development', in Marketing into the eighties: proceedings, European Academy of Advanced Research in Marketing and Marketing Education Group in the UK (eds M.J. Baker and M.A. Saren), Edinburgh, March.

Sasaki, N. (1984) In Mark Barrett's 'Quality management: Japan's secret weapon' *Purchasing and Supply Management*, November, p. 30.

Scanlon, B. (1980) 'Fight off the recession with a lab. full of new products', *Chief Executive*, May.

Scheer, I. (1976) 'Design's key role in selling UK products', *The Times*, Tuesday, 20 April.

Schmookler, J. (1966) *Invention and Economic Growth*, Harvard University Press, Cambridge, Massachusetts.

Schon, D.A. (1967) *Technology and Change*, Delacorte Press, New York.

Schott, K. (1981) *Industrial Innovation in the United Kingdom, Canada and the United States*, British–North American Committee, Contemprint Ltd, London.

Schumpeter, J. (1939) *Business Cycles*, McGraw-Hill, New York.

Schwarz, K.K. (1984) 'Design and innovation', *IEE Proceedings*, vol. 131, no. 8, November.

Science and Engineering Research Council (SERC) (1983) A report of the engineering design working party, SERC, London.

Senker, P., Sciberras, E., Swords, N., and Huggett, C. (1977) 'Forklift trucks: a study of a sector of the UK engineering industry', Science Policy Research Unit, University of Sussex.

Shankleman, (1975) 'Britain's post war export performance', in D. Stout's 'International price competitiveness, non-price factors and export performance', NEDO, April, 1977, p. 37.

Sharpe, B. (1984) in B. Dale and R.J. Mortiboys 'Management of quality and quality of management: seven case histories', *Purchasing and Supply Management*, July.

Smith, Adam (1776) in E. Cannon (ed.) (1950) '*Wealth of Nations:* an inquiry into the nature and causes of the wealth of nations', vol. 2, 6th edn, Methuen, London.

Smith, B.P. (1978) 'The morality and management of design', *Journal of Royal Society of Arts*, March.

Smith, W.R. (1956) 'Product differentiation and market segmentation as alternative marketing strategies', *Journal of Marketing*, vol. 21, July.

Spandler, R. (1980) 'Foreign mopeds rule', *Marketing*, June.

Stanton, W.J. (1971) *The Fundamentals of Marketing*, 3rd edn, McGraw-Hill, New York.

Steur, Ball and Eaton (1966) 'The effect of waiting times on foreign orders for machine tools', *Economica*, London.

Sutherland, A. (1984) 'Getting the right product at the right time', *IEE Proceedings*, vol. 131, no. 8, November.

Thackara, J. (1983) 'Mass propaganda for design: no. 10 seminar, when the Prime Minister met representatives of industry and the design profession', *Design*, January.

Thatcher, Margaret (1984) '10 Downing Street: The Prime Minister', a foreward written by the Prime Minister and appended to a report on 'Managing design: an initiative in management education', CNAA, London, June.

Thompson, H.U. (1962) *Product Strategy*, Business Publications Ltd, London.

Tinnesand, B. (1973) 'Toward a general theory of innovation', PhD thesis, University of Wisconsin, Madison.
Topalian, A. (1980) *The Management of Design Projects*, Associated Business Press, London.
Trafford, D.B. (1984) 'Computer aid for design', *IEE Proceedings*, vol. 131, no. 8, November.
Turnbull, G.F. (1984) 'Managing for profit — by product design', *IEE Proceedings*, vol. 131, no. 8, part A, November.
Turner, B.T. (1968) 'Management of design', Industrial and Commercial Techniques Ltd, December.
Twiss, B. (1980) *Managing Technological Innovation*, 2nd edn, Longman, London.
Udell, J.G. (1968) 'Toward a theory of marketing strategy', *British Journal of Marketing*, Winter.
Ughanwa, D.O. (1986) 'The role of new product design in international competitiveness', PhD Thesis, University of Strathclyde, Glasgow, vols 1 and 2, October.
Unilever Ltd (1981) 'Research and development in Unilever', Unilever Publication, Unilever Ltd, London.
US Department of Defense (1969) 'Principles and applications of value engineering', The US Department of Defense, vol. 1.
US National Academy of Science (1966) Report to the Ad Hoc Committee of Prinicipals of Research — engineering interaction, Washington, Materials Advisory Board, National Research Council.
Utterback, J.M. (1971) 'The process of innovation: a study of the organisation and development of ideas for new scientific instruments', *IEE Transactions* on Engineering Management, Em-18.
Utterback, J.M., Thomas, J.A., Holloman, J.H. and Sirbu, M.H. (1976) 'The process of innovation in five industries in Europe and Japan', IEEE Transactions on Engineering Management, no. 1, February, pp. 3–9.
von Hippel, E. (1977) 'The dominant role of the user in semi-conductor and electronic sub-assembly process innovation', *IEE Transactions* on Engineering Management, May.
von Hippel, E. (1979) 'A customer-active paradigm for industrial product idea generation', in M.J. Baker (ed), *Industrial Innovation, Technology, Policy, Diffusion*, Macmillan, London.
Walker, N. (1976) 'Technical change and economic performance in the UK mechanical engineering industry: a review of the literature', Mimeo, Science Policy Research Unit, University of Sussex.
Walker, W. (1976) 'International trade in portable power tools', Science and Policy Research Unit, University of Sussex, Brighton.
Walker, W. (1980) 'Britain's industrial performance 1850–1950: a failure to adjust', in K. Pavritt (ed.) *Technological Innovation and British Economic Performance*, Macmillan, London.
Walsh, V. and Roy, R. (1983) *Plastics Products: Good Design, Innovation and Business Success*, Design Innovation Group, the Open University,

Milton Keynes.
Watkin, B. (1984) in H. Mara's 'Hardware for the handicapped', *Financial Times*, Tuesday, 29 May.
Watt, A.W. (1980) 'Managing the design function in the UK textile industry or I'm no expert but I know what I like', in Marketing into the eighties: proceedings, European Academy of Advanced Research in Marketing and the Marketing Education Group of the UK, Edinburgh (eds M.J. Baker and M.A. Saren), March.
Wells, S.J. (1964) *British Export Performance: A Comparative Study*, Cambridge University Press, Cambridge.
Wentz, W.B., Eyrich, G.I., and Stevenson, D.K. (1973) 'Marketing of products', in H. Britt (ed.) *Marketing Managers Handbook*, Dartnell Corp.
White, D. (1980) in J. Lott 'What makes a market leader, what makes a design?', *Design*, August, pp. 34–43.
Whitfield, A. and Wiltshire, T. (1983) 'Colour', in C.H. Flurscheim (ed). *Industrial Design in Engineering*, The Design Council, London.
Wicks, I. (1984) 'Organisational structure' *IEE Proceedings*, vol. 131, no. 8, November.
Williams, M. (1984) 'Testing time', *Purchasing and Supply Management*, September.
Willcock, J. (1981) 'The design triangle', *Designer*, April.
Wilmshurst, J. (1978) *The fundamentals and practice of marketing*, Heinemann, London, p. 1.
Wilson, A. (1984) 'Innovation in the marketplace', *Management Today*, June.
Witcher, B. (1985) 'Innovation and marketing', *The Quarterly Review of Marketing*, Winter.
Wooderson, T.T. (1966) *Introduction to engineering design*, McGraw-Hill, New York.
Zeid, E.H.A. (1981) 'Marketing and export success', PhD Thesis, University of Strathclyde.

Name Index

Subject Index